Applying AutoCAD 2006

Terry T. Wohlers

Applying AutoCAD 2006 is a textbook for those who wish to learn how to use AutoCAD software. AutoCAD is a computer-aided drafting and design package produced by Autodesk, Inc. For information on how to obtain the AutoCAD software, contact Autodesk.

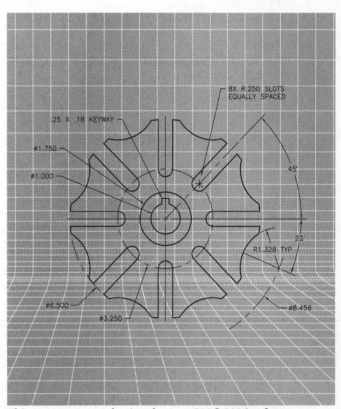

This text was created using the AutoCAD® 2006 software.

McGraw Hill **Glencoe**

New York, New York   Columbus, Ohio   Chicago, Illinois   Peoria, Illinois   Woodland Hills, California

Reviewers & Contributors:

John M. Felser
Broward County Schools
Ft. Lauderdale, FL

Timothy J. Caffrey
Caffrey Consulting
Fayetteville, AR

Don Grout, Ed.D.
Professor, retired, Champlain College
Burlington, VT

Christina Hammond
Tennessee Technology Center
Whiteville, TN

Gary J. Hordemann
Professor and Chair
Department of Mechanical Engineering
Gonzaga University
Spokane, WA

Ronald W. Koyle
Chaffey College
Rancho Cucamonga, CA

Wayne Samuelson
Chippewa Valley Technical College,
Retired
Eau Claire, WI

James F. Shaum
Madison Comprehensive High School
North Central Ohio College Tech Prep
Consortium
Mansfield, OH

Stuart Soman, Ed.D.
Educational Consultant
Long Island, NY

Thanks to **Jim Quanci and the members of the AutoCAD 2006 team at Autodesk, Inc.**, for answering questions and providing assistance.

Cover design: Squarecrow Creative Group.

Products carrying the Autodesk logo are not necessarily Autodesk products and as such are not warranted by Autodesk. Autodesk, the Autodesk logo, and AutoCAD are registered trademarks of Autodesk, Inc.

Applying AutoCAD 2006 is not an Autodesk product and is not warranted by Autodesk. Autodesk, the Autodesk logo, and AutoCAD are registered trademarks of Autodesk, Inc.

Word and Excel are trademarks of Microsoft, Inc.

 **Glencoe**

The **McGraw·Hill** Companies

Send all inquiries to:
Glencoe/McGraw-Hill
3008 W. Willow Knolls Drive
Peoria, IL 61614

ISBN 0-07-873837-7 (Student text)
Printed in the United States of America.
1 2 3 4 5 6 7 8 9 10   009   10 09 08 07 06 05

Contents in Brief

(Continues)

Contents in Brief (Continued)

What's New in 2006

In AutoCAD 2006, Autodesk made important changes to almost every aspect of the AutoCAD software. You will notice the following differences in *Applying AutoCAD 2006:*

- New **DIMARC** and **DIMJOGGED** commands make dimensioning arcs and large radii faster and easier.

- **Dynamic blocks** allow parametric block editing, greatly increasing the flexibility of blocks.

- Enhanced **attribute extraction** allows you to create bills of materials within AutoCAD—no need for external software.

- Improved **TABLE** command permits the use of formulas and styles.

About the Author

For more than two decades, Terry Wohlers has focused his education, research, and practice on design and manufacturing. He has authored more than 275 books, articles, reports, and technical papers on engineering and manufacturing automation. He has presented to thousands of engineers and managers and has been a keynote speaker at major industry events in Asia, Europe, the Middle East, the Americas, and South Africa.

Wohlers is a renowned CAD educator. He developed and taught the first graduate and undergraduate courses on CAD at Colorado State University in 1983. Since then, he has taught many courses and given countless lectures on CAD and related subjects. In 1986, he founded Wohlers Associates, Inc., an independent consulting firm that provides organizations with technical and strategic advice on the new developments and trends in CAD/CAM, rapid prototyping, and manufacturing.

He was also the author and instructor of the first self-instructional university independent study courses on AutoCAD and CADKEY. An estimated 2,000 practicing design professionals from around the world have enrolled in them.

In May 2004, Wohlers received an Honorary Doctoral Degree of Mechanical Engineering from Central University of Technology, Free State (Bloemfontein, South Africa).

User's Guide

A Practical Approach

Applying AutoCAD 2006 presents each feature of the AutoCAD®
2006 software in a logical, sequential format. You will build skills
as you read about and apply techniques, solve problems, and
practice computer-aided drafting and design.

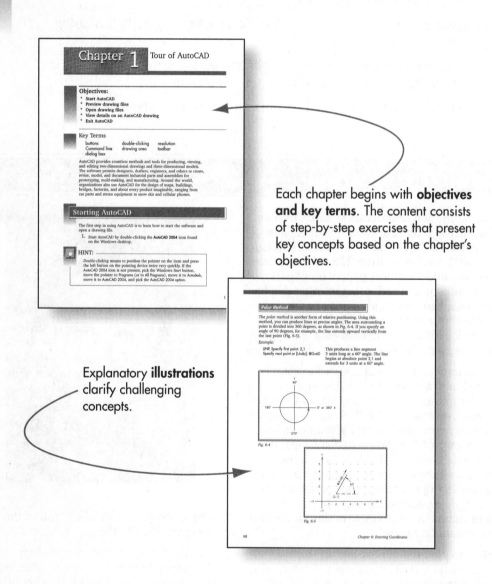

Each chapter begins with **objectives and key terms**. The content consists of step-by-step exercises that present key concepts based on the chapter's objectives.

Explanatory **illustrations** clarify challenging concepts.

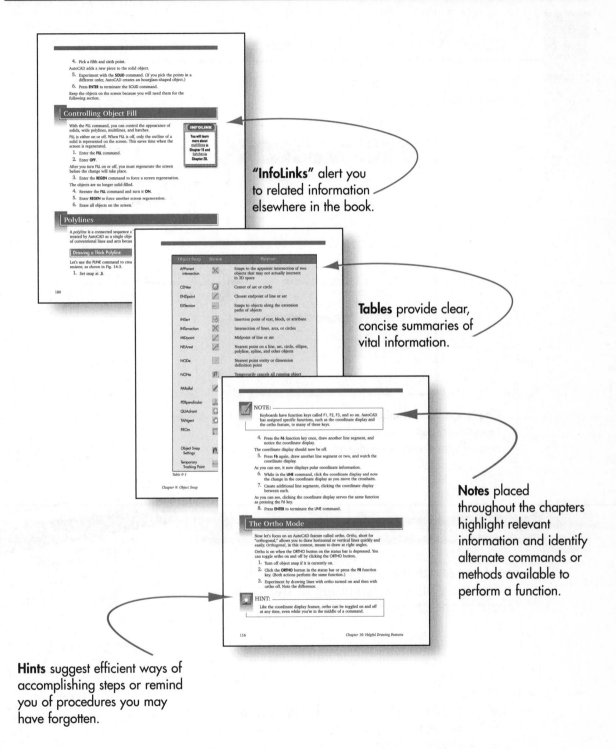

"InfoLinks" alert you to related information elsewhere in the book.

Tables provide clear, concise summaries of vital information.

Notes placed throughout the chapters highlight relevant information and identify alternate commands or methods available to perform a function.

Hints suggest efficient ways of accomplishing steps or remind you of procedures you may have forgotten.

User's Guide (Continued)

Chapter Review & Activities

Each chapter concludes with activities that review, reinforce, and expand learning.

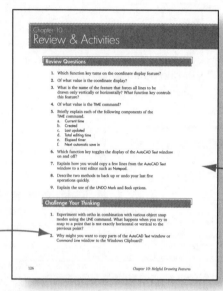

Review Questions at the end of each chapter allow you to check your comprehension of basic chapter content.

Challenge Your Thinking questions require you to reason, research, and explore concepts in further detail.

Applying AutoCAD Skills

Problems at the end of each chapter help you gauge your understanding of the skills taught in the chapter.

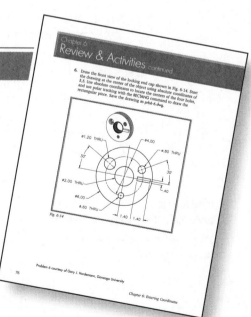

Problems include **real-world objects and concepts** to help you practice and apply chapter skills and concepts in a work-related context.

Using Problem-Solving Skills

Problem-solving activities use representative tasks that you might encounter in industry. These problems require you to synthesize the AutoCAD skills presented throughout the text to arrive at practical solutions.

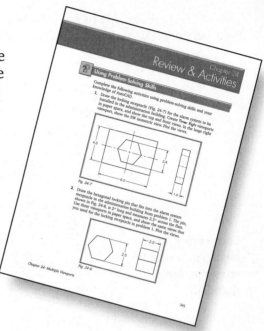

Part Projects

A real-world project is located at the end of each of the nine parts in this book. These projects tie together key concepts and skills and help you apply the skills and techniques you have learned in a realistic manner.

Careers Using AutoCAD

Careers Using AutoCAD pages help you explore the different types of careers open to people with AutoCAD knowledge and skills.

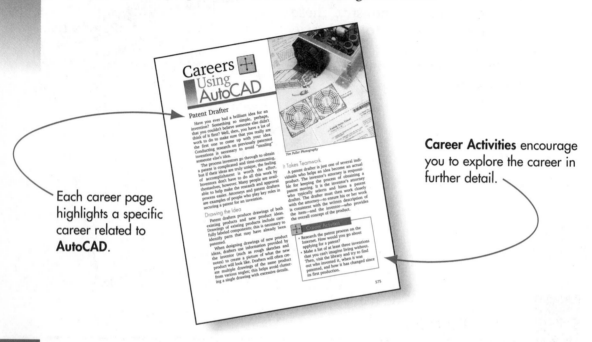

Each career page highlights a specific career related to **AutoCAD**.

Career Activities encourage you to explore the career in further detail.

Advanced Projects

Each advanced project presents a problem that you might encounter on the job. Little instruction is given—you are required to plan a solution and then carry it out. This is an opportunity for you to show initiative and creativity on a long-term or group project.

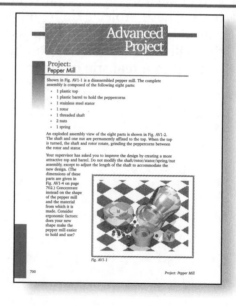

Additional Problems

Additional Problems provide further opportunities to put your
AutoCAD skills to work.

Problems provide
**specific instructions
and guidelines** to
help you get started.

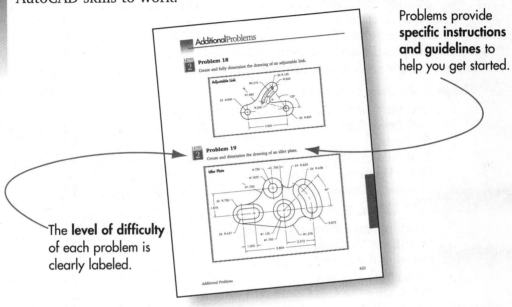

The **level of difficulty**
of each problem is
clearly labeled.

Instructor Resource CD

Includes:

- Instructional Plans
- PowerPoint® Presentations
- *ExamView® Pro Test Generator*
- Student Practice Files
- Answer Keys
- Additional Instructor Resources

Table of Contents

Part 2 – Drawing Aids and Controls

Table of Contents — continued

Part 3 – Drawing and Editing

Part 4 – Text and Tables

Part 5 – Preparing and Printing a Drawing

Table of Contents — *continued*

Part 6 – Dimensioning and Tolerancing

Part 7 – Groups and Details

Table of Contents — *continued*

Part 8 – 3D Drawing and Modeling

Part 9 – Solid Modeling

Careers Using AutoCAD

Brand Disclaimer

Publisher does not necessarily recommend or endorse any particular company or brand name product that may be discussed or pictured in this text. Brand name products are used because they are readily available, likely to be known to the reader, and their use may aid in the understanding of the text. Publisher recognizes that other brand name or generic products may be substituted and work as well or better than those featured in the text.

Chapter 1 — Exploring AutoCAD

Objectives:

- **Start AutoCAD**
- **Preview drawing files**
- **Open drawing files**
- **View details on an AutoCAD drawing**
- **Exit AutoCAD**

Key Terms

buttons	drawing area	resolution
Command line	dynamic input	thumbnail
dialog box	icons	toolbar
double-clicking		

AutoCAD provides countless methods and tools for producing, viewing, and editing two-dimensional drawings and three-dimensional models. The software permits designers, drafters, engineers, and others to create, revise, model, and document industrial parts and assemblies for prototyping, mold-making, and manufacturing. Around the world, organizations also use AutoCAD for the design of maps, buildings, bridges, factories, and about every product imaginable, ranging from car parts and stereo equipment to snow skis and cellular phones.

Starting AutoCAD

The first step in using AutoCAD is to learn how to start the software and open a drawing file.

1. Start AutoCAD by double-clicking the **AutoCAD 2006** icon found on the Windows desktop.

HINT:

Double-clicking means to position the pointer on the item and press the left button on the pointing device twice very quickly. If the AutoCAD 2006 icon is not present, pick the Windows Start button, move the pointer to Programs (or to All Programs), move it to Autodesk, move it to AutoCAD 2006, and pick the AutoCAD 2006 option.

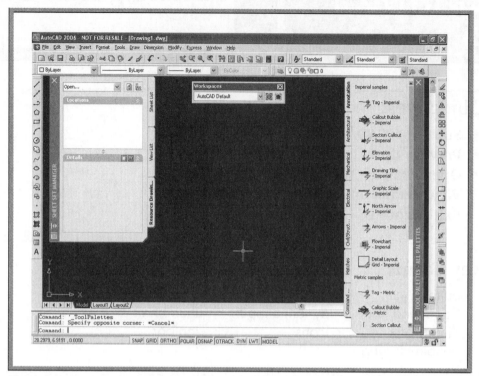

Fig. 1-1

The AutoCAD window appears as shown in Fig. 1-1.

NOTE:

If this window does not appear and you instead see a gray screen with a dialog box named Startup in the foreground, click the Cancel button in the dialog box. The AutoCAD window will then appear. You will learn more about this dialog box later in this book.

The exact configuration may vary depending on how it was left by the last person to use the software. The display hardware can cause the appearance of screen elements to vary. At lower resolutions, the buttons at the top may extend across the entire window. The *resolution* is the number of pixels per inch on a display system; this determines the amount of detail that you can see on the screen.

Notice the Sheet Set Manager, Workspaces, and Tool Palettes windows in the AutoCAD window. These windows appear by default in AutoCAD 2006. However, many users choose to close them when they are not in use to gain valuable drawing space and to reduce clutter. In this textbook, the windows will not be shown unless they are being used.

Buttons in AutoCAD are the small areas that contain pictures, called *icons*, normally found along the top and sides of the AutoCAD drawing area. Pressing a button enters its associated command or function. The *drawing area* is the main portion of the window where the drawing appears.

Previewing Drawing Files

1. From AutoCAD's **File** pull-down menu, select the **Open...** item.

This displays the Select File dialog box as shown in Fig. 1-2. A *dialog box* provides information and permits you to make selections and enter information. In this case, the dialog box enables you to open a drawing file, as well as select other options. The folders that you see in Fig. 1-2 are those stored in the main AutoCAD 2006 folder.

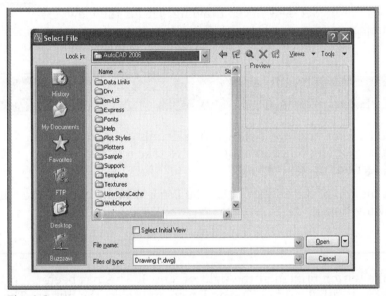

Fig. 1-2

2. Find and double-click AutoCAD's folder named **Sample**.

This displays all of the files and folders contained in the Sample folder.

3. Rest the pointer on any of the files in the list. If nothing displays in the Preview area of the dialog box, single-click the file. (Your computer may be configured one way or the other.)

Notice that a small picture of the drawing, or *thumbnail*, appears at the right in the area labeled Preview. This gives you a quick view of the graphic content of the selected file.

4. Preview each of the remaining sample drawing files by resting the pointer or single-clicking on each of them.

5. Double-click the file named **Welding Fixture Model.dwg** or single-click it and pick the **Open** button. (The DWG extension may not be shown as part of the file name. This depends on the configuration of your computer.)

A three-dimensional (3D) model of a welding fixture assembly appears. It may take a few seconds to load. Notice the tabs named Model and Layout1 at the bottom of the drawing area.

INFOLINK

See Chapter 25 for more about model space, paper space, and layouts.

6. Pick the **Layout1** tab.

This displays the drawing as it would appear on a printed or plotted sheet.

7. Pick the **Model** tab.

This displays the model view of the drawing. This is the view for most drafting and design work in AutoCAD.

Viewing Details

AutoCAD offers powerful tools for viewing details on a drawing. For example, the zooming function allows you to take a closer look at very small sections of the drawing.

1. Near the top of the window, pick the **Zoom Window** button. (See the following hint.)

Standard

 HINT:

Notice the button printed at the right of Step 1. This has been provided as a visual aid to help you locate and select the button. The name of the toolbar in which the button is located appears above the button for reference. A *toolbar* is the strip that holds the buttons.

Buttons appear in the margin throughout the chapter in which they are introduced and in the subsequent three or four chapters. They also appear occasionally in later chapters, when the button has not been used recently, to refresh your memory.

In the lower left area of the window, you will see AutoCAD's *Command line*, which allows you to key in commands and information. AutoCAD also displays important information in this area. Notice that it is now asking you to specify the first corner of the window.

Also notice the specification box that attaches to the cursor. This box provides *dynamic input*—a new feature for AutoCAD 2006. It allows you to launch commands, enter values, and read prompts right at the pointer, instead of at the Command line.

INFOLINK

You will learn more about the Command line and dynamic input in Chapter 3.

Imagine a small window surrounding the upper area of the drawing. This will become the zoom window.

2. Using the pointing device, pick a point (using the left button) to specify one of the four corners of the imaginary window. See Fig. 1-3 for the location of the window.

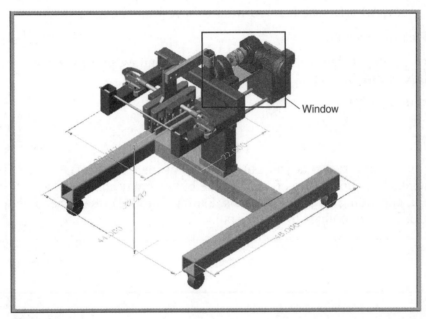

Window

Fig. 1-3

3. Move the pointing device to the opposite corner to create the window and pick another point.

INFOLINK

You will learn more about zooming in Chapter 12.

AutoCAD zooms in to fill the drawing area with the area represented by the zoom window.

4. Move the pointing device slowly around the drawing. Notice that each part of the assembly becomes highlighted as the cursor touches it.

5. Pick the **Zoom Previous** button, which is located to the right of the Zoom Window button.

This zooms the window to the previous zoom magnification.

Exiting AutoCAD

1. Type the **CLOSE** command using upper- or lowercase letters and press **ENTER.**

Notice that the letters you type appear near the cursor as well as at the Command line. This is part of the dynamic input feature. When you press ENTER, AutoCAD asks whether you want to save your changes.

2. Pick the **No** button.

3. In the upper right corner of the AutoCAD window, pick the small box containing the **x.**

This exits AutoCAD.

NOTE:

It is very important that you exit AutoCAD at the end of each chapter. The appearance of some of the buttons may change as you work, depending on your selections. Exiting AutoCAD resets the buttons in the toolbars to their default settings.

Chapter 1
Review & Activities

Review Questions

1. Name five examples of products or industries that use AutoCAD.

2. How do you start AutoCAD?

3. Explain the process of previewing AutoCAD drawing files prior to opening one.

4. Explain how it is possible to view small details on a complex AutoCAD drawing.

5. How do you exit AutoCAD?

Challenge Your Thinking

1. Explore the concept behind the screen resolution. Why is this resolution important, and how do you change it?

2. In AutoCAD's Select File dialog box, explore a second method of previewing thumbnails of AutoCAD drawing files.

Applying AutoCAD Skills

Work the following problems to practice the commands and skills you learned in this chapter.

1. Start AutoCAD.

2. Preview the drawing files located in AutoCAD's Support folder.

3. Open the drawing file named Hummer Elevation.dwg located in AutoCAD's Sample folder. Pick each of the tabs located at the bottom of the drawing area.

4. Open the drawing file named Welding Fixture-1.dwg in the Sample folder. Pick the Zoom Window button on the docked Standard toolbar and make a window to take a closer look at the title block, parts list, and other details in the drawing.

5. Exit AutoCAD. Do not save your changes.

Using Problem-Solving Skills

Complete the following activities using problem-solving skills and your knowledge of AutoCAD.

1. Start AutoCAD. Open Welding Fixture-1.dwg, which is located in AutoCAD's Sample folder. Notice the tabs at the bottom of the screen. Pick each tab, noticing the name of the tab and the contents that are displayed. Zoom in on different areas of the drawing. Why did the drafter create so many layout views? Explain the advantage of AutoCAD's multiple layout capabilities. Exit AutoCAD without saving.

2. Start AutoCAD. Using AutoCAD's preview feature, find and open each of the following files in AutoCAD's Sample folder:

   Lineweights.dwg

   Plot screening and Fill Patterns.dwg

   Tablet.dwg

   TrueType.dwg

   Explore the drawings, including the model and layout tabs used. What is the purpose of these files? Explain how knowing that these files exist can help you as you use AutoCAD in the future. Exit AutoCAD without saving.

Careers
Using
AutoCAD

The Image Bank/John Humble

Civil Drafter

Before a construction project begins, a great deal of information must first be collected. Distances are measured, the earth's topography is observed, and directions are noted. This information is used when planning the construction of bridges, highways, pipelines, and water and sewage systems.

In addition to the architects and engineers (who conceptualize these types of projects) and the surveyors (who collect preliminary data), every project needs another type of workers. These individuals are responsible for creating topographical relief maps and other detailed drawings used for construction projects: the civil drafters.

Using Technology

As in many areas of drafting today, civil drafters often use CAD (computer-aided drafting) programs to help create drawings. While some freehand skills are also helpful, the ability to work with computers is important as design technology becomes commonplace. Aspiring civil drafters will benefit from learning graphic design programs such as AutoCAD or Pro/Engineer. Classes providing basic instruction for computer-aided drafting programs are offered at many high schools, junior colleges, and vocational schools.

Analyzing Data

Civil drafters must be able to process the many different types of information they receive. They are expected to decipher notes and sketches from surveyors, engineers, architects, and builders. The drafters use this information collectively to help them prepare drawings for builders and manufacturers. The ability to analyze data from a variety of different sources is a necessary skill for anyone involved in civil drafting.

Work Environment

Civil drafting can involve extensive outdoor work. Drafters are often asked to travel to potential sites for on-site analysis and dimensions. They usually work closely with civil engineers and construction personnel to prepare ongoing documentation as modifications are made to original designs.

Career Activities

- Think of several situations in which you use analytical skills. What kinds of analytical skills do you use on a regular basis?
- In what types of "real life" situations do you use these skills?

9

Chapter 2 — User Interface

Objectives:

- Identify the parts that make up the AutoCAD user interface and describe the function of each part
- Navigate AutoCAD's pull-down menus
- Display and reorganize docked and floating toolbars
- Describe the function of the **Command Line** window, status bar, and scrollbars
- Display and make transparent the **Tool Palettes** window

Key Terms

cascading menu	floating toolbar
crosshairs	pull-down menus
docked toolbar	status bar

AutoCAD's user interface includes many parts that are important to the efficient operation of the software. For example, the proper use of docked and floating toolbars improves the speed with which you create, edit, and dimension drawings. Productive users of AutoCAD use a combination of toolbars, pull-down menus, dialog boxes, and the keyboard to enter commands and interact with AutoCAD. Each of these elements is a part of the AutoCAD user interface.

Overview of the AutoCAD Window

Figure 2-1 describes the parts that make up the AutoCAD window.

1. Start AutoCAD.

The AutoCAD window appears.

NOTE:

If this window does not appear and you instead see a gray screen with a dialog box named Startup in the foreground, click the Cancel button in the dialog box. The AutoCAD window will then appear.

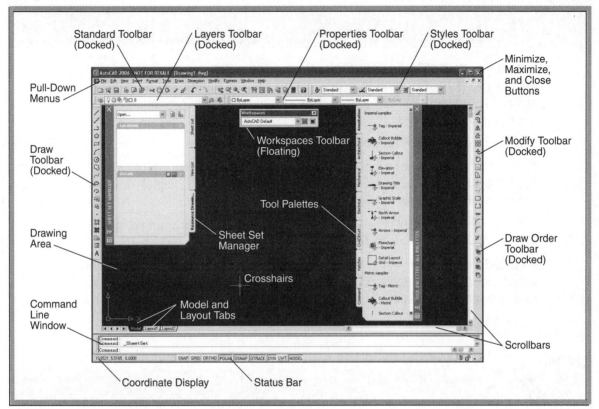

Fig. 2-1

Note the locations of the parts labeled pull-down menus, docked and floating toolbars, Command Line window, status bar, and scrollbars. Most of the rest of the AutoCAD window is the drawing area.

As you move the pointing device in the drawing area, notice that the pointer is a special cursor called the *crosshairs*. You will use the crosshairs to pick points and select objects in the drawing area. When you move outside the drawing area, the normal Windows pointer appears. You will learn more about the features in the drawing area in future chapters.

NOTE:

As explained in Chapter 1, the exact configuration of the AutoCAD window may vary depending on the screen resolution and on how it was left by the last person to use the software. If all of the toolbars are not present, that's okay. You will learn how to display and position them later in this chapter, in the section titled "Docked Toolbars."

Pull-Down Menus

Notice the words File, Edit, View, Insert, etc., that appear in the upper left area of the AutoCAD window. These words are the names of *pull-down menus*. These menus allow you to perform many drawing, editing, and file manipulation tasks.

1. Select the **View** pull-down menu.

Notice that several menu items contain a small arrow pointing to the right. This means that when you point to them, they will display another menu called a *cascading menu*.

2. Move the pointer to any one of the menu items containing a small arrow and allow it to rest for a moment.

A cascading menu appears, displaying another set of menu options. Menu items that contain ellipsis points (...), as in the Toolbars... item, display a dialog box when picked.

3. Press the **ESC** key to remove each open menu.

Toolbars

Toolbars contain collections of buttons that can save you time when working with AutoCAD. Under the menu bar (File, Edit, View, etc.) you should see the Standard toolbar docked horizontally across the top of the screen. A *docked toolbar* is one that appears to be a part of the top, bottom, left, or right border of the drawing area. Below the Standard toolbar, you should see the Layers toolbar, also docked horizontally. Located at the right of the Standard and Layers toolbars are the Styles and Properties toolbars. When a toolbar is docked, the name of the toolbar does not appear.

Floating Toolbars

Let's focus on displaying and positioning toolbars. Toolbars that are not docked are called *floating toolbars*. The Workspaces toolbar in the previous illustration is an example.

1. If any floating toolbars are present, close them by picking the **x** (close button) in the upper right corner of the toolbar.

2. Rest the pointer on any one of the docked toolbars and right-click.

This displays the list of toolbars that AutoCAD makes available to you, as shown in Fig. 2-2. By right-clicking on any docked or floating toolbar on the screen, you can display this list of toolbars. Those that are checked are the ones that are present on the screen. To display a toolbar from the list, single-click its name.

3. Single-click the word **Shade** to display the Shade toolbar.

4. Close the Shade toolbar.

It is easy to move toolbars to a new location. As you work with AutoCAD, you should move floating toolbars to a location that is convenient, yet does not interfere with the drawing.

5. Open the **Solids** toolbar.

6. Move the Solids toolbar by clicking the bar located at the top of the toolbar and dragging it. (This bar contains the word Solids.)

NOTE:

AutoCAD allows users to lock toolbars in place. If the Solids toolbar does not move when you try to drag it, right-click on any open toolbar, go to the bottom of the pop-up menu, and rest the cursor on Lock Location. From the secondary menu, pick the check mark next to Floating Toolbars to unlock the toolbar's position.

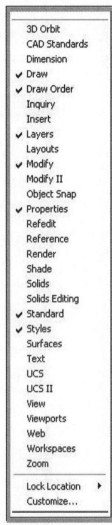

Fig. 2-2

Changing a Toolbar's Shape

AutoCAD makes it easy to change the shape of toolbars. You might want to change the shape of a toolbar, for example, to make it fit more readily in the area of the drawing in which you are working.

1. Move the pointer to the bottom edge of the Solids toolbar, slowly and carefully positioning it until a double arrow appears.

2. When the double arrow is present, click and drag downward until the toolbar changes to a vertical shape and then release the button.

3. Move the pointer to the right edge of the Solids toolbar, positioning it until a double arrow appears.

4. When the double arrow is present, click and drag to the right until the toolbar changes back to a horizontal shape.

5. Close the Solids toolbar.

Docked Toolbars

1. If the Standard, Styles, Layers, and Properties toolbars are not present on the screen, skip to Step 5.

2. Move the pointer to the Standard toolbar and click and drag the outer left edge (border) of the toolbar, moving the toolbar down into the drawing area.

3. Repeat Step 2 with the remaining three toolbars that are docked along the top of the AutoCAD window.

4. Close all four of these floating toolbars.

5. Right-click in any of the three remaining toolbars to display the list of toolbars and select **Layers** to display the Layers toolbar.

6. Repeat Step 5 to display the Properties, Standard, and Styles toolbars.

7. Click and drag the Standard toolbar and carefully dock it under the menu bar (File, Edit, View, etc.).

The toolbar locks into place.

8. Click and drag the other three toolbars and carefully dock them into place. (Refer to page 11 if you are unsure where they belong.)

9. If the Draw, Modify, and Draw Order toolbars are not present and locked into place, position them now as shown on page 11.

You should now see seven docked toolbars on the screen.

 NOTE:

In future chapters, *Applying AutoCAD* assumes that these seven toolbars are present on the screen in a configuration similar to the one on page 11. This book has been written to work with AutoCAD straight from the installation. It assumes that the screen color, crosshairs, and other elements of the AutoCAD window have not been customized.

10. If you did not complete Steps 2 through 4, do so now. Also, complete Steps 5 through 8.

Tool Palettes Window

The Tool Palettes window provides access to drawing content. It may or may not be present in the AutoCAD window, depending on how the last user of the AutoCAD software left it. Figure 2-3 shows the Tool Palettes window. It is a long, vertically oriented window with tabs along its left edge.

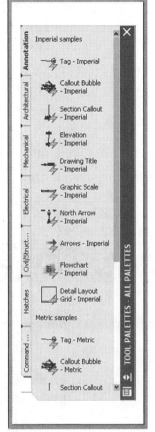

Fig. 2-3

1. If it is present, close it by picking the **x** located in the upper right corner of the Tool Palettes window.

2. From the **Tools** pull-down menu, find and select **Tool Palettes Window**.

This causes the window to display. Pressing the CTRL and 3 keys simultaneously also causes it to display, or to disappear if it is present.

Auto-Hide

1. Rest the pointer inside the Tool Palettes window (but not over a button or text) and right-click.

This displays a menu of options.

2. Select **Auto-hide**.

3. Move the pointer away from the Tool Palettes window.

This causes the window to disappear, except for its title bar.

4. Move the pointer to the title bar.

The window reappears.

5. Move the pointer away from the window and its title bar.

The window disappears. This feature helps to preserve screen real estate, which is especially helpful when the AutoCAD window becomes busy.

6. Rest the pointer on the title bar of the Tool Palettes window, right-click, and uncheck **Auto-hide**.

Transparency

Transparency is another feature that helps with screen real estate.

1. Rest the pointer inside the Tool Palettes window, right-click, and select the **Transparency...** item.

This displays a small Transparency dialog box.

2. Click and drag the slider in the dialog box to the far right and pick the **OK** button.

This causes the Tool Palettes window to become semi-transparent.

3. From the File pull-down menu, select Open... and open the file named **Hotel Motel.dwg**, which can be found in AutoCAD's Sample folder.

You should be able to see the drawing through the Tool Palettes window.

4. Right-click the **Tool Palettes** window, select **Transparency...**, move the slider to the middle position, and pick **OK**.

The Tool Palettes window becomes less transparent.

5. Right-click the **Tool Palettes** window, select **Transparency...**, move the slider to the far left, and pick **OK**.

The Tool Palettes window is no longer transparent.

Other Features

The AutoCAD window offers other parts that you will use many times throughout this book. The Command Line window and the status bar are key parts of AutoCAD, so pay close attention to them as you work.

Command Line Window

Notice the Command Line window near the bottom of the screen. As you learned in Chapter 1, you can enter AutoCAD commands at the Command line. However, in many cases, it is faster and easier to use dynamic input.

It is possible to click and drag the Command Line window into the drawing area, making it a floating window, similar to a floating toolbar. It is also possible to right-click the Command Line window, select Transparency, and make it semi-transparent, similar to the Tool Palettes window.

Status Bar

The *status bar* at the bottom of the screen tells you the coordinates of the screen crosshairs and the status of various AutoCAD modes and settings. The more you work with AutoCAD, the more you will appreciate the availability of the buttons in the status bar. They will save time and effort as you create drawings.

Review Questions

1. What is a cascading menu?

2. Describe how to display a toolbar that is not currently displayed on the screen.

3. How do you move a toolbar? *use the = at the top of the toolbar*

4. How do you make a floating toolbar disappear? *click on the X or Drag it to the side*

5. Explain how to make the Tool Palettes window semi-transparent.

6. Write the names of items A through Q in Fig. 2-4.

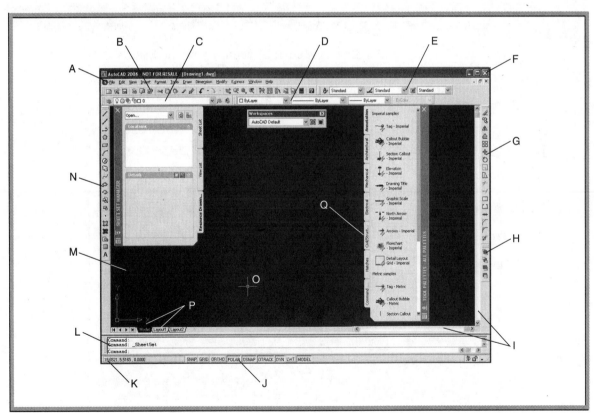

Fig. 2-4

Challenge Your Thinking

1. AutoCAD provides ways to change the appearance and colors of the AutoCAD window. For example, if you prefer to work on a gray background, you can change the drawing area from black to gray. Find out how to customize AutoCAD's appearance and colors. Then write a short paragraph explaining how to customize the graphics screen and why it might sometimes be necessary to do so.

2. The AutoCAD window includes two sets of Windows-standard Maximize, Minimize, and Close buttons. Explain why there are two sets. What is the function of each?

Applying AutoCAD Skills

Work the following problems to practice the commands and skills you learned in this chapter.

1–3. Rearrange the toolbars so that they are similar to the ones in Figs. 2-5 through 2-7.

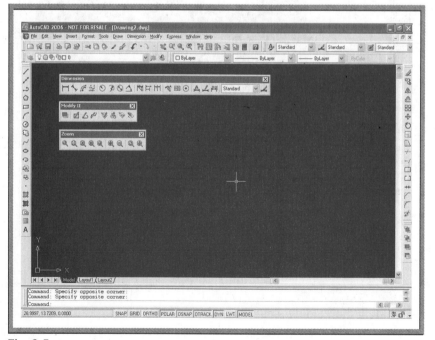

Fig. 2-5

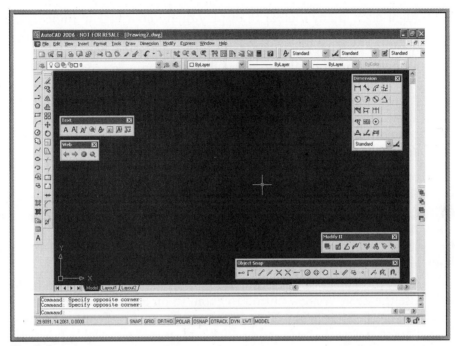

Fig. 2-6

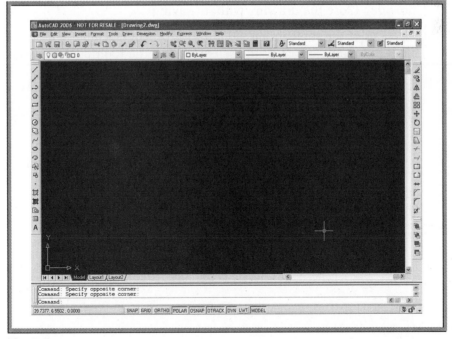

Fig. 2-7

Chapter 2
Review & Activities continued

4. Right-click in any toolbar to display the toolbar list, and uncheck a toolbar. Repeat until all toolbars are unchecked. Notice that the drawing area has increased, but you have lost the convenience of toolbar command selection. Use dynamic input to type –toolbar, then Standard, then Show. The Standard toolbar reappears on the screen. Right-click in the Standard toolbar and reopen the other toolbars that appear by default in AutoCAD. If some of the toolbars are out of position or docked on top of others, move the toolbars so that they are docked in their default position.

 ## Using Problem-Solving Skills

Complete the following activities using problem-solving skills and your knowledge of AutoCAD.

1. Open the Taisei Detail Plan.dwg file found in AutoCAD's Sample folder. Use the Zoom Window button to zoom in on a small area near the center of the drawing. Then use the horizontal and vertical scrollbars to navigate the drawing. Pick and drag the movable box within the scrollbars to move quickly around the drawing. Experiment to find other ways to navigate. What other methods does AutoCAD provide? Close the drawing without saving.

2. Open all of the toolbars provided in AutoCAD. Dock them along the sides, top, and bottom of the screen. Using the tooltips, investigate the buttons that make up the toolbars. (Tooltips are the words that appear when the pointer rests on a button.) What two problems might you encounter if you left all the toolbars open on the screen? When you have finished, close all the toolbars except the seven that are docked by default in AutoCAD.

Chapter 3 Command Entry

Objectives:

- Create, save, and open an AutoCAD drawing file
- Enter commands using the keyboard and toolbars
- Apply shortcut methods to enter commands
- Reenter commands

Key Terms

command alias   shortcut menu
context-sensitive   tooltips
rubber-band effect

AutoCAD offers several methods of entering and reentering commands. When entering them frequently, as you will practice in this chapter, you will want to use methods that minimize the time it takes. Command aliases and other shortcuts permit you to enter commands with as little effort as possible. If you take advantage of these time-saving features of AutoCAD, you will require less time to complete a drawing. Consequently, you become more attractive to prospective companies looking for highly productive individuals.

In the steps that follow, you will practice entering commands by doing some basic operations such as creating a file and drawing lines. You will apply the concepts you learn in this chapter nearly every time you use AutoCAD.

Creating and Saving a Drawing File

1. Start AutoCAD.

The AutoCAD window appears.

NOTE: ────────────────────────────

If the AutoCAD window does not appear and you instead see the Startup dialog box, click the Cancel button in the dialog box.

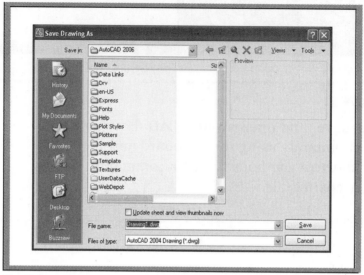

Fig. 3-1

It is important that you start (or restart) AutoCAD at the beginning of each chapter, unless instructed otherwise. This ensures that the buttons in the toolbars are set to their default settings.

2. Pick either **Save** or **Save As...** from the **File** pull-down menu.

The Save Drawing As dialog box appears as shown in Fig. 3-1. If you use other Windows software, the dialog box should look familiar to you.

Standard

3. Pick the **Create New Folder** button (located in the upper right area of the dialog box).

This creates a new folder named New Folder.

4. Right-click the new folder, select **Rename** from the menu, type your first and last name, and press **ENTER**.

5. Double-click this new folder to make it current.

6. In the File name text box, highlight the name of the drawing file (e.g., **Drawing1.dwg**) by double-clicking it, and enter **stuff**.

Standard

7. Pick the **Save** button in the dialog box to create a file named stuff.dwg.

 NOTE:

If you do not see the file extension (dwg) after the names of drawing files, this feature may be turned off on your computer. Your work with AutoCAD will be much easier if you turn on the extension display.

Notice that the drawing file name now appears in the title bar of the AutoCAD window.

8. From the **File** pull-down menu, pick **Close** to close the stuff.dwg file.

The stuff.dwg file disappears, and the drawing area becomes gray because no drawing file is open. Notice also that many of the pull-down menus and all of the toolbars disappear except for an abbreviated version of the Standard toolbar.

Opening a Drawing File

If you are already working in AutoCAD, you can open a file easily using the Open button.

1. Pick the **Open** button on the docked Standard toolbar, or select **Open...** from the **File** pull-down menu.

Standard

The Select File dialog box appears.

AutoCAD defaults to the folder you named after yourself.

2. Double-click **stuff.dwg** in the selection window.

You have reopened the stuff.dwg file. Currently, the drawing area is blank.

Entering Commands

For the purpose of learning to enter commands, we will use the LINE command. You will learn more about using various AutoCAD commands in the chapters that follow.

Using Dynamic Input

The newest way to enter commands is with dynamic input.

1. With the pointer anywhere in the drawing area, type **LINE** using upper- or lowercase letters and press the **ENTER** key.

Notice that the command appears near the crosshairs as you type.

2. Press **ENTER**.

AutoCAD asks you to specify the first point. AutoCAD also shows the coordinates (current position) of the crosshairs.

INFOLINK

You will learn more about coordinates in Chapter 6.

3. Pick a point anywhere in the drawing area.

AutoCAD's dynamic input prompt now requests the next point. It also displays the distance and angle of the crosshairs from the last point, as shown in Fig. 3-2.

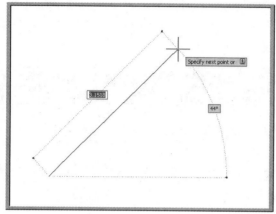

Fig. 3-2

4. Move the crosshairs on the screen and notice how the distance and angle values change dynamically.

5. Instead of picking the next point, type **10** and press the **TAB** key.

Notice that the value you typed appears in the distance box of the dynamic input. Also, a padlock appears to indicate that the value of 10 is locked into place.

6. Type **45** and press **ENTER**.

AutoCAD creates a line segment that is 10 units long at an angle of 45°.

7. Produce the second and third line segments to create a triangle and then press **ENTER** to terminate the LINE command.

Dynamic input can be toggled on and off by clicking the DYN button on the status bar. You will use dynamic input more in the chapters that follow.

NOTE:

Depending on the current AutoCAD settings, you may notice that temporary dotted lines and yellow symbols (or another color) appear on the screen as you complete the triangle in Step 7. These are produced by the AutoCAD features known as object snap and AutoTrack™. You will use these features extensively in the chapters that follow. For now, you can ignore them.

Using the Command Line

The original way to enter a command in AutoCAD was to enter the command name on the Command line. This method is still available, although other methods have been added in recent years to give you a choice.

1. Single-click the **DYN** button on the status bar to toggle dynamic input off.

2. Type **LINE** and press the **ENTER** key.

In the Command window, notice that AutoCAD asks you to specify the first point.

3. Pick a point anywhere in the drawing area.

AutoCAD requests the next point.

4. Pick a second point anywhere on the screen.

AutoCAD requests another point.

5. Move the crosshairs away from the last point and notice the stretching effect of the line segment.

This is known as the *rubber-band effect*.

6. Produce second and third line segments to create a triangle and then press **ENTER** to terminate the LINE command.

7. Pick **DYN** on the status bar to reactivate dynamic input.

NOTE:

You can use the Command line method of entry even when dynamic input is active. Simply single-click at the Command line before typing the command.

Picking Buttons on a Toolbar

It can be faster to pick a button than to enter a command at the keyboard or from a pull-down menu.

Draw

1. Rest the pointer on the **Line** button located at the top of the docked Draw toolbar, but do not pick it.

As discussed briefly in Chapter 1, one or more words appear after about one second. These words, called *tooltips*, help you understand the purpose of the button. Notice also the status bar at the bottom of the AutoCAD window. When a tooltip appears, the status bar changes to provide more detailed information about the button.

2. Slowly position the pointer on top of other buttons and read the information that AutoCAD displays.

3. Pick the **Line** button and move the pointer into the drawing area.

Draw

You have just entered the LINE command. You can see that AutoCAD is again asking where you want the line to begin.

4. Pick a point anywhere in the drawing area; then pick a second point to form a line.

5. Press the **ENTER** key to terminate the LINE command.

6. Pick another button (any one of them) from the Draw toolbar.

7. Press the **ESC** key to cancel the last entry at the Command line.

HINT:

As you work with AutoCAD, you will be entering commands by picking buttons, selecting them from menus, or entering them at the keyboard. Occasionally you might accidentally select the wrong command or make a keyboarding error. It's easy to correct such mistakes.

If you catch a keyboarding error *before* you press ENTER...	use the backspace key to delete the incorrect character(s). Then continue keyboarding.
If you select the wrong button or pull-down menu item...	press the ESC key to clear the Command line.
If you key in the wrong command...	press the ESC key to clear the Command line.
If you accidentally pick a point or object on the screen...	press the ESC key.

Command Aliases

AutoCAD permits you to issue commands by entering the first few characters of the command. Command abbreviations such as these are called *command aliases*.

1. Type **L** in upper- or lowercase and press **ENTER**.

This enters the LINE command. You can also enter other commands using aliases. Some of the more common command aliases are listed in Table 3-1.

Command	Entry	Command	Entry
ARC	A	LINE	L
CIRCLE	C	MOVE	M
COPY	CO	PAN	P
DONUT	DO	POLYGON	POL
ELLIPSE	EL	SCALE	SC
ERASE	E	ZOOM	Z

Table 3-1

2. Press the **ESC** key to cancel the LINE command.

3. Enter each of the commands listed in Table 3-1 using the alias method. For now, press **ESC** to cancel each command.

You will learn more about these commands in future chapters.

4. Reenter the **LINE** command using the command alias method and create another line.

5. Press **ENTER** to terminate the LINE command.

Even when you do not know a command alias, AutoCAD can help you enter the command without typing the entire command. We will use the POLYGON command as an example.

6. Type **PO** in upper- or lowercase letters, but do not press ENTER.

7. Press the **TAB** key repeatedly until the POLYGON command is displayed on the screen.

AutoCAD cycles through an alphabetical list of all commands that begin with the letters PO.

8. Press **ENTER** to enter the POLYGON command.

9. Press **ESC** to terminate the command.

 NOTE:

Command aliases are defined in the AutoCAD file acad.pgp. You can view this file or create additional command aliases by modifying it using a text editor. The file is located in the following folder:

C:\Documents and Settings\username\Application Data\Autodesk\AutoCAD 2005\R16.2\enu\Support

Substitute your user name for username.

Reentering the Last Command

After you have entered a command, you have several choices for repeating it.

1. Enter the **LINE** command and create a line segment anywhere in the drawing area. Press **ENTER** to end the command.

2. Press the spacebar or **ENTER.**

This enters the last-entered command, in this case the LINE command. You will find that this saves time over alternative methods of entering the command.

3. Create a line segment and then press **ENTER** to end the LINE command.

4. Press the right button on the pointing device.

Right-clicking in AutoCAD displays a context-sensitive *shortcut menu.* The top item on the menu, when selected, repeats the last command you entered.

5. Pick **Repeat LINE.**

6. Draw two of the polygons shown in Fig. 3-3.

The Repeat option is *context-sensitive.* This means that the software remembers the last command you used.

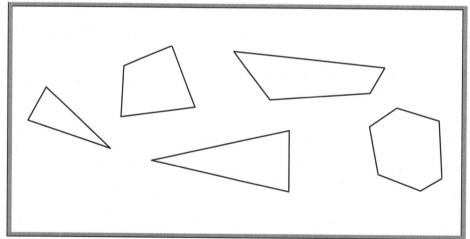

Fig. 3-3

HINT:

After you have entered at least two line segments, the prompt changes to include the Close option. This option allows you to accurately close a polygon automatically and terminate the LINE command. To choose the Close option, type the letter C and press ENTER.

7. Draw the remaining polygons shown in Fig. 3-3 using the **LINE** command. Be sure to use one of the shortcut methods of reentering the last command.

8. Enter the **LINE** command and draw two connecting line segments, but do not terminate the command.

9. In reply to Specify next point or [Close/Undo], enter **U** for Undo.

AutoCAD backs up one segment, undoing it so that you can recreate it.

10. Practice the **Undo** option with additional polygons.

11. Pick the **Save** button from the docked Standard toolbar to save your work.

Standard

As you work with the different methods of entering commands, you may prefer one method over another. Picking the Line button from the toolbar may or may not be faster than entering L at the keyboard. Keep in mind that there is no right or wrong method of entering commands. Experienced users of AutoCAD use a combination of methods.

12. Exit AutoCAD.

Chapter 3
Review & Activities

Review Questions

1. Describe the purpose of the Open button located on the docked Standard toolbar and the Open... item on the File pull-down menu.

2. When a file has not yet been saved for the first time, what appears when you pick either Save or Save As... from the File pull-down menu?

3. What appears when you rest the pointer on a button contained in a toolbar?

4. What is the fastest and simplest method of reentering the previously entered command?

5. How can you enter commands such as LINE, CIRCLE, and ERASE quickly at the keyboard?

6. What is a fast method of closing a polygon when using the LINE command?

7. Explain the use of the LINE Undo option.

Challenge Your Thinking

1. Discuss the advantages and disadvantages of having more than one way to enter a command. Which method of entering a command is most efficient? Collect opinions from other AutoCAD users and consider ergonomic factors.

2. Describe a way to open a drawing file in AutoCAD when AutoCAD is not currently running on the computer.

Applying AutoCAD Skills

Work the following problems to practice the commands and skills you learned in this chapter.

1. Create a new drawing file named prb3-1.dwg and store it in the folder named after yourself. Create the bookcase shown in Fig. 3-4. For practice, enter the LINE command in various ways: using dynamic input, using the Command line, or picking a button. Use various shortcut methods. Save the drawing.

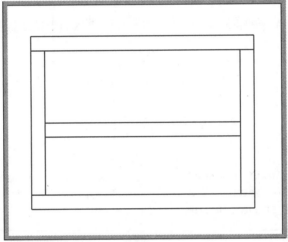

Fig. 3-4

2. Create a new drawing file named prb3-2.dwg and store it in the folder named after yourself. Draw the concrete block shown in Fig. 3-5 using the most efficient method(s) of entering the LINE command.

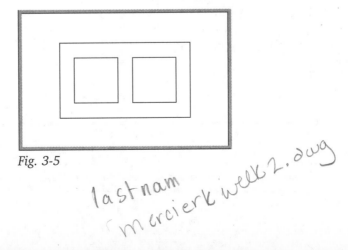

Fig. 3-5

lastnam mercierk week2.dwg

3. Create a new drawing file named prb3-3.dwg and store it in the folder named after yourself. Draw the simple house elevation drawing shown in Fig. 3-6 by picking the Line button from the docked Draw toolbar. When reentering the command, use a shortcut method.

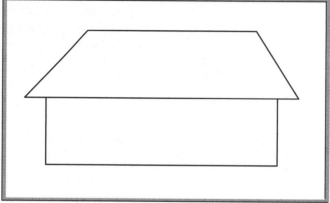

Fig. 3-6

4. Create a new drawing file named prb3-4.dwg and store it in the folder named after yourself. Draw the front and side views of the sawhorse shown in Fig. 3-7 by entering the LINE command's alias at the keyboard. Be sure to use shortcut methods when reentering the command.

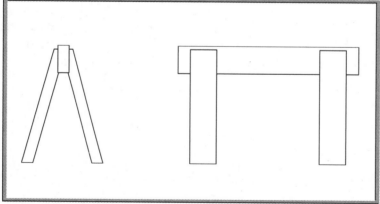

Fig. 3-7

Problem 4 courtesy of Joseph K. Yabu, Ph.D., San Jose State University

Chapter 3: Command Entry

Using Problem-Solving Skills

Complete the following problems using problem-solving skills and your knowledge of AutoCAD.

1. A boat trailer manufacturer needs a new design for the rollers that facilitate raising a boat onto the trailer. The engineering division of your company, which designs the rollers, has proposed the design shown in Fig. 3-8. Create a drawing to be submitted to the trailer manufacturer for approval. Notice that the drawing is dimensioned. Experiment with the buttons on the status bar and see if you can figure out a way to draw the roller approximately to size. Do not dimension the drawing. Save the drawing as roller1.dwg.

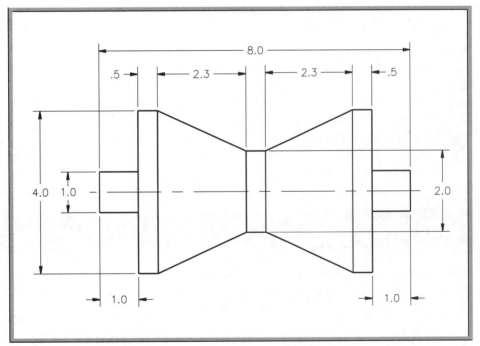

Fig. 3-8

2. The trailer manufacturer accepted the roller design with the following modification: the 4-inch overall diameter is to be increased to 5 inches. All other dimensions are to remain the same. Open roller1.dwg and make the change. Save the drawing as roller2.dwg.

Chapter 4 — Basic Objects

Objectives:

- Draw two-dimensional views of holes, cylinders, and rounded and polygonal features
- Create curved objects such as circles, arcs, ellipses, and donuts
- Create rectangles and regular polygons

Key Terms

concentric
diameter
donuts
ellipse

radial
radius
regular polygon
tangent

two-dimensional (2D) representation

Most products, buildings, and maps contain holes, rounded features, and other shapes that are circular in nature. Consequently, when drawing them with AutoCAD, you must use commands that produce circular and radial features. A *radial* feature is one in which every point is the same distance from an imaginary center point. Figure 4-1, a racecar engine, contains many examples of circular and radial features. Technical drawings also include rectangles and other polygonal shapes as shown in this and other drawings in this book.

Creating Circles

AutoCAD makes it easy to draw *two-dimensional (2D) representations* of holes, cylinders, and other round shapes. A 2D representation is a single profile view of an object seen typically from the top, front, or side.

1. Start AutoCAD. If the AutoCAD window does not appear and you instead see the Startup dialog box, click the **Cancel** button in the dialog box.

 NOTE: ───────────────────────

As a reminder, be sure to exit and restart AutoCAD if it was loaded prior to Step 1.

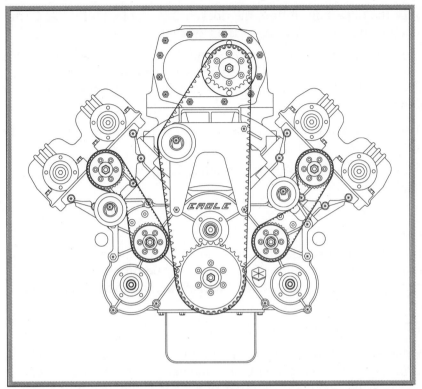

Fig. 4-1   Courtesy of Joe Schubeck, Eagle Engine Manufacturing

2. Pick the **Save** button from the Standard toolbar, or select **Save As...** from the **File** pull-down menu.

This causes the Save Drawing As dialog box to appear.

3. Open the folder with your name and highlight **Drawing1.dwg** located to the right of File name.

4. Type **engine** in upper- or lowercase letters and pick **Save** or press **ENTER**.

This creates and stores a new drawing file named engine.dwg in your named folder.

The detail drawing in Fig. 4-2 shows one of the engine's cylinder heads. Let's create the center portion of the cylinder head. (For now, disregard the bolt at the center of the head. You will add it later in this chapter.)

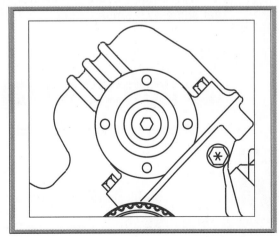

Fig. 4-2

5. Pick the **Circle** button from the docked Draw toolbar, or enter **C** (the command alias for **CIRCLE**) at the keyboard.

Draw

Notice the instruction on the Command line: Specify center point for circle. Other options are in brackets. These same instructions are displayed on the dynamic input prompt. Other options are displayed if you press the down arrow key.

6. Use the pointing device to pick a center point near the center of the drawing area.

7. Move the crosshairs and notice that you can "drag" the radius of the circle.

NOTE:

If the radius of the circle does not change as you move the crosshairs, an AutoCAD feature called DRAGMODE may be turned off on your system. To turn it back on, enter DRAGMODE and then enter A (for Auto).

Notice that AutoCAD is requesting the radius or diameter of the circle. The *radius* of a circle is the length of a line extending from its center to one side of the circle. The *diameter* of a circle is the length of a line that extends from one side of a circle to the other and passes through its center.

8. Pick a point a short distance from the center as shown in Fig. 4-3.

This completes the command and forms the innermost circular feature of the cylinder head.

Now let's draw a concentric circle to represent the next feature. *Concentric* circles are ones that share a common center.

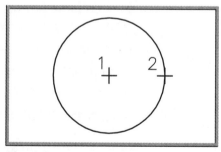

Fig. 4-3

Concentric Circles

Draw

9. Reenter the **CIRCLE** command.

HINT:

Use one of the shortcuts to reenter the command.

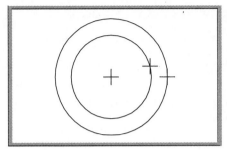

Fig. 4-4

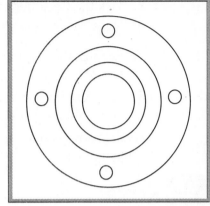

Fig. 4-5

10. Move the crosshairs to the center of the circle you created.

A small yellow circle appears at the center of the circle. This is the Center object snap. Object snaps are like magnetic points that lock onto a specific point on an object easily and accurately.

11. While the small yellow circle is present, pick a point.

AutoCAD snaps to the center point of the existing circle.

12. Drag the radius into place to create a second circle (Fig. 4-4). Review the detail of the cylinder head to approximate the size of the circle.

13. Repeat Steps 9 through 12 to place the remaining concentric circles as shown in the detail of the cylinder head.

14. Reenter the **CIRCLE** command and create the four small circles to represent bolt holes. Approximate their size and location.

Your drawing should now look similar to the one in Fig. 4-5.

15. Save and close the drawing file.

INFOLINK

You will learn more about object snaps in Chapter 9

Draw

Tangent Circles

Tangent circles are circles that meet at a single point. The Ttr option allows you to create a circle with a specified radius that is tangent to two other circles or arcs.

1. Select **New...** from the **File** pull-down menu to begin a new drawing file.

This displays the Select template dialog box. This dialog box permits you to select from a number of template files. A template file contains pre-established settings and values that speed the creation of new drawings. You will learn more about them later in the book.

2. Pick the **Open** button to select the default acad.dwt template file.

3. Create two circles and position them relatively close to one another.

Draw

4. Reenter the **CIRCLE** command, and enter **T** to select the Ttr (tan, tan, radius) option.

5. Read the instruction on the Command line and then move the crosshairs over one of the two circles.

A small yellow symbol appears, showing that the center of the crosshairs is tangent to the circle.

NOTE: ————————————————————————————

If the yellow symbol does not appear, pick the OSNAP button on the status bar to depress it, and then try Step 5 again.

6. Pick a point on the first circle.

7. Move the crosshairs to the other circle and pick a point.

8. For the radius of the new circle, enter **2**.

AutoCAD creates a new circle that is tangent to the two circles. It's okay if the circle extends off the screen.

Creating Arcs

The person who created the engine drawing shown in Fig. 4-1 used arcs to produce the radial features on some of the individual parts. In AutoCAD, the ARC command is often used to produce arcs.

Draw

1. Pick the **Arc** button from the Draw toolbar, or enter **A** at the keyboard.

This enters the ARC command.

2. Pick three consecutive points anywhere on the screen.

3. Reenter the **ARC** command and produce two additional arcs using the same method.

Options for Creating Arcs

AutoCAD offers several options for creating arcs. The option you choose depends on the information you know about the arc. Suppose you want to create an arc for an engine part and you know the start point, the endpoint, and the radius of the arc.

1. Enter the **ARC** command and pick a start point.

2. Enter **E** for End and pick an endpoint a short distance from the start point.

3. Move the crosshairs and watch the arc form dynamically on the screen.

4. Enter **R** for Radius and enter **3**.

An arc forms on the screen.

Suppose you don't know the radius of the arc or where the endpoint should be. However, you know where the center point of the arc should be, and you know how long the arc must be.

Draw

5. Reenter the **ARC** command and pick a point.

6. Enter **C** for Center and pick a point.

7. Enter **A** for Angle and enter a number (positive or negative) up to 360. (The number specifies the angle in degrees.)

An arc forms on the screen.

8. Experiment with the other options on your own.

As you can see, AutoCAD's options allow you to create an arc accurately using the information that is available to you. Understanding these options can save time when you need to produce accurate engineering drawings that contain circular or radial features.

Series of Tangent Arcs

In some cases, you may need to draw a series of arcs that are connected to one another. The Continue option of the ARC command provides an efficient way to draw such a series.

The line in Fig. 4-6 is really a series of tangent arcs. This line was developed for a map. It represents a stretch of highway 34 through Rocky Mountain National Park, west of Estes Park, Colorado. Let's reproduce the map line to practice creating tangent arcs.

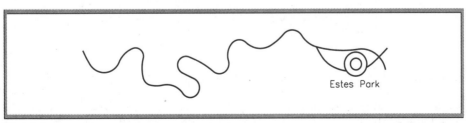

Fig. 4-6

1. Enter the **ARC** command and create an arc using any of the options.

2. Reenter the **ARC** command by pressing the spacebar or **ENTER**; press the spacebar or **ENTER** a second time.

Notice how another arc segment develops as you move the crosshairs.

3. Create the next arc segment.

Notice that it is tangent to the first. (The point of tangency is the point at which the two arcs join.)

4. Repeat Steps 2 and 3 until you have finished the road.

5. Create the concentric circles to show Estes Park, but omit the text.

Drawing Ellipses

The ELLIPSE command enables you to create mathematically correct ellipses. An *ellipse* is a regular oval shape that has two centers of equal radius. Ellipses are used to construct shapes that become a part of engineering drawings. For example, the racecar engine includes a partial ellipse, as shown in Fig. 4-7.

Fig. 4-7

1. From the Draw toolbar, pick the **Ellipse** button to enter the ELLIPSE command.

2. Pick a point for Axis endpoint 1 as shown in Fig. 4-8.

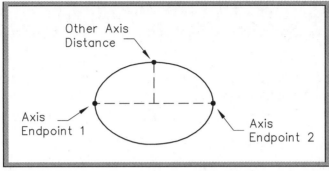

Other Axis
Distance

Axis
Endpoint 1

Axis
Endpoint 2

Fig. 4-8

3. Pick a second point for Axis endpoint 2 directly to the right or left of the first point.

HINT:

As noted in the previous chapter, temporary dotted lines appear when you move the crosshairs over the top of existing objects on the screen. This feature allows you to align and create new points from the existing objects. Take advantage of the dotted horizontal line that appears when creating axis endpoint 2.

4. Move the crosshairs and watch the ellipse develop. Pick a point, or enter a numeric value such as **.3**, and the ellipse will appear.

5. Experiment with the **Center** option on your own. How might this option be useful?

Elliptical Arcs

Many parts that include elliptical shapes do not use the entire ellipse. To draw these shapes, you can use the ELLIPSE Arc option.

Draw

1. Pick the **Ellipse Arc** button.

2. Pick three points similar to those shown in Fig. 4-8.

A temporary ellipse forms.

3. In reply to Specify start angle, pick a point anywhere on the ellipse.

4. As you slowly move the crosshairs counterclockwise, notice that an elliptical arc forms.

5. Pick a point to form an elliptical arc.

Donuts

The DONUT command allows you to create thick-walled or solid circles, known in AutoCAD as *donuts*. Drafters commonly use donuts to represent features on a machine part, architectural drawing, or map.

1. From the Draw pull-down menu, pick **Donut**, or enter **DO** for DONUT.

2. Specify an inside diameter of **.4** ...

3. ... and an outside diameter of **.75**.

The outline of a small donut locks onto the crosshairs and is ready to be dragged and positioned by its center.

4. Place the donut anywhere in the drawing by picking a point.

5. Move the crosshairs away from the new solid-filled donut and notice the Command line at the bottom of the screen.

6. Place several additional donuts in the drawing.

7. Press **ENTER** to terminate the command.

That's all there is to the DONUT command.

Rectangles

One of the most basic shapes used by drafters and designers is the rectangle. You can create rectangles using the LINE command, but doing so has some disadvantages. For example, you would have to take the time to make sure that the corner angles are exactly 90°. Also, each line segment would be a separate object. Therefore, AutoCAD provides the RECTANG command, which allows you to create a rectangle with perfect corners and as a single object. Clicking anywhere on a rectangle created with the RECTANG command selects the entire rectangle.

Draw

1. From the Draw toolbar, pick the **Rectangle** button.

As you can see, AutoCAD is asking for the first corner of the rectangle. Other options appear in brackets at the Command line.

2. In reply to Specify first corner point, pick a point at any location.

3. Move the pointing device in any direction and notice that a rectangle begins to form.

4. Pick a second point at any location to create the rectangle.

5. Create a second rectangle. Since the RECTANG command was just entered, reenter it by pressing the spacebar or **ENTER**, or right-click and pick **Repeat Rectangle** from the shortcut menu.

6. Create a third rectangle.

7. Close the drawing without saving it.

Polygons

The POLYGON command enables you to create regular polygons with 3 to 1024 sides. A *regular polygon* is one with sides of equal length. Using the POLYGON command, let's insert the bolt head into the engine drawing you started earlier in this chapter.

1. Open the **engine.dwg** file by picking the **Open** button from the abbreviated Standard toolbar.

Draw

2. From the Draw toolbar, pick the **Polygon** button, or enter the **POL** alias.

Notice the <4> at the end of the AutoCAD prompt. This is the default value, meaning that if you were to press ENTER now, AutoCAD would enter 4 in reply to Enter number of sides. To represent the bolt head shown in Fig. 4-2 you will need a hexagon (six-sided polygon).

3. Enter **6**.

AutoCAD now needs to know if you want to define an edge of the polygon or select a center point. Let's specify a center.

4. Move the pointing device over any of the larger circles in the drawing. When the small yellow symbol appears at the center of the cylinder head, pick a point to select the center of the circle as the center point of the hexagon. (The OSNAP button on the status bar must be depressed.)

AutoCAD allows you to create a polygon by inscribing it in a circle of a specified diameter or by circumscribing it around the specified circle. Figure 4-9 shows the difference.

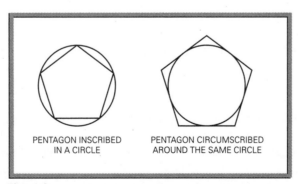

Fig. 4-9

5. Press **ENTER** to select the I (Inscribed) default value.

AutoCAD now wants to know the radius of the circle within which the polygon will appear.

6. With the pointing device, move the crosshairs from the center of the polygon and notice that a hexagon begins to form.

7. Pick a point to create the hexagonal bolt head at an appropriate size relative to the cylinder head. Refer to Fig. 4-2 if necessary.

 NOTE: ───────────────────────────────

To create the polygon with a more accurate size, you could have entered a specific numeric value, such as .5, at the keyboard. (Entering a 0 before the decimal point is optional.)

The drawing of the cylinder head component is now complete. Your drawing should look similar to the one in Fig. 4-10.

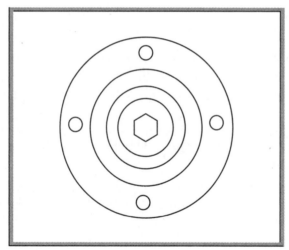

Fig. 4-10

Standard

8. Save your work and exit AutoCAD.

 HINT: ───────────────────────────────

Back up your files regularly. The few seconds required to make a backup copy may save you hours of work. Experienced users back up faithfully because they know the consequences of not doing so.

Review Questions

1. Briefly describe the following methods of producing circles.
   a. 2 points
   b. 3 points
   c. tan tan radius

2. In which AutoCAD toolbar are the ARC and CIRCLE buttons found?

3. What function does the ARC Continue feature serve?

4. Explain the purpose of the DONUT command.

5. When you create a polygon using the Inscribed in circle option, does the polygon appear inside or outside the imaginary circle?

6. What are the practical differences between creating a rectangle using the LINE command and creating it using the RECTANG command? Why might you choose one command instead of the other?

7. In addition to the LINE and RECTANG commands, which other command is capable of creating a rectangle? When might you choose this method?

Challenge Your Thinking

1. Review the information in this chapter about specifying an angle in degrees. How might you be able to create an arc in a clockwise direction? Try your method to see if it works, and then write a paragraph describing the method you used.

2. Experiment further with the ARC command. Is it possible to create a single arc that has a noncircular curve using the options available for this command? (A noncircular curve is one in which not all the points are exactly the same distance from a common center point.) Explain your answer.

Applying AutoCAD Skills

Work the following problems to practice the commands and skills you learned in this chapter.

1. Using the commands you just learned, complete the drawings shown in Figs. 4-11 and 4-12. Don't worry about text matter or exact shapes, sizes, or locations, but do try to make your drawings look similar to the ones in the figures. Save the drawings as prb4-1a.dwg and prb4-1b.dwg.

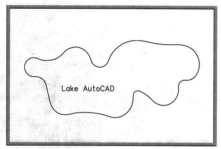

Fig. 4-11

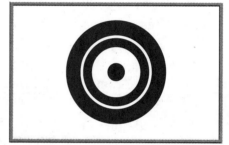

Fig. 4-12

2. Use the LINE, POLYGON, CIRCLE, and ARC commands to create the hex bolt shown in Fig. 4-13. Save the drawing as prb4-2.dwg.

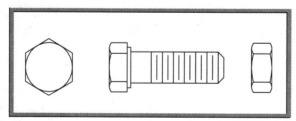

Fig. 4-13

Problem 2 courtesy of Joseph K. Yabu, Ph.D., San Jose State University

3. Create the object shown in Fig. 4-14 by first drawing a six-sided polygon with the POLYGON command. Then draw the six-pointed star using the LINE command. Save the drawing as prb4-3.dwg.

4. In the same drawing, create the same object again, but draw the star using two three-sided polygons. Save your work.

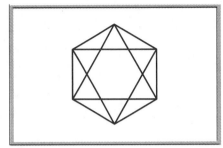

Fig. 4-14

5. How would you draw a five-pointed star? Draw it in the same drawing file and save your work.

6. Draw a block with a rectangular cavity like the one shown in Fig. 4-15 using the RECTANG and LINE commands. Save the drawing as prb4-6.dwg.

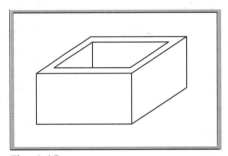

Fig. 4-15

7. The two objects shown in Fig. 4-16 are composed entirely of equal-sided and equal-sized polygons with common edges surrounding a central polygon. Can this be done with equal-sided and equal-sized polygons of any number of sides? Answer this question by trying to draw such objects using the POLYGON command with polygons of five, six, seven, and eight sides.

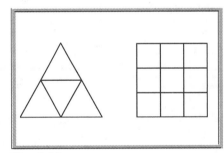

Fig. 4-16

8. Draw the screw heads shown in Fig. 4-17. Save the drawing as prb4-8.dwg.

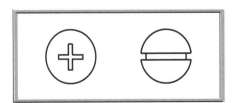

Fig. 4-17

Problems 3 through 7 courtesy of Gary J. Hordemann, Gonzaga University

9. Draw the eyebolt in Fig. 4-18 using all of the commands you have learned. Save the drawing as prb4-9.dwg.

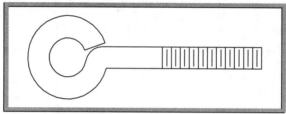

Fig. 4-18

10. Using the LINE, ARC, and CIRCLE commands, draw the two views of a hammer head as shown in Fig. 4-19. Save the drawing as prb4-10.dwg.

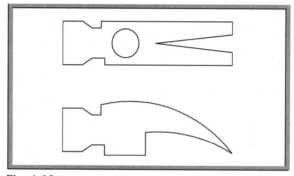

Fig. 4-19

11. Use the ARC and LINE commands to create the screwdriver shown in Fig. 4-20. Save the drawing as prb4-11.dwg.

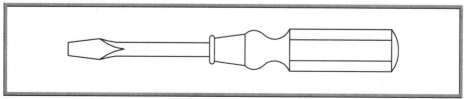

Fig. 4-20

Problems 9 through 11 courtesy of Joseph K. Yabu, Ph.D., San Jose State University

Using Problem-Solving Skills

Complete the following activities using problem-solving skills and your knowledge of AutoCAD.

1. Draw the ski lift rocker arm for a design study by a gondola manufacturer (Fig. 4-21). Do not include dimensions. Save your drawing as rockerarm.dwg.

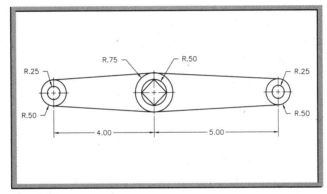

Fig. 4-21

2. The traditional mantle clock body drawing shown in Fig. 4-22 is necessary for the reconstruction of antique and collectible clocks. Draw the clock body. Do not include dimensions. Save your drawing as mantle.dwg.

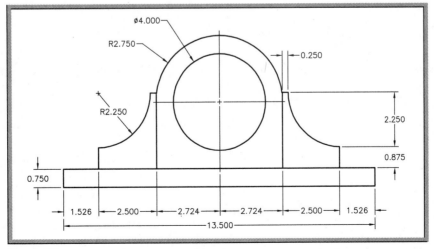

Fig. 4-22

Careers Using AutoCAD

Ship Designer

If you have seen the movie "Titanic," you will surely remember the scene when the ship's designer shared his fears about the Titanic's fate. He knew there were not enough dinghies (small boats) to take all the passengers to safety. He had been meticulous in every other aspect of ship design, but his superiors had not thought that adding more dinghies was necessary. The designer was only allowed to plan room enough for a few of these small vessels. As a result, many people were left behind as the massive Titanic slowly sank into the ocean.

Giant water vessels, such as the Titanic, aren't the only types of ships that require careful planning during the initial design phase. Other marine vessels, including tankers, barges, submarines, recreational yachts, aircraft carriers, and others, also require the expertise of drafters to help build safe and functional vessels.

Who Designs Ships?

Currently, twenty-two major shipyards are operational in the United States. Drafters specializing in ship design are employed at these locations. They work closely with naval architects and marine engineers to design the machinery ships need to operate properly, such as engine parts, navigation equipment, boilers, and gauges. Hull designers prepare drawings that provide exact measurements of ship dimensions. Ship detail drafters create plans for electrical and plumbing fixtures. Because the people who specialize in designing and building ships exhibit expertise in a variety of areas, the planning process is an excellent example of teamwork.

Corbis/Vittorio Rastelli

Special Skills

Drafters who want to specialize in one of the many drafting specialties related to shipbuilding must acquire other skills in addition to basic drafting skills. For example, they must develop an understanding of basic nautical safety. They must also understand how various building materials interact with fresh and salt water.

Where Can I Learn More?

The Shipbuilders Council of America website (www.shipbuilders.org) provides information about the shipbuilding industry. It also includes a listing of major shipyards in the United States and provides various employment descriptions.

Career Activities

- If you were considering becoming a drafter with a focus on shipbuilding, what type of experience do you think would be beneficial to your goals?
- How might you become familiar with ships and ship design? Think of some areas near your city that might provide opportunities to become familiar with ships.

Chapter 5 Object Selection

Objectives:

- **Identify AutoCAD objects**
- **Select objects to create a selection set**
- **Add and remove objects from the selection set**
- **Erase and restore objects in AutoCAD**
- **Resize objects parametrically**

Key Terms

entity	pickbox
grips	selection previewing
noun/verb selection	selection set
object	selection window
parametric shape editing	verb/noun selection

When using AutoCAD, you will find yourself selecting objects so that you can move, copy, erase, or perform some other operation. Because drawings can become very dense with lines, dimensions, and text, it is important to use the most efficient methods of selecting these objects. Usually, the quickest way to move or copy an object is to click and drag. This is true whether you are using an inexpensive shareware program on a low-cost computer or sophisticated CAD software on an expensive computer. This chapter covers this and alternative methods of selecting objects and performing basic editing operations, such as resizing objects and erasing.

Objects

An *object*, also called an *entity*, is an individual predefined element in AutoCAD. The smallest element that you can add to or erase from a drawing is an object.

The following list gives examples of object types in AutoCAD.

3Dface	leader	solid
3Dsolid	line	spline
arc	mline	text
attribute	multiline text	tolerance
body	point	trace
circle	polyline	vertex
dimension	ray	viewport
ellipse	region	xline
image	shape	

You'll learn more about these objects and how to use them in future chapters.

Selecting Objects

Basic object selection in AutoCAD is straightforward—you select the object by picking it with the left button on the pointing device. However, AutoCAD also offers more advanced object selection techniques. This chapter illustrates many of these techniques.

1. Start AutoCAD. If the AutoCAD window does not appear, click the **Cancel** button in the Startup dialog box to display it.

NOTE: —————————————————————————————

In this book, all new drawings use Imperial (English) units unless otherwise specified.

Draw

Draw

2. Using the **LINE** and **CIRCLE** commands, draw the triangle and circle shown in Fig. 5-1.

Fig. 5-1

Window Selection

3. Enter the ERASE command by picking the **Erase** button from the Modify toolbar or by entering **E** at the keyboard.

The AutoCAD prompt changes to Select objects. Notice that the crosshairs have changed to a small box called the *pickbox*. The pickbox is used to pick objects. At this point, you can also create a window to select points.

4. Pick a point anywhere to the lower left of the triangle.

5. Move the pointing device and notice that a blue box forms.

This box defines the *selection window*.

6. Move the pointing device so that the triangle fits completely inside the window and pick a point.

The objects to be erased are highlighted with broken lines as shown in Fig. 5-2. The highlighted objects make up the *selection set*.

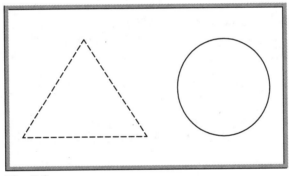

Fig. 5-2

7. Press **ENTER** to make the triangle disappear. (This also terminates the ERASE command.)

Restoring Erased Objects

What if you have erased an object by mistake and you want to restore it? Let's restore the triangle.

Standard

8. Pick the **Undo** button on the docked Standard toolbar.

The triangle reappears. The Undo button conforms to standard Windows functionality, so you may be familiar with it from other Windows applications. Picking this button repeatedly undoes your previous actions, one at a time, beginning with the most recent.

Using a Crossing Window

The Crossing object selection option selects all objects that cross the window boundary as well as those that lie completely within it.

Modify

1. Enter the **ERASE** command.

2. Type **C**, press **ENTER**, and pick a point to the right of the triangle.

3. Move the crosshairs to the left to form a green box and notice that the box is made up of broken lines.

4. Form the box so that it crosses over at least one side of the triangle and press the pick (left) button in reply to Specify opposite corner.

Notice that AutoCAD selected those objects that cross over the box as well as any objects that lie completely within it.

5. Press **ENTER** to erase the object(s).

HINT: ———————————————————————————

You can also select objects using a crossing window without formally entering the Crossing option. When the Select objects prompt appears, simply pick two points to form a window, being sure to pick the *right-most* point first.

Polygonal Selection Options

The WPolygon and CPolygon object selection options are similar to the Window and Crossing options. However, they offer more flexibility because they are not restricted to the rectangular window shape used by the Window and Crossing options.

Modify

1. Fill the screen with several objects, such as lines, circles, arcs, rectangles, polygons, ellipses, and donuts.

2. Enter the **ERASE** command.

3. In reply to Select objects, enter **WP** for WPolygon. (W is for Window.)

4. Pick a series of points that form a polygon of any shape around one or more objects. As you create the polygon, notice that the area inside the polygon is blue.

5. Press **ENTER** to close the polygon.

NOTE: ———————————————————————————

When forming the polygon, AutoCAD automatically connects the last point to the first point.

AutoCAD selects those objects that lie entirely within the polygon.

6. Press **ENTER** to erase the objects.

The CPolygon option is similar to WPolygon.

Modify

7. Enter the **ERASE** command or press the spacebar.

8. In reply to Select objects, enter **CP** for CPolygon. (C is for Crossing.)

9. Pick a series of points to form a polygon. Make part of the polygon cross over at least one object and press **ENTER**.

As you create the polygon, notice that it is made up of broken lines, and that the area inside the polygon (the selection window) is green. The broken lines and green selection window indicate that the Crossing option is in effect. Solid lines around a blue selection window indicate a normal (not crossing) selection.

AutoCAD selects those objects that cross the polygon, as well as those that lie completely within it. As you can see, CPolygon is similar to the Crossing option, whereas WPolygon is similar to the basic window option.

10. Press **ENTER** to complete the erasure.

Using a Fence

The Fence option is similar to the CPolygon option except that you do not close a fence as you do a polygon. When you select objects using the Fence option, AutoCAD looks for objects that touch the fence. Objects "inside" the fence are not selected unless they actually touch the fence.

Modify

1. Enter **ERASE**.

2. Enter **F** for Fence.

3. Draw a line that crosses over one or more objects and press **ENTER**.

4. Press **ENTER** to complete the erasure.

Selecting Single Objects

Suppose you want to select a single object, such as a line segment.

1. Recreate the triangle and circle you drew previously.

Modify

2. Pick the **Erase** button or enter **E** at the keyboard.

3. Using the pickbox, pick two of the line segments in the triangle.

NOTE:

AutoCAD automatically highlights objects as the pickbox rolls over them. This is called *selection previewing*. This feature is helpful when you need to select individual objects in a crowded area of the drawing. It allows you to see in advance which object will be selected when you click the mouse button.

4. Press **ENTER** to complete the command and to make the highlighted objects disappear.

HINT:

If you are using a pointing device that has more than one button, pressing one of them, usually the one on the right, is the same as pressing the ENTER key. This is normally faster than pressing ENTER at the keyboard.

Selecting the Last Object Drawn

The Last object selection option automatically selects the last object you drew. This option works whenever the Select objects prompt is present.

1. Pick the **Undo** button to restore the line segments.

Standard

Notice that because you erased both line segments in a single ERASE operation, you only have to pick the Undo button once to make both lines reappear.

2. Enter the **ERASE** command.

3. Type **L** (for Last) and press **ENTER**.

Modify

AutoCAD highlights the last object you drew.

NOTE:

The last object you drew may be different from the last object you selected in the drawing. If you want to reselect the last object you selected, enter P (for Previous) instead of Last. The Previous option is especially useful when you need to reselect an entire group of objects.

4. Press **ENTER**.

If you continue to enter ERASE and Last, you will erase objects in the reverse order from which you created them. However, it is usually faster to use the Undo button for this purpose.

Standard

5. Pick the **Undo** button to restore the object you deleted using the Last option.

Selecting the Entire Drawing

The All object selection option permits you to select all the objects in the drawing quickly.

Modify

1. Enter the **ERASE** command.

2. In reply to Select objects, type **All** and press **ENTER**.

AutoCAD selects all objects in the drawing.

3. Press **ENTER** again to erase all the objects.

Standard

4. Pick **Undo** to restore the objects.

Editing Selection Sets

AutoCAD allows you to add and remove objects from a selection set while you are in the process of creating it. For example, if you accidentally pick an object that you did not mean to select, you can remove it without affecting the rest of the selected objects. This is especially helpful when you have many objects in a selection set on a complex drawing.

Removing Objects from a Selection Set

To remove an unwanted object from a selection set, you must enter the Remove objects mode.

Modify

1. Enter the **ERASE** command.

2. Create a selection window to select the triangle and the circle.

Both the triangle and the circle should now be highlighted. Suppose you have decided not to erase the circle.

3. Type **R** (for Remove) and press **ENTER**.

Notice that the AutoCAD prompt changes to Remove objects. You can now remove one or more objects from the selection set. The object(s) you remove from the selection set will not be erased.

4. Remove the circle from the selection set by picking it.

Note that it is no longer highlighted.

5. Press **ENTER** to complete the ERASE operation.

6. Pick the **Undo** button to restore the triangle.

Standard

Adding to the Selection Set

After you have used the Remove option to remove objects from a selection set, you can select additional objects to include in the set.

7. Repeat Steps 1 through 4 in the previous section ("Removing Objects from a Selection Set").

8. Instead of pressing ENTER, type **A** for Add and press **ENTER**.

Notice that the AutoCAD prompt changes back to Select objects.

9. Select the circle.

10. Press **ENTER** to complete the ERASE operation.

So you see, you can add and remove objects as you wish until you are ready to perform the operation. The objects selected are indicated by broken lines. These selection procedures work not only with the ERASE command, but also with other commands that require object selection, such as MOVE, COPY, MIRROR, ARRAY, and many others.

Standard

11. Pick the **Undo** button to restore the objects to the screen.

Using Grips

Another method of selecting objects is to pick them using the pickbox located at the center of the crosshairs. When you pick an object without first entering a command, small boxes called *grips* appear at key points on the object. Using these grips, you can perform several basic operations such as moving, copying, or changing the shape or size of an object.

Draw

1. Enter the **RECTANG** command and draw a rectangle of any size and at any location.

2. Select any point on the rectangle. (No command should be entered.)

Notice that blue grips appear at each corner of the rectangle.

Editing Size and Shape with Grips

3. Move the crosshairs until it locks over the lower right grip, but do not pick the grip yet.

Chapter 5: Object Selection

AutoCAD displays the current length of the two lines that meet at the lower right grip, as shown in Fig. 5-3.

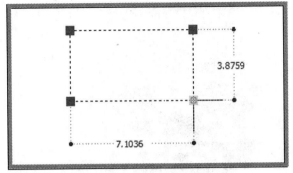

4. Move the crosshairs to each of the other grips to see the current length of the other sides of the rectangle.

5. Pick the lower right grip to select it.

Fig. 5-3

The grip turns red, showing that you have selected it. The length of the right side of the rectangle is now displayed in an editable text box, as shown in Fig. 5-4.

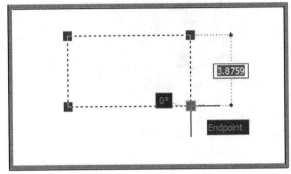

6. Move the crosshairs on the screen and notice how the lengths update automatically as you move the grip around. See Fig. 5-5.

7. At the keyboard, enter a new length of 3.

Fig. 5-4

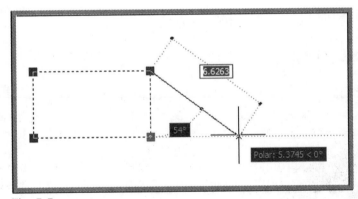

Fig. 5-5

The length of the right side of the rectangle changes to a length of exactly 3 units, and the shape is no longer rectangular. This method of editing a CAD object by entering new size values is known as *parametric shape editing*.

8. Press **ESC** to deselect the object.

9. Pick the **Undo** button to restore the original shape and size of the rectangle.

Moving Objects with Grips

1. Select the rectangle again and select any one of the four grips.

2. Press the right button on the pointing device (right-click) to display a shortcut menu.

3. Pick **Move**, drag the rectangle to a new location, and pick a point.

4. Display the shortcut menu again and select one of the other editing options, such as **Rotate**, and complete the operation. Follow the instructions at the AutoCAD prompt.

Copying Objects with Grips

You can also copy objects using grips.

1. Select the circle and pick one of the four grips that lie on it.

2. Move the pointing device and notice that you can adjust the circle's radius.

3. Pick a point to change the circle's radius.

4. Pick the circle's center grip, right-click, and pick **Copy**.

5. Copy the circle three times by picking three points anywhere on the screen.

6. Press **ENTER** to terminate the copying.

Modifying the Grips Feature

For most AutoCAD users, the standard grips characteristics, such as color and size of the grip boxes, are adequate. In some companies or circumstances, however, it may be necessary to adjust the grips so that you can work with them more easily. For example, if your company has customized the background of the drawing area to dark gray, you may want to change the color of unselected grips boxes from blue to yellow so that you can see them more easily. You may need to adjust the size of the grip boxes depending on your display resolution. Grip boxes viewed at 1600 × 1200 resolution are twice as small as those viewed at 800 × 600. AutoCAD allows you to make changes such as these using the Options dialog box.

1. Select the **Tools** pull-down menu and select the **Options...** item.

The Options dialog box includes nine tabs with important information and settings in each one.

2. Pick the **Selection** tab.

Information and settings related to the grips feature appear in the right half of the dialog box, as shown in Fig. 5-6.

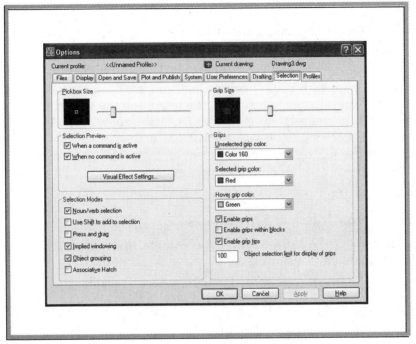

Fig. 5-6

3. In the area labeled Grip Size, move the slider bar and watch how it changes the size of the box to the left of it.

4. Increase the size of the grips box and pick the **OK** button.

5. Select one of the objects and notice the increased size of the grips.

6. Pick one of the grips and perform an editing operation.

7. Display the Options dialog box and adjust the size of the grips box to its original size, but do not close the dialog box.

The Selection tab also permits you to enable and disable the grips feature, as well as control the assignment of grips within blocks. You can also make changes to the color of selected and unselected grips. For now, do not change these settings.

8. Pick the **OK** button to close the Options dialog box.

INFOLINK

You will learn more about blocks in Chapter 32.

Using Grips with Commands

The traditional AutoCAD method of entering a command and then selecting the object to be edited is sometimes called *verb/noun selection*. First you tell the software what to do (using a verb such as *erase*), and then you select the object to be acted upon (such as a line).

When you use grips with editing commands, you are using *noun/verb selection*. First you select the object, and then you tell the software what to do with the object. You may find the grips (noun/verb) method convenient to use, especially if you are familiar with other graphics programs that use this technique. However, either method is acceptable. The method you choose is a matter of personal preference.

1. Select one of the circles.

2. Pick the **Erase** button or enter **E** for ERASE.

Modify

As you can see, AutoCAD erased the circle without first asking you to select objects. You have just used the noun/verb technique.

3. Pick another circle and press the **Delete** key on the keyboard.

As you can see, pressing the Delete key is the same as entering the ERASE command.

Standard

4. Pick the **Undo** button from the docked Standard toolbar.

5. Exit AutoCAD without saving.

Review Questions

1. In AutoCAD, what is an object? Give three examples.

2. After you enter the ERASE command, what does AutoCAD ask you to do?

3. Experiment with and describe each of these object selection options.

   a. Last
   b. Previous
   c. Crossing
   d. Add
   e. Remove
   f. Undo
   g. WPolygon
   h. CPolygon
   i. Fence
   j. All

4. How do you place a window around one or more objects during object selection?

5. If you erase an object by mistake, how can you restore it?

6. How can you retain part of what has been selected for erasure while remaining in the ERASE command?

7. What is the fastest way of erasing the last object you drew?

8. When you use a graphics program, such as AutoCAD, what usually is the quickest way to edit an object?

9. Describe how you would remove one or more objects from a selection set.

10. Explain the primary benefit of using the grips feature.

11. What editing functions become available to you when you use grips?

12. Explain how to copy an object using grips.

13. Explain the difference between the noun/verb and verb/noun selection techniques.

1. As you have seen, AutoCAD's grips feature is both easy and convenient to use. With that in mind, discuss possible reasons AutoCAD also includes specific commands that you must enter to perform some of the same functions you can do easily with grips.

2. Describe a situation in which the Remove option can be useful for more than removing an object that was added to a selection set by mistake.

Applying AutoCAD Skills

1. Create a new drawing from scratch. Draw a circle and a rectangle as illustrated in Fig. 5-7. Use the RECTANG command to create the rectangle. Enter the ERASE command and try to erase one line of the rectangle. At another location in the drawing, create another rectangle using the LINE command. Try to erase one line of the second rectangle. What is the difference? Do not save the drawing.

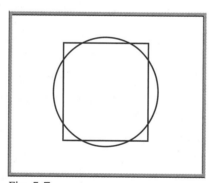

Fig. 5-7

2. Create a new drawing. Use the LINE command to draw polygons *a* through *e* as shown in Fig. 5-8.

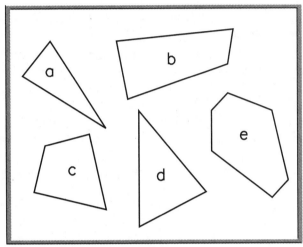

Fig. 5-8

Enter the ERASE command and use the various object selection options to accomplish the following without pressing ENTER.

- Place a window around polygon *a*.

- Pick two of polygon *b*'s lines for erasure.

- Use the Crossing option to select polygons *c* and *d*.

- Remove one line selection from polygon *c* and one from polygon *d*.

- Pick two lines from polygon *e* for erasure, but then remove one of the lines so it won't be erased.

After you have completed the five operations listed above, press ENTER. You should have nine objects (line segments) left on the screen. Save the drawing as prb5-2.dwg in the folder with your name, but do not close the file.

3. From the File pull-down menu, select Save As... and enter prb5-2.dwg for the file name. Restore the line segments that you deleted in problem 2. Then use grips to rearrange the polygons in order of size from smallest to largest. Save your work.

Using Problem-Solving Skills

Complete the following activities using problem-solving skills and your knowledge of AutoCAD.

1. The computer sketch in Fig. 5-9 was sent to the project engineer for comments and has now been routed to your desk with the attached note. Use the ELLIPSE, ARC, and LINE commands to create the sketch of the bushing. Because this is only a rough sketch, you may approximate the dimensions. Make the change requested by the engineering department. Save the drawing as ch5-bushing.dwg.

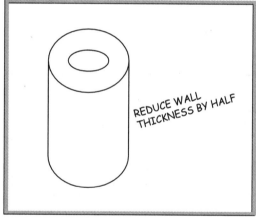

Fig. 5-9

2. Your architectural firm's customer wants the gable-end roof changed to a hip roof (Fig. 5-10). Draw the front elevation with both roof lines. Since this is only a roof line change, approximate the size and location of the windows and door. Save the drawing as ch5-roof.dwg.

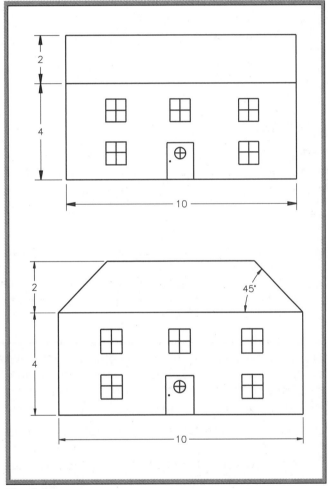

Fig. 5-10

Objectives:

- Describe several methods of entering *x,y* coordinates
- Locate points and draw objects using the Cartesian coordinate system
- Apply the absolute, relative, polar, polar tracking, and direct distance methods of entering coordinates

Key Terms

absolute points
alignment paths
Cartesian coordinate system
coordinate pair
coordinates
direct distance method

origin
polar method
polar tracking method
relative method
world coordinate system (WCS)

Coordinates are sets of numbers used to describe specific locations. In AutoCAD, coordinates are used to specify the location of objects in a drawing. In previous chapters, when you used the pointing device to pick the endpoints of lines, etc., AutoCAD recorded the coordinates of the points that you picked on the screen.

You can also use the keyboard to enter coordinates. The keyboard method allows you to specify points and draw lines of any specific length and angle. This method also applies to creating arcs, circles, and other object types. By specifying the coordinates of objects, you can achieve the accuracy needed for engineering drawings.

AutoCAD uses a *Cartesian coordinate system* similar to that used in geometry. In AutoCAD, this system is called the *world coordinate system (WCS)*. In the two-dimensional version of this system, two axes are used. (See Fig. 6-1.) This is the version used for most drafting work. The X axis is an infinite, imaginary line that runs horizontally. The Y axis is an infinite, imaginary line that runs vertically. Each axis is numbered with sequential positive and negative numbers. Each combination of one *x* value and one *y* value (a *coordinate pair*) specifies a unique point in the coordinate system.

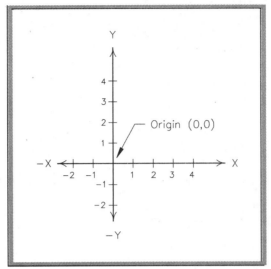

Fig. 6-1

The point at which the axes cross is known as the *origin*. At the origin, the values of *x* and *y* are both zero. The values to the right and above the origin are positive. Those below and to the left of the origin are negative. In AutoCAD's drawing area, the origin is located at the lower left corner by default.

In three-dimensional modeling, a third axis—the Z axis—is added at right angles to the other two axes to allow you to specify depth.

Methods of Entering Coordinates

The Cartesian system does not limit you to entering coordinate pairs to specify points. Consider the following ways to specify points using the LINE command.

Absolute Method

When you enter specific *x* and *y* values, AutoCAD places the points according to the Cartesian coordinate system. Points entered in this way are considered *absolute points* (Fig. 6-2).

Example:

LINE Specify first point: 2,1   This begins the line at absolute
Specify next point or [Undo]: 6,3   point 2,1 and ends it at absolute
                                       point 6,3.

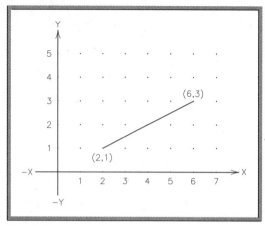

Fig. 6-2

Relative Method

The *relative method* allows you to enter points based on the position of a point that has already been defined. AutoCAD determines the position of the new point relative to a previous point (Fig. 6-3). Notice the use of the @ symbol in the following example. Use of this symbol tells AutoCAD that the following coordinate pair should be read relative to the previous point.

Example:

LINE Specify first point: 2,1
Specify next point or [Undo]: @2,0

This draws a line 2 units in the positive X direction and 0 units in the Y direction from absolute point 2,1. In other words, the distances 2,0 are relative to the location of the first point.

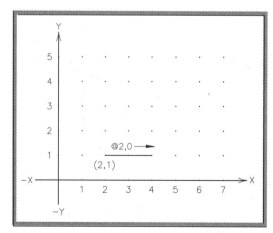

Fig. 6-3

Polar Method

The *polar method* is another form of relative positioning. Using this method, you can produce lines at precise angles. The area surrounding a point is divided into 360 degrees, as shown in Fig. 6-4. If you specify an angle of 90 degrees, for example, the line extends upward vertically from the last point (Fig. 6-5). Notice the use of the < symbol to specify the angle.

Example:

LINE Specify first point: 2,1
Specify next point or [Undo]: @3<60

This produces a line segment 3 units long at a 60° angle. The line begins at absolute point 2,1 and extends for 3 units at a 60° angle.

∠ = Angle of

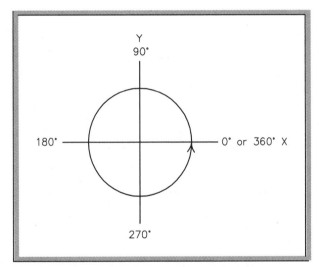

Fig. 6-4

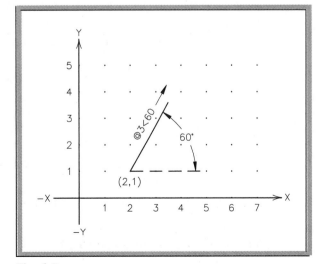

Fig. 6-5

Polar Tracking Method

The *polar tracking method* is similar to the polar method, except it's faster and easier. With this method, you can produce lines at precise angles and lengths with minimal keyboard entry (Fig. 6-6).

Polar tracking causes AutoCAD to display temporary *alignment paths* at prespecified angles to help you create objects at precise positions and angles. Polar tracking is on when the POLAR button on the status bar is depressed.

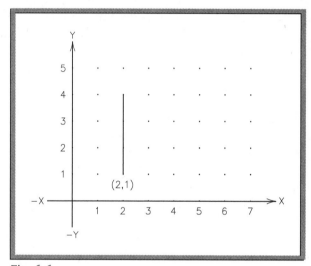

Fig. 6-6

Example:

LINE Specify first point: 2,1
Specify next point or [Undo]: Use the 90° alignment path that appears automatically on the screen and enter 3.

This produces a line segment 3 units long at a 90° angle. The line begins at absolute point 2,1 and extends 3 units upward vertically.

The alignment paths appear at 90° angles by default. You can change the angle of the alignment paths to meet your needs for specific drawing tasks. The following steps guide you through this process.

1. Start AutoCAD and display the AutoCAD window.

2. From the **Tools** pull-down menu, select the **Drafting Settings...** item and pick the **Polar Tracking** tab.

The Drafting Settings dialog box appears as shown in Fig. 6-7.

3. In the Polar Angle Settings area of the dialog box, pick the down arrow and change the increment angle to **30**.

4. Pick the **OK** button to close the dialog box.

5. Enter the **LINE** command and pick a point anywhere on the screen.

6. Move the crosshairs away from and around the point.

At 30° increments around the point, AutoCAD displays an alignment path and a small box showing the distance and angle from the point.

7. Position the crosshairs so that they are at a 60° (1 o'clock) position from the point and enter **2**.

AutoCAD produces a line 2 units long at a 60° angle.

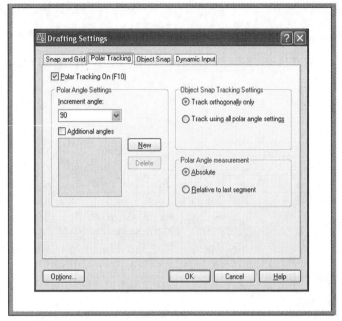

Fig. 6-7

8. Use the polar tracking feature to produce additional line segments.

9. Erase everything in the drawing file.

10. Display the Drafting Settings dialog box, set the polar tracking increment angle to **90**, and pick the **OK** button.

Direct Distance Method

AutoCAD also offers a *direct distance method* of entering points. This method is similar to polar tracking. After you specify a point, you can "point" at any angle using the pointing device (without using alignment paths) and enter a number at the keyboard to specify the distance. This method is fast, but it is not as accurate as polar tracking because you cannot specify an exact angle.

Example:

LINE Specify first point: 2,1
Specify next point or [Undo]: Use the pointing device to point toward the upper right corner of the screen and enter 2.

This produces a line segment 2 units long at the specified angle.

Applying Methods of Coordinate Entry

Some methods of coordinate entry work better than others in a given situation. The best way to become familiar with the various methods of coordinate entry, and which to use when, is to practice using them.

1. Enter the **LINE** command.

2. Using the methods presented in the chapter, create the drawing of the gasket shown in Fig. 6-8. Don't worry about the exact locations of the holes, and don't dimension the drawing. However, do make the drawing precisely this size. The top and bottom edges of the gasket are perfectly horizontal. (You may need to enter **Z** and **A** to ZOOM All. This makes the gasket larger on the screen. You will learn more about ZOOM All later in the book.)

HINT:

As you draw the holes in the gasket, keep in mind that the ø (*theta*) symbol indicates that the dimensions are given in diameters.

NOTE:

You can enter negative values for line lengths and angles.

LINE Specify first point: 5,5
Specify next point or [Undo]: @2<–90

This polar point specification produces a line segment 2 units long downward vertically from absolute point 5,5. This is the same as entering @2<270. Try it and then erase the line.

3. Save your work in a file named gasket.dwg and exit AutoCAD.

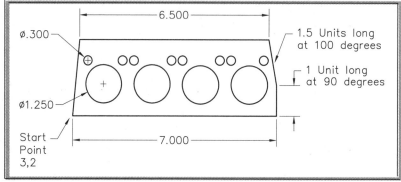

Fig. 6-8

Review Questions

1. What is the relationship between AutoCAD's world coordinate system and the Cartesian coordinate system?

2. Briefly describe the differences among the absolute, relative, polar, and polar tracking methods of point specification.

3. Explain the advantage of specifying endpoints from the keyboard rather than with the pointing device.

4. What is the advantage of specifying endpoints with the pointing device rather than the keyboard?

5. What happens if you enter the @ character at the Specify first point prompt after entering the LINE command?

6. What is the effect on the direction of the new line segment when you specify a negative number for polar coordinate entry?

7. What is the difference between direct distance entry and the polar tracking method of entering coordinates?

Challenge Your Thinking

1. Why may entering absolute points be impractical much of the time when completing drawings?

2. Consider the following command sequence:

   LINE Specify first point: 4,3
   Specify next point or [Undo]: 5,1

   What polar coordinates could you enter at the Specify next point or [Undo] prompt to achieve exactly the same line?

3. When might the polar tracking feature be useful?

Applying AutoCAD Skills

Work the following problems to practice the commands and skills you learned in this chapter. ZOOM All at the beginning of each drawing.

1. Figure 6-9 shows the end view of a structural member for an architectural drawing. Create the rectangle using the LINE command and the following keyboard entries.

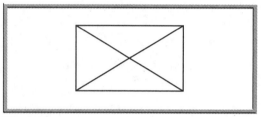

Fig. 6-9

Specify first point: 1,1

Specify next point or [Undo]: 10.5,1

Specify next point or [Undo]: @0,5.5

Specify next point or [Close/Undo]: @9.5<180

Specify next point or [Close/Undo]: c

Draw a line from the intersection at the lower left corner of the rectangle to the upper right corner of the rectangle. Use coordinate pairs to place the endpoints of the diagonal line exactly at the corners of the triangle. The new line defines the rectangle's diagonal distance, or the hypotenuse of each of the two triangles you have created. Then draw a line from the lower right corner to the upper left corner of the rectangle to create the other diagonal. Save the drawing as prb6-1.dwg.

2-3. List on a separate sheet of paper exactly what you would enter when using the LINE command to produce the drawings of sheet metal parts in Figs. 6-10 and 6-11. Then step through the sequence in AutoCAD. Try to use four methods—absolute, relative, polar, and polar tracking—to enter the points. (Note that the horizontal lines in the drawings are perfectly horizontal.) Be sure to ZOOM All at the beginning of each drawing. Save the drawings as prb6-2.dwg and prb6-3.dwg.

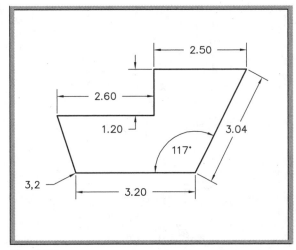

Fig. 6-10

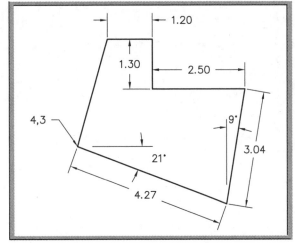

Fig. 6-11

4. Draw the 3½″ disk shown in Fig. 6-12. Save the drawing as prb6-4.dwg.

5. Draw the front view of the spacer plate shown in Fig. 6-13. Start the drawing at the lower left corner using absolute coordinates of 3,3. Determine and use the best methods of coordinate entry to draw the spacer plate. Save the drawing as prb6-5.dwg.

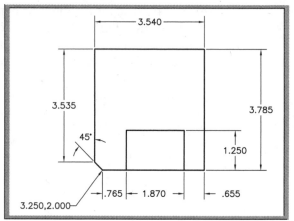

Fig. 6-12

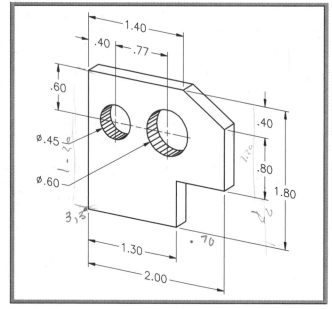

Fig. 6-13

Problem 5 courtesy of Gary J. Hordemann, Gonzaga University

6. Draw the front view of the locking end cap shown in Fig. 6-14. Start the drawing at the center of the object using absolute coordinates of 5,5. Use absolute coordinates to locate the centers of the four holes, and use polar tracking with the RECTANG command to draw the rectangular piece. Save the drawing as prb6-6.dwg.

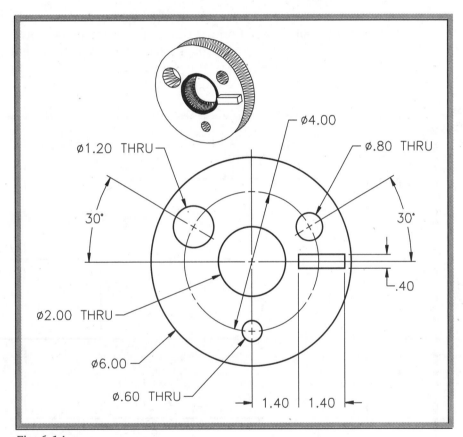

Fig. 6-14

Problem 6 courtesy of Gary J. Hordemann, Gonzaga University

Using Problem-Solving Skills

Complete the following activities using problem-solving skills and your knowledge of AutoCAD.

1. Draw the shim (Fig. 6-15). Decide on the best method of coordinate entry for each point, based on the dimensions shown and your knowledge of the various methods. Position the drawing so that the lower left corner of the shim is at absolute coordinates 1,1. Do not include the dimensions. Save the drawing as ch6-shim.dwg.

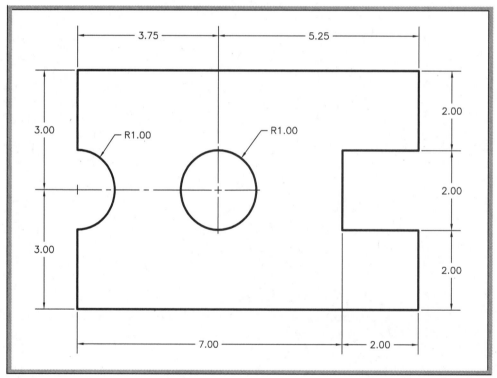

Fig. 6-15

2. Draw the front view of the slider block shown in Fig. 6-16.
Assume the spacing between dots to be .125. Start the drawing
at the lower left corner using absolute coordinates of 1,3. Use
the RECTANG command to draw the rectangular slot. Use relative
polar coordinates to locate the endpoints of the lines, and absolute
rectangular coordinates to locate the centers of the three holes.
Save the drawing as ch6-block.dwg.

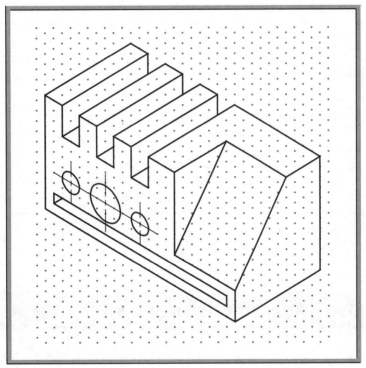

Fig. 6-16

Chapter 7   Getting Help

Objectives:

- Browse a group of files
- Obtain help on AutoCAD commands and topics
- Use AutoCAD's context-sensitive help features

Key Term

context-sensitive help

AutoCAD provides a rich mix of on-line help and file searching capabilities. Its digital versions of the *User's Guide, Command Reference, Driver and Peripheral Guide, Installation and Licensing Guides*, and *Customization Guide* are no more than a few clicks away no matter what you are doing in AutoCAD. Several of these guides are also available to you in PDF form. Adobe's® Portable Document Format (PDF) files open with the free Acrobat Reader® software. The *Quick Reference* and *Frequently Asked Questions and Answers* are also available as PDF files. Autodesk, the maker of AutoCAD, also provides on-line assistance on how you can obtain consulting, training, and support.

Browsing and Searching

When you open a drawing, AutoCAD provides visual help, making the file selection process faster.

1. Start AutoCAD and display the AutoCAD window.

2. Pick the **Open** button from the Standard toolbar.

The Select File dialog box appears.

3. From the **Views** pull-down menu, located in the upper right area of the dialog box, select **Details**, unless it is grayed out, which means it is the current selection.

4. Rest the pointer over the name of each drawing file. If this does not produce a thumbnail in the Preview area, single-click the file.

As you have seen in previous chapters, the drawings appear in a preview box in the right area of the dialog box. This makes it easier to find drawings that you or others have created and saved.

5. Pick the **Cancel** button.

Finding Drawing Files

Using the Select File dialog box, AutoCAD provides other options for reviewing and finding files.

1. Pick the **Open** button on the Standard toolbar.

2. From the **Tools** pull-down menu, located in the upper right area of the dialog box, select the **Find...** item.

This displays the Find dialog box, as shown in Fig. 7-1.

3. In the text box after Named, enter **stadium north elevation** and pick the **Find Now** button.

AutoCAD searches and attempts to find the file named Stadium North

Fig. 7-1

Elevation.dwg. If AutoCAD successfully finds it and displays the file details in the lower area of the dialog box, go to Step 5. If the search is unsuccessful, proceed to Step 4.

4. Click the down arrow located at the right of Look in, select the disk drive in which AutoCAD was installed, and pick the **Find Now** button.

5. Pick the **OK** button to open the folder containing this file.

AutoCAD opens the folder containing the file.

6. Select **Find...** again from the **Tools** pull-down menu.

7. Pick the down arrow located at the right of Type.

This permits you to specify other file types to search.

8. Pick the **Browse** button.

It enables you to select a specific folder or disk drive to search.

9. Pick the **Cancel** button.

10. Pick the **Date Modified** tab.

This gives you the option of specifying dates within which the file you are seeking was created or modified.

11. Pick the **Cancel** button.

12. Pick the **Cancel** button again to remove the Select File dialog box.

You will learn more about the Select File dialog box in the next chapter.

Obtaining Help

The AutoCAD software offers many forms of help. Most of these can be obtained by entering the HELP command.

1. Find and open the **engine.dwg** file.

The cylinder head component you created in Chapter 4 should now be visible on the screen.

Standard

2. Pick the **Help** button from the Standard toolbar, or enter **HELP** or **?** at the keyboard.

HINT:

You can also obtain help by pressing the F1 function key.

The AutoCAD 2006 Help window appears.

Searching by Content Area

At this point, you have several options for obtaining help.

3. Pick the **Contents** tab, if it is not already selected.

4. Double-click **Command Reference** or click the small box next to it containing the plus (+) character.

5. Click the small box containing the plus character next to Commands.

6. Click the small box containing the plus character next to C Commands.

AutoCAD displays all of the commands and subtopics that begin with the letter C.

7. Click on the **COPY** command.

This displays help on the COPY command.

8. In the upper left area of the AutoCAD 2006 Help window, pick the **Back** button.

This displays the previous screen of information.

9. Close the AutoCAD 2006 Help window by clicking the **x** in the upper right corner of the window.

Using the Help Index

Another way to search for help in AutoCAD is to use the help index. Using this method, you can find entries in all of the on-line help documents by entering just the first few letters of the command or topic for which you need help.

1. Reopen the **AutoCAD 2006 Help** window.

HINT:

To display the AutoCAD 2006 Help window, you may need to click the AutoCAD 2006 Help button on the Windows taskbar located at the bottom of the screen. This will bring it to the foreground.

2. Pick the **Index** tab.

3. Use the scrollbar to scroll down through the list of topics.

Notice a blinking cursor inside the empty entry box.

4. Type **M** in upper- or lowercase, but do not press ENTER.

AutoCAD very quickly finds the first topic in the list box that begins with the letter M.

5. After the M, type **O** and then **V**, but do not press ENTER.

AutoCAD finds the first topic that begins with MOV.

6. Locate and double-click the item named **MOVE command**.

This displays the topics related to the MOVE command on the right side of the AutoCAD 2006 Help window.

7. Click on the Concepts, Procedures, and Commands tabs to display more information on this item.

Using the Search Tab

You can also use the Search tab to search for specific topics in AutoCAD.

1. Pick the **Search** tab.

2. Click in the edit box and type **move 3D objects**.

3. Pick the **List Topics** button to see a list of related topics.

4. Use the scrollbar to move through the topics.

5. Locate and double-click the item named **Move or Rotate Objects – Procedures**.

AutoCAD lists procedures that contain the any or all of the words *move*, *3D*, or *objects*.

6. Pick in the right side of the AutoCAD 2006 Help dialog box and scroll down to see the complete list of procedures. Notice that the search words are highlighted in blue so that you can see them easily.

Using the Ask Me Option

This option permits you to phrase a specific question that you want answered.

1. Pick the **Ask Me** tab.

2. Read the message that displays and pick the **ASK** button.

3. In the text box located after Type a question and press Enter, type **How do I use the MOVE command?** and press **ENTER**.

AutoCAD displays the topics that may answer your question.

4. Review the list of topics and select **To move an object using two points – Procedures**.

AutoCAD provides step-by-step instructions on how to move an object using two points.

5. Close the AutoCAD 2006 Help window.

Other Sources of Help

The Help pull-down menu provides several sources of help. Selecting the first item on the menu, Help, is the same as entering the HELP command. The remaining items on the pull-down menu offer other types of assistance, as shown below.

Info Palette	Displays a list of procedures relevant to the current command. Its guidance is often enough to get you started with unfamiliar or rarely used tasks.
New Features Workshop	Helps you learn what is new in AutoCAD 2006.
Subscription e-Learning Catalog	A subscription service that provides self-paced, interactive lessons that are organized into product catalogs.

Create Support Request	Provides direct, one-on-one support communication with Autodesk support technicians.
View Support Requests	Allows you to track and manage your requests and their responses.
Edit Subscription Center Profile	Allows you to view and edit your subscription profile.
Additional Resources	Displays the following five options, each of which requires an Internet connection and Web browser.
Support Knowledge Base	Offers information on current issues and permits you to search the Support Knowledge Base.
Online Training Resources	Gives instructor-led and self-paced training options, how-to articles, and learning tips.
Online Developer Center	Provides software developers with tools and technologies to develop third-party software applications.
Developer Help	A Help window specifically for third-party AutoCAD developers.
Autodesk User Group International	Describes the user group.

1. Experiment with the **Info Palette** and **New Features Workshop**.

2. Close any help-related windows before proceeding to the next section.

Context-Sensitive Help

AutoCAD allows you to obtain help in the middle of a command, when you're most likely to need it. If another command is active when you enter the HELP command, AutoCAD automatically displays help for the active command. This is known as *context-sensitive help*. When you close the help window, the command resumes.

1. Enter the **GRID** command at the keyboard.

2. Pick the **Help** button.

Standard

AutoCAD displays help information on the GRID command.

3. After closing the help window, press the **ESC** key to cancel the GRID command.

4. Close any help windows that might be open and exit AutoCAD without saving.

Chapter 7
Review & Activities

Review Questions

1. Suppose you want to open a specific drawing file in the folder you named after yourself, but you've forgotten the name of it. Explain how to browse through the files in that folder to find the file you want to open.

2. When you pick the HELP button, what does AutoCAD display?

3. How do you obtain a listing of all AutoCAD topics?

4. Suppose you have entered the MIRROR command. At this point, what is the fastest way of obtaining help on the command?

5. Name and briefly describe each of the options available in the Help pull-down menu.

6. What is context-sensitive help? Explain how to use it in AutoCAD.

Challenge Your Thinking

1. For some commands, the help information may be long and complicated. Discuss ways to make a permanent hard copy of commonly used help information to keep for reference.

2. Explain the benefit of using the Ask Me capability in the AutoCAD 2006 Help window. When would you use it instead of the Index?

Applying AutoCAD Skills

Work the following problems to practice the commands and skills you learned in this chapter.

1. You know you have an elevation drawing of a public building somewhere on your hard drive, but you don't remember the full name of the drawing. You recall that it starts with the letters "hu," and because it's a drawing file, you know that it has a DWG file extension. Find the drawing. What is its name and what does it contain?

Chapter 7
Review & Activities

2. Obtain AutoCAD help on the following commands and state the basic purpose of each.

   a. EXPORT
   b. SNAP
   c. TIME
   d. SCALE
   e. TRIM

3. Locate each of the above commands in the on-line *Command Reference* and read what it says about each of them.

 ## Using Problem-Solving Skills

Complete the following activities using problem-solving skills and your knowledge of AutoCAD.

1. Your company has purchased a new plotter for the AutoCAD network. You have been asked to configure it, but since you don't do many configurations, you need some help. Find the instructions in the AutoCAD help system for configuring a network nonsystem plotter, and write down the path for future reference.

2. You need to draw two concentric circles (circles that share the same center point). You have already drawn the inner circle. The outer circle needs to be .5″ larger. Enter the OFFSET command from the Modify toolbar and access context-sensitive help to offset the diameter of the first circle by .5″.

Chapter 7: Getting Help

Chapter 8   File Maintenance

Objectives:
- **Copy, rename, move, and delete drawing files**
- **Create and delete folders**
- **Audit and recover damaged drawing files**

Key Terms
audits
file attributes

When working with AutoCAD, there will be times when you want to review the contents of a folder or delete, rename, copy, or move files. This chapter gives you practice in using these functions, as well as commands that attempt to repair drawing files containing errors.

File and Folder Navigation

AutoCAD's Select File dialog box can help you organize and manage AutoCAD-related files.

1. Start AutoCAD and display the AutoCAD window.

2. Pick the **Open** button from the Standard toolbar.

AutoCAD displays the Select File dialog box (Fig. 8-1). Notice the buttons along the top of the dialog box.

3. Find and open the folder with your name—the one you created in Chapter 3.

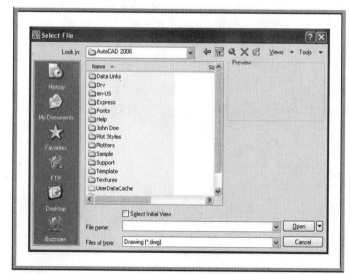

Fig. 8-1

Viewing File Details

4. Pick the **Views** drop-down menu and select **Details**, unless it is grayed out, which means it is the current selection.

This provides a listing of each file, along with its size, type, and the date and time it was last modified. These items are known as *file attributes*. In folders that contain many drawing files, it is sometimes easier to find files if they are listed in a certain order. By default, files are listed in alphabetical order by name. You can change the order in which the files and folders are listed by clicking the column names, which are actually buttons.

5. Click the gray bar containing **Name**.

This reverses the order of the files. They are now in reverse alphabetical order (from Z to A).

6. Click the same gray **Name** bar again.

The files are once again listed in alphabetical order.

7. Click the gray bar containing **Size**.

This places the files in order of size from smallest to largest. Clicking it again would reorder the files from largest to smallest. Clicking the gray bars containing Type and Date Modified organizes the files according to type and date, respectively.

8. Click the gray bar containing **Date Modified**; pick it again.

9. From the **Views** drop-down menu, pick **List** to produce a basic listing of the files.

Copying Files

The Select File dialog box permits you to copy files easily from within AutoCAD.

1. Right-click and drag the **stuff.dwg** file to an empty area within the list box area and release the right button on the pointing device.

A shortcut menu appears.

2. Pick **Copy Here**.

A new file named Copy of stuff.dwg appears.

Renaming Files

You can also rename files using the Select File dialog box.

1. Right-click the new file named **Copy of stuff.dwg**.

2. Select **Rename** from the shortcut menu.

You should now see a blinking cursor at the end of the file name with a box around it.

3. Type the name **junk.dwg** and press **ENTER**.

 NOTE: ————————————————————

Instead of entering an entirely new name, you can also edit the name of the file by picking a new cursor location and typing new text. This method of renaming also permits you to use the backspace key and the spacebar.

The new file named junk.dwg contains the same contents as stuff.dwg.

Creating New Folders

It's also fast and easy to create new folders.

1. Pick the **Create New Folder** button.

A new folder named New Folder appears.

2. To rename the folder, right-click the new folder, select **Rename** from the menu, type **Junk Files**, and press **ENTER**.

Moving Files

Let's move a file to the new folder.

1. Right-click **junk.dwg** and drag and drop it into the **Junk Files** folder.

A shortcut menu appears.

2. Pick **Move Here**.

The junk.dwg file is now located in the Junk Files folder.

Deleting Files and Folders

AutoCAD also allows you to delete unwanted files and folders using the Select File dialog box.

1. Open the **Junk Files** folder.

2. Right-click **junk.dwg** and select **Delete** from the shortcut menu.

3. Pick the **Yes** button in the Confirm File Delete dialog box if you are sure you have selected junk.dwg.

4. Pick the **Up one level** button.

5. Right-click the **Junk Files** folder and select **Delete** from the shortcut menu.

6. If you are sure that you selected the Junk Files folder, pick the **Yes** button in the Confirm Folder Delete dialog box to send this folder to the Windows Recycle Bin.

7. Close the Select File dialog box.

Diagnosing, Repairing, and Recovering Files

Occasionally, AutoCAD drawing files may become corrupted. AutoCAD supplies two commands to help fix corrupted files.

Auditing a Drawing File

The AUDIT command is available as a diagnostic tool. It *audits*, or examines the validity of, the current drawing files. It can also correct some of the errors it finds. The AUDIT command automatically creates an audit report file containing an ADT file extension when the AUDITCTL system variable is set to 1 (On). The default setting is 0.

Probable causes of damaged files include:

• An AutoCAD system crash

• A power surge or disk error while AutoCAD is writing the file to disk

When a file becomes damaged, AutoCAD may refuse to edit or plot the drawing. In some cases, a damaged file may cause an AutoCAD internal error or fatal error.

1. Open the file named **stuff.dwg**.

2. From the **File** pull-down menu, select **Drawing Utilities** and **Audit**.

This enters the AUDIT command.

3. In reply to Fix any errors detected? press **ENTER** to accept the No default.

4. Press the **F2** function key to view the text window and read all of the message.

Information similar but not identical to that shown in Fig. 8-2 appears.

If you had entered Yes in reply to Fix any errors detected? the report would have been the same if this file has no errors.

```
Command: _audit
Fix any errors detected? [Yes/No] <N>:
1      Blocks audited
Pass 1 27      objects audited
Pass 2 27      objects audited
Total errors found 0 fixed 0
```

Fig. 8-2

NOTE:

AutoCAD offers a command similar to AUDIT named RECOVER. This command attempts to open and repair damaged drawing files. The "recover" process has been embedded in the OPEN command, so when you try to open a damaged file, AutoCAD automatically executes the RECOVER command and attempts to repair it.

Permanently Damaged Files

A drawing file may be damaged beyond repair. If so, drawing recovery with the AUDIT and RECOVER commands will not be successful.

Each time you save an AutoCAD file, AutoCAD saves the changes to the current DWG file. AutoCAD also creates a second file with a BAK (backup) file extension. This file contains the previous version of the file—the version prior to saving. If the DWG file becomes damaged beyond repair, you can rename the BAK file to a DWG file and use it instead. To prevent the loss of data, save often and produce a backup copy of your DWG files frequently.

The Drawing Recovery Manager

When a hardware problem, power failure, or software problem causes AutoCAD to terminate unexpectedly, the Drawing Recovery Manager opens the next time you start AutoCAD. It displays a list of the drawing files that were open at the time the program terminated.

If you pick a file, AutoCAD lists all the available versions of the drawing that may be recovered, including the DWG file, the BAK file, and the SV$ (auto-save) file. These file types are listed in the order of their time stamps. You can double-click the most recent file, and AutoCAD will attempt to repair and open it. This allows you to recover the most recently saved version of the drawing that is available.

5. Exit AutoCAD without saving unless the file was damaged and recovered. In that case, save the changes to the file.

Managing Drawing Files with Windows

Microsoft Windows makes it possible to copy, rename, move, and delete AutoCAD drawing files. Also, it enables you to view thumbnails of the drawing files.

1. On your Windows desktop, find and double-click **My Computer**.

2. Browse until you find the folder with your name.

HINT:

If your folder is located on the Local Disk (C:), double-click it. If your folder is located in the folder named Documents and Settings, double-click it. Continue to find and select folders until the folder with your name appears.

When the folder with your name appears, you can copy, rename, move, and delete AutoCAD drawing files, the same as you did earlier with the Select File dialog box in AutoCAD.

3. Right-click the **engine.bak** file and delete it; then delete **gasket.bak**, if it exists. Be sure to select the BAK files and not the DWG files.

It is also possible to view the drawing files as thumbnails in this window.

4. From the **View** drop-down menu, select **Thumbnails**.

Thumbnail images of the drawing files appear.

NOTE:

If you do not see the thumbnails, it could be one of two problems: 1) the RASTERPREVIEW system variable was set to 0 (instead of 1) when the drawing file was saved, or 2) the video hardware or driver is not permitting the display of the thumbnails. Possible solutions are to decrease the video hardware acceleration or to update the video driver software from the driver manufacturer's Web site. If you do not know how to proceed, consult your instructor or system administrator.

5. From the **View** drop-down menu, select **Details**.

The files appear in detailed format.

6. Close the window.

Chapter 8
Review & Activities

Review Questions

1. Explain the purpose of each of the five items located in the upper right area of the Select File dialog box.

2. How do you copy a file in the Select File dialog box?

3. In the Select File dialog box, what information can you display by picking Details from the Views drop-down menu?

4. Explain how to create a folder from within AutoCAD.

5. What is the primary purpose of the AUDIT command?

Challenge Your Thinking

1. Is it possible to move and copy files from the Select File dialog box to folders on the Windows desktop? Is it possible to move and copy files from folders on the Windows desktop to the Select File dialog box? Explain why this might be useful.

2. AutoCAD gives you the option of creating a report file when you run the AUDIT command. Explain why such a report might be useful.

Applying AutoCAD Skills

Using the features described in this chapter, complete the following activities.

1. Display a list of drawing files from one of AutoCAD's folders.

2. Display a list of template files found in one of AutoCAD's folders.

3. Rename one of your drawing files. Then change it back to its original name.

4. Make a copy of the file Colorwh.dwg in the AutoCAD Sample folder. Rename it Mycolorwh.dwg, and move it to your named folder.

5. Delete Mycolorwh.dwg from the folder with your name.

6. Diagnose one of your drawing files and fix any errors detected. Display the text of the AutoCAD message. How many passes did the diagnostic tool run? How many objects were included?

Using Problem-Solving Skills

Complete the following activities using problem-solving skills and your knowledge of AutoCAD.

1. It's time to build the bookcase you designed in Chapter 3. Make a copy of the bookcase you saved in the folder with your name as prb3-1.dwg. Modify the drawing as shown in Fig. 8-3, and save the modification as bookcase.dwg. Delete the copy of prb3-1.dwg.

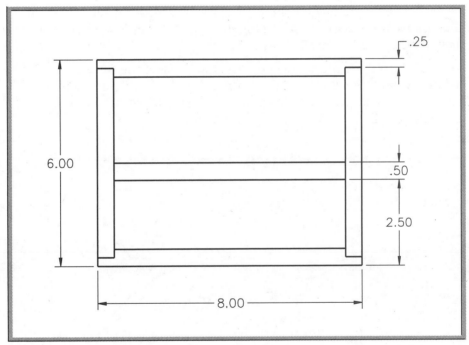

Fig. 8-3

2. To keep track of projects, you need to organize them into folders. Make a new folder called Build within your named folder. Move bookcase.dwg into the new folder and rename it shelves.dwg. Check the file for corruption.

Part 1 Project

Rocker Arm

Engineering drawing refers to the creation of highly accurate drawings using drafting instruments or, more commonly today, CAD software. Engineering drawings are used, among other things, to describe a part so that the manufacturer can build it correctly.

Description

The rocker arm shown in Fig. P1-1 was submitted to your manufacturing company as a hand sketch. The basic design has been approved, and now the marketing department needs an electronic version of the drawing for use in a new product brochure. The drawing must reflect the actual dimensions of the rocker arm, although for this use, it is not necessary to show the dimensions on the drawing.

Your task is to draw the rocker arm accurately using the AutoCAD commands and procedures you learned in Part 1. Be sure to read the "Hints and Suggestions" for this project before you begin. Save the drawing as **rockerarm.dwg** in your drawing file.

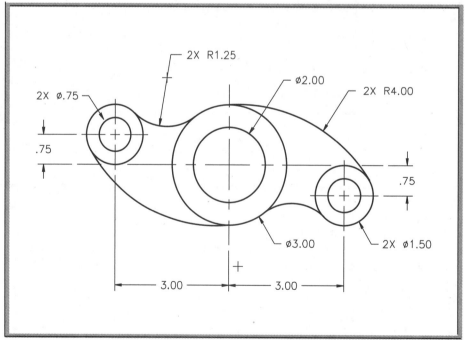

Fig. P1-1

Hints and Suggestions

1. The lines made up of short and long dashes are center lines (lines that intersect at the exact center of a feature). Use these lines and your knowledge of coordinates to draw the rocker arm accurately.

2. Study the drawing to determine the best order in which to draw the parts of the rocker arm. Keep in mind that it is *not* necessary to construct the part linearly, so that each line connects to the next. For example, you may choose to create all the holes and cylindrical features before you draw the other parts of the rocker arm.

3. If you have trouble visualizing how to draw the part in AutoCAD, try sketching it first on graph paper.

Summary Questions

Your instructor may direct you to answer these questions orally or in writing. You may wish to compare and exchange ideas with other students to help increase efficiency and productivity.

1. Pinpoint any difficulties you had in drawing the rocker arm and explain how you solved them.

2. Describe the order in which you drew the parts to make up the rocker arm. Why did you choose this order?

3. Study the drawing in P1-1 once more. Does this drawing provide everything needed to manufacture the actual part? If not, what is missing?

4. In this project, you have created an electronic drawing file to be used by the marketing department because marketing needed the file in electronic format. However, there are other advantages to creating the file electronically. How else can this file be used? How would it need to be modified to be useful to other departments? Explain your answer.

Careers Using AutoCAD

Mechanical Drafter

Mechanical devices are all around you. At the grocery store, conveyer belts help move shoppers' selections from their carts to the cashier. Mechanics use hydraulic lifts to raise cars in order to work on parts underneath. Military personnel use all sorts of equipment—gauges, tracking devices, etc.—to perform their jobs. Have you ever wondered how mechanical devices are able to perform certain functions? Have you ever taken a mechanical object apart just to see how it works? If so, you might be interested in learning more about mechanical drafting.

On the Job

Mechanical drafters make up the largest category of drafters in the United States. They concentrate on designing different types of machinery and mechanical tools. Accurate, detailed drawings of mechanical equipment must be provided for manufacturers to use as guides during the building process. Today, most mechanical drafters use AutoCAD or a similar CAD program to assist with the overall design process.

A Mechanical Mindframe

Curiosity about how mechanical devices operate and good visual aptitude (the ability to visualize how an object might look from all angles and dimensions) are necessary skills for a mechanical drafter to possess. The ability to communicate and work well with others is also important.

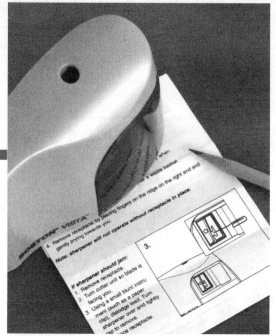

Tim Fuller Photography

Mechanical drafters often consult with the individuals who initially thought up product ideas; they also confer regularly with other drafters and mechanical engineers. Perhaps most importantly, mechanical drafters must have a knack for understanding exactly how mechanical items work. Attention to detail is important when studying existing devices in order to improve future designs.

Career Activities

- Look around the room. What types of mechanical devices do you see? How much planning do you think went into the design phase of the objects at which you are looking?
- Try to sketch a rough example of how you think a design drawing might look. For example, you probably know that an electric clock has gears, and a pencil sharpener contains rotating blades. Share your drawings with a classmate and compare ideas.

Chapter 9 — Object Snap

Objectives:
- Set and use running object snap modes
- Apply object snaps that are not currently turned on
- Change object snap settings to increase productivity
- Use the **Quick Setup** wizard

Key Terms

aperture box	quadrant points
object snaps	running object snap modes
object snap tracking	running object snaps
perpendicular	system variable
quadrant	wizard

Object snaps are like magnets that permit you to pick endpoints, midpoints, center points, points of tangency, and other specific points easily and accurately. *Object snap tracking*, which is used in conjunction with object snap, provides alignment paths that help you produce objects at precise positions and angles. Combined with polar tracking, these features speed the precision drawing process.

Running Object Snaps

You can use object snaps in different ways to make a drawing task easier. When you know that you will be needing one or more specific object snaps several times, you can preset them to "run" in the background as you are working. Object snaps that have been preset in this way are known as *running object snaps*.

Quick Setup Wizard

In preparation for using object snaps, we will use the Quick Setup wizard to establish basic settings quickly in AutoCAD. Quick Setup is based on the acad.dwt template file.

1. Start AutoCAD.
2. Enter **STARTUP** at the AutoCAD prompt and enter **1**.

STARTUP is a system variable. A *system variable* is a setting that controls a specific function in AutoCAD. The default value of the STARTUP system variable is 0. When it is set at 1, AutoCAD displays the Startup dialog box when you start the software.

Standard

3. Enter the **NEW** command or pick the **QNew** button from the Standard toolbar.

The Create New Drawing dialog box appears. The four buttons at the top permit you to open a drawing, start a drawing from scratch, choose a template file, or use a wizard. A *wizard* is a prompted sequence of steps that streamlines a task or set of tasks. Using the Start from Scratch button is equivalent to beginning a new drawing with STARTUP set at 0. In either case, all of the settings come from the acad.dwt template file.

4. Pick the **Use a Wizard** (fourth) button.

5. Select **Quick Setup** and pick **OK**.

AutoCAD displays a dialog box that focuses on the unit of measurement. Decimal is the default selection. Notice the sample in the right area of the box.

6. Accept the Decimal default selection and pick the **Next** button.

The dialog box concentrates on the drawing area. As you can see, the default drawing area is 12 × 9 units.

7. For Width, enter **95**, and for Length, enter **70**.

8. Pick the **Finish** button and save your work in a file named **bike.dwg**.

The drawing area now represents 95 × 70 units. These units can represent millimeters, kilometers, feet, miles, or whatever we want them to be. We will use them as inches.

9. Enter **Z** for ZOOM and **A** for All.

This zooms the drawing area to the full 95 × 70 inches.

> **INFOLINK**
>
> See Chapter 12 for more about the ZOOM command.

Specifying Object Snap Modes

1. Use the pick button on the pointing device to depress the **POLAR**, **OSNAP**, and **OTRACK** buttons on the status bar, unless they are already depressed.

The MODEL button should also be depressed, but the SNAP, GRID, and ORTHO buttons should not be.

2. Enter the **OSNAP** command.

Chapter 9: Object Snap

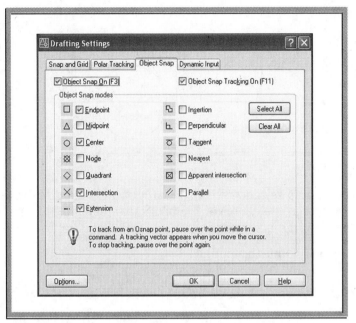

Fig. 9-1

This displays the Drafting Settings dialog box. Entering the OSNAP command is equivalent to selecting Drafting Settings... from the Tools menu and selecting the Object Snap tab in the dialog box (Fig. 9-1).

3. Review the object snap modes.

The ones that are checked are called *running object snap modes*. These object snaps are "on" and function automatically as you work with drawing and editing commands in AutoCAD.

4. Check **Endpoint**, **Center**, **Intersection**, and **Extension**, and uncheck all the others. (Some or all of them may be checked already.)

Notice that Object Snap On and Object Snap Tracking On are checked. This is because the OSNAP and OTRACK buttons on the status bar are on (depressed). A third way to turn them on and off is to press the F3 and F11 function keys, as indicated in the dialog box.

5. Pick the **OK** button.

Using Preset Object Snaps

Let's use AutoCAD's object snaps to create a basic drawing of a new mountain bike frame design (Fig. 9-2 on page 104). Be sure to save your work every few minutes as you work through this chapter.

Draw

1. In the lower left area of the screen, draw a 22″-diameter circle to represent the front wheel as shown in Fig. 9-2.

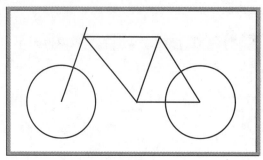

Fig. 9-2

The Center object snap selects the exact center of a circle or arc. Let's use it to start a 24″ line at the-center of the bike's front wheel. This line will begin the outline of the bike frame.

Draw

2. Enter the **LINE** command.

3. At the Specify first point prompt, move the crosshairs so that it touches the circle or move the crosshairs to the circle's center.

A small yellow marker appears at the circle's center. The yellow marker indicates that AutoCAD has selected the point at the center of the circle.

4. While the yellow marker is present, pick a point.

AutoCAD snaps to the circle's center.

5. For the next endpoint, use polar coordinates to extend the line **24″** up and to the right at a **70°** angle.

HINT:

Type 24, press the TAB key, type 70, and press ENTER, or enter @24<70 at the keyboard.

6. Press **ENTER** to terminate the LINE command.

Object Snap Tracking

Next, we are going to draw the horizontal line using AutoCAD's object snap tracking feature.

1. Reenter the **LINE** command and move the crosshairs to the second endpoint of the line until the Endpoint object snap marker appears.

The purpose of doing this is to "acquire" the Endpoint object snap. When you acquire an object snap, a small yellow + (plus sign) appears at the snap point. An acquired object snap also displays a small white x near the crosshairs. To remove an acquired object snap, pass over the snap point a second time.

2. Slowly move the crosshairs around the endpoint and notice the temporary alignment paths that appear.

As you may recall, alignment paths are temporary lines that display at specific angles from an acquired object snap. Object snap tracking is active when the alignment paths appear from one or more acquired object snaps. This feature is a part of AutoCAD's AutoTrack™. You can toggle AutoTrack on and off with the POLAR and OTRACK buttons on the status bar. (OTRACK stands for "object snap tracking.") Because object snap tracking works in conjunction with object snaps, one or more object snaps must be set in advance before you can track from an object's snap point.

3. Move the crosshairs along the line that you created earlier so that a small x appears on the line.

4. Enter **3**.

This starts a new line 3″ from the endpoint of the first line.

5. Draw a horizontal line segment **24″** to the right.

HINT:

When creating the second point, use polar tracking to force the line horizontal and enter 24.

6. In reply to Specify next point, click on the Command line and enter **@21<250** to create the next line segment.

NOTE:

The dynamic input prompt can also be used, but the result will depend on the position of the crosshairs at the time you type the angle value of 250. This is because AutoCAD measures the angle either clockwise or counterclockwise from the horizontal, depending on where you position the crosshairs.

7. In reply to Specify next point, move the crosshairs to the intersection of the first and second line segments.

NOTE:

The intersection of the two lines is also an endpoint of the second line. Therefore, either the Intersection or Endpoint object snap marker will appear. A yellow x-shaped marker depicts the Intersection object snap, and a square marker depicts the Endpoint object snap.

8. Snap to this point and press **ENTER** to terminate the LINE command.

9. Using the **Endpoint** object snap and polar tracking, create the second horizontal line. Make it **20″** long.

10. Create the final segment of the bike frame, as shown in Fig. 9-2.

11. Create the second wheel for the bike. Use the **Endpoint** object snap to place the center of the wheel at the right endpoint of the lower horizontal line. Make it the same size as the front wheel.

The drawing of the new bike design is now complete.

12. Save your work.

Specifying Object Snaps Individually

Another way to use object snaps is to specify them individually as you need them. You can do this either from the keyboard or by displaying and using the Object Snap toolbar. Using this toolbar streamlines the selection of object snaps.

1. Right-click on any of the six docked toolbars and select **Object Snap** from the shortcut menu.

The Object Snap toolbar appears as shown in Fig. 9-3.

Fig. 9-3

2. Rest the pointer on each of the buttons and read the tooltips.

Table 9-1 contains a list of object snap modes and their functions. To enter an object snap mode at the keyboard, enter only the capitalized letters.

Snapping to Quadrant Points

1. Enter the **LINE** command.

2. From the Object Snap toolbar, pick the **Snap to Quadrant** button.

Object Snap

Object Snap	Button	Purpose
APParent intersection		Snaps to the apparent intersection of two intersection objects that may not actually intersect in 3D space
CENter		Center of arc or circle
ENDpoint		Closest endpoint of line or arc
EXTension		Objects along the extension paths of objects
INSert		Insertion point of text, block, or attribute
INTersection		Intersection of lines, arcs, or circles
MIDpoint		Midpoint of line or arc
NEArest		Nearest point on a line, arc, circle, ellipse, polyline, spline, and other objects
NODe		Nearest point entity or dimension definition point
NONe		Temporarily cancels all running object snaps (for one operation only)
PARallel		Snaps to a point on an alignment path that is parallel to the selected object
PERpendicular		Perpendicular to line, arc, or circle
QUAdrant		Quadrant point of arc or circle
TANgent		Tangent to arc or circle
FROm		Used in combination with other object snaps to establish a temporary reference point as a base point
Object Snap Settings		Displays the Object Snap tab of the Drafting Settings dialog box
Temporary Tracking Point		Allows you to set a temporary OTRACK point while a drawing command is active

Table 9-1

Quadrant points on a circle are the points at 0°, 90°, 180°, and 270°. If you were to connect these points with horizontal and vertical lines, as shown in Fig. 9-4, you would create four equal parts. Each part represents a *quadrant* of a circle.

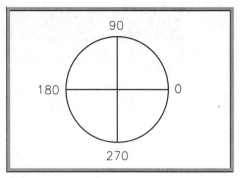

Fig. 9-4

3. Move the crosshairs over one of the two wheels until a small yellow diamond marker appears, along with the tooltip Quadrant.

4. Move to each of the four quadrant points and pick one of them.

5. Enter **ESC** to cancel the LINE command.

Snapping Perpendicularly

The Perpendicular object snap allows you to snap to a point perpendicular to an object such as a line, multiline, polyline, ray, spline, construction line, or even a point on a circle or arc. (A line is *perpendicular* to another line when the two form a right angle.)

1. Enter the **LINE** command and pick a point anywhere above the top horizontal segment of the bike.

Object Snap

2. Pick the **Snap to Perpendicular** button from the Object Snap toolbar.

3. Move the crosshairs along the top horizontal segment of the frame until a right angle marker appears, along with the tooltip Perpendicular, and pick a point.

AutoCAD snaps the line perfectly perpendicular to the horizontal segment of the frame.

Standard

4. Press **ENTER** to terminate the LINE command and pick the **Undo** button to remove the line.

Snapping Parallel Lines

You can use the Parallel object snap to produce parallel line segments.

Object Snap

1. Reenter the **LINE** command, pick another point above the bike, and pick the **Parallel** object snap from the toolbar.

2. Touch the crosshairs on the fork of the frame (the first line you drew) and then move the crosshairs until a parallel marker appears, and then back and forth until a temporary line is parallel to the fork.

3. Pick a point anywhere to form the parallel line.

4. Press **ENTER** and pick the **Undo** button.

Snapping to Points of Tangency

The Tangent object snap permits you to snap to a point on an arc or circle that, when connected to the last point, forms a line tangent to that object.

Object Snap

1. Enter the **LINE** command and pick the **Snap to Tangent** button from the Object Snap toolbar.

2. Move the crosshairs along the bottom half of the front wheel until a Snap to Tangent marker appears, along with the tooltip Deferred Tangent.

AutoCAD defers the point of tangency until after you've picked both endpoints of the line segment.

3. Pick a point anywhere along the bottom half of the front wheel.

Object Snap

4. Pick the **Snap to Tangent** button again.

5. Move the crosshairs along the bottom half of the rear wheel until the Snap to Tangent marker appears and pick a point.

AutoCAD draws a line tangent to both wheels.

6. Using polar tracking, extend the line **12″** to the right and press **ENTER** to end the LINE command.

Snapping to the Nearest Point

The Nearest Point object snap mode snaps to a location on the object that is closest to the crosshairs. This is particularly useful when you want to make certain that the point you pick lies precisely on the object and not a short distance away from it.

Object Snap

1. Pick the **Object Snap Settings** button from the Object Snap toolbar.

This also displays the Object Snap tab of the Drafting Settings dialog box. (You entered the OSNAP command to display it the last time.)

2. Pick the **Select All** button and pick **OK**.

All object snaps are now on.

3. Enter the **LINE** command and slowly move the crosshairs over the snap points of the objects you've drawn.

A yellow hourglass-shaped marker depicts the Nearest object snap.

4. Snap to a Nearest point on one of the objects.

Cycling Through Snap Modes

Pressing the TAB key cycles through all of the running snap modes associated with the object. This is useful when an area is densely covered with lines and other objects, making it difficult to snap to a particular point.

1. Rest the crosshairs on the rear wheel.

2. Press the **TAB** key.

3. Press **TAB** several more times, stopping for a moment between each.

As you can see, AutoCAD moves from one object snap to the next.

4. Press **ESC** to cancel the LINE command.

5. Pick the **Object Snap Settings** button from the Object Snap toolbar.

6. Pick the **Clear All** button, check **Endpoint, Center, Intersection**, and **Extension**, and pick **OK**.

Object Snap

Object Snap Settings

Other settings allow you to tailor the behavior of object snaps to meet your individual needs.

Changing the Settings

1. Pick the **Object Snap Settings** button from the Object Snap toolbar.

2. Pick the **Options...** button from the dialog box.

Marker, Magnet, and Display AutoSnap tooltip should be checked. When Marker is checked, AutoCAD geometric markers appear at each of the snap points. (These are the yellow markers you have used throughout this chapter.) When Magnet is checked, AutoCAD locks the crosshairs onto the snap target. When Display AutoSnap tooltip is checked, AutoCAD displays a tooltip that describes the name of the snap location.

3. Check the **Display AutoSnap aperture box** check box.

This causes an aperture box to appear at the center of the crosshairs when you snap to objects. The *aperture box* defines an area around the center of the crosshairs within which an object or point will be selected when you press the pick button.

4. Under **AutoSnap marker color**, pick the down arrow to display a list of colors and select green.

5. In the **AutoSnap Marker Size** area, move the slider bar slightly to the right, increasing the size of the marker.

In the AutoTrack Settings area of the dialog box, all three check boxes should be checked. When Display polar tracking vector is checked, AutoCAD displays an alignment path for polar tracking. When Display full-screen tracking vector is checked, AutoCAD displays alignment vectors as infinite lines. When Display AutoTrack tooltip is checked, AutoCAD displays AutoTrack tooltips.

The Alignment Point Acquisition area controls the method of displaying alignment vectors. When Automatic is picked, AutoCAD displays tracking vectors automatically when the aperture moves over an object snap. When Shift to acquire is picked, AutoCAD displays the tracking vectors only if you press the SHIFT key while you move the aperture over an object snap.

6. In the **Aperture size** area, slightly increase the size of the aperture using the slider bar.

7. Pick the **Apply** and **OK** buttons.

8. Pick the **OK** button in the Drafting Settings dialog box.

Applying the New Settings

Now let's examine how the changes made in the dialog box affect the drawing process.

Draw

1. Enter the **ARC** command and move the crosshairs over the drawing.

Notice that the object snap markers appear when the aperture box contacts the object.

2. Press **ESC** to cancel the **ARC** command.

The AutoSnap settings are stored with AutoCAD, not with the drawing. This means that AutoCAD remembers the changes. Likewise, running object snap modes are stored with AutoCAD.

3. Pick the **Object Snap Settings** button from the Object Snap toolbar and pick the **Options...** button.

Object Snap

4. Uncheck **Display AutoSnap aperture box**, change the marker color to yellow, and reduce the size of the marker and aperture box.

5. Pick **Apply** and **OK**; pick **OK** again.

As you work with polar tracking, object snap, and object snap tracking, keep in mind that you can toggle them on and off at any time using the POLAR, OSNAP, and OTRACK buttons on the status bar. Most of the time, you will want to use these features, but there will be times when you will want to disable them.

6. Practice using the remaining object snaps on your own.

7. Close the Object Snap toolbar and exit AutoCAD.

8. Enter **STARTUP** at the AutoCAD prompt and enter **0**.

9. Save your work and exit AutoCAD.

Chapter 9
Review & Activities

Review Questions

1. How is the Quick Setup wizard useful?

2. Explain the purpose of the object snaps.

3. Describe two methods of using AutoCAD's object snap modes.

4. In order to snap a line to the center of a circle, what part of the circle must the crosshairs touch?

5. What is the benefit of using running object snap modes?

6. Describe a situation in which you would want to change the aperture box size.

7. Briefly describe the use of each of the following object snap modes.
   a. Apparent Intersection
   b. Center
   c. Endpoint
   d. Extension
   e. Insertion
   f. Intersection
   g. Midpoint
   h. Nearest
   i. Node
   j. None
   k. Parallel
   l. Perpendicular
   m. Quadrant
   n. Tangent

8. Explain the benefit of object snap tracking.

9. Explain the relationship between polar tracking, object snap, and object snap tracking. Can you use object snap tracking when object snap is turned off?

Challenge Your Thinking

1. Imagine a new object snap that does not currently exist in AutoCAD. The new snap mode must provide a useful service. Write a short paragraph describing the new object snap. Explain what it does and why it is useful.

2. Investigate the Apparent Intersection object snap. Of what use is an object snap that allows you to snap to a point where two objects only *seem* to intersect? Give an example.

Applying AutoCAD Skills

Work the following problems to practice the commands and skills you learned in this chapter.

1. Draw the square on the left in Fig. 9-5. Then use object snaps to make the additions shown on the right to produce a top view of a jeweler's new design of a cut gemstone.

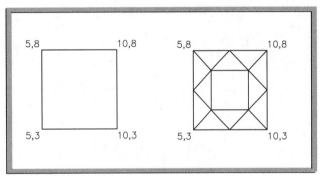

Fig. 9-5

2. Create a new drawing and set running snap modes for Center and Endpoint. Then create the shaft bearing in Fig. 9-6 using the dimensions shown. Start the lower left corner of the drawing at absolute coordinates 1,1, and work counterclockwise. Save the file as shaftbearing.dwg.

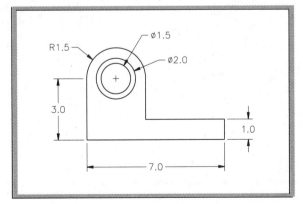

Fig. 9-6

3. Draw the top view of the video game part shown in Fig. 9-7.

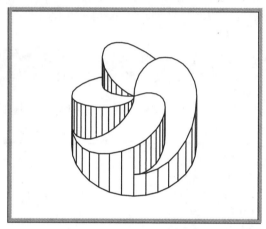

Fig. 9-7

Start by drawing a circle of any size at any location. Then use the Quadrant and Center object snaps to draw the four lines to divide the circle into quadrants, as shown on the left of Fig. 9-8. (You must draw four lines, not two.) Use the ARC command with the Start, Center, End option and Quadrant, Midpoint, and Center snap modes to draw the arcs. Erase the quadrant lines. Save the drawing as gamepart.dwg.

Fig. 9-8

Problem 3 courtesy of Gary J. Hordemann, Gonzaga University

Chapter 9: Object Snap

Using Problem-Solving Skills

Complete the following activities using problem-solving skills and your knowledge of AutoCAD.

1. Suppose your company provides special machine parts for a plastic injection molding company. Draw the new design for a pivot arm to swing the mold halves apart (Fig. 9-9). Use the Quick Setup wizard and select the appropriate drawing unit of measurement and drawing area. Set the appropriate object snap modes. Do not include dimensions. Save the drawing as injection.dwg.

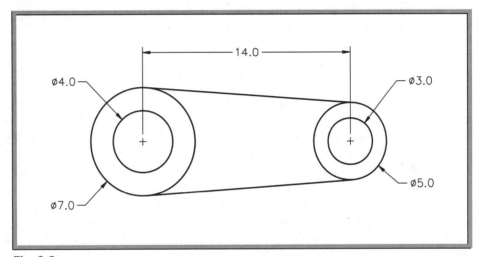

Fig. 9-9

2. The drawing in Fig. 9-10 represents a variation of a magnet for a new motor design. Draw the magnet. Select the appropriate drawing unit of measurement, drawing area, and object snap modes. Do not include dimensions, but save the drawing as magnet.dwg.

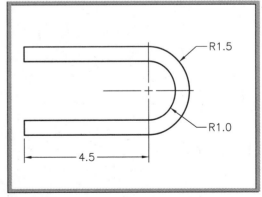

Fig. 9-10

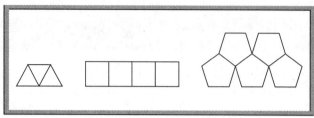

Fig. 9-11

3. The three figures in Fig. 9-11 are constructed of polygons according to the following rules:

   a. There are as many polygons as there are sides to the polygons.

   b. All polygons, except the two on the ends, share a single edge with each of two other polygons.

   If we imagine these figures to be groups of blocks sitting on a horizontal surface, then all three groups are stable (assuming a fair amount of friction between the surfaces). Using the Edge option of the POLYGON command with either the Intersection or the Endpoint object snap, draw similar groupings of hexagons and heptagons. Draw them as though they are sitting on a horizontal surface, and be sure to follow the two rules stated above. Is either grouping unstable? Can you formulate a general rule about the stability of such groupings that would apply to polygons with any number of sides?

Problem 3 courtesy of Gary J. Hordemann, Gonzaga University

Chapter 10   Helpful Drawing Features

Objectives:

- Use and change the display of coordinate information
- Restrict lines to vertical or horizontal orientations
- Track time spent on a drawing file
- Use the **AutoCAD Text** and **Command Line** windows
- Undo and redo multiple operations

Key Terms

ortho
orthogonal

Several features in AutoCAD provide information to help users become more productive. The coordinate display, for example, gives size information on the drawing area. The ortho feature helps you draw horizontal and vertical lines quickly and accurately. The TIME command tracks the time you spend working with AutoCAD. The AutoCAD Text and Command Line windows provide historical information on your interactions with the software. With AutoCAD's undo and redo features, you don't have to worry about making a wrong move.

Coordinate Display

Coordinate information, in digital form, is located in the lower left corner of the AutoCAD window and is part of the status bar. The coordinate display tracks the current position (or coordinate position) of the crosshairs as you move the pointing device. It also gives the length and angle of line segments as you draw.

1. Start AutoCAD and display the AutoCAD window.

2. Move the crosshairs in the drawing area by moving the pointing device and note how the coordinate display changes with the movement of the crosshairs.

3. Enter the **LINE** command and draw the first line segment of a polygon. Note the coordinate display as you draw the line segment.

NOTE: ─────────────────────────────────────

Keyboards have function keys called F1, F2, F3, and so on. AutoCAD has assigned specific functions, such as the coordinate display and the ortho feature, to many of these keys.

4. Press the **F6** function key once, draw another line segment, and notice the coordinate display.

The coordinate display should now be off.

5. Press **F6** again, draw another line segment or two, and watch the coordinate display.

As you can see, it now displays polar coordinate information.

6. While in the **LINE** command, click the coordinate display and note the change in the coordinate display as you move the crosshairs.

7. Create additional line segments, clicking the coordinate display between each.

As you can see, clicking the coordinate display serves the same function as pressing the F6 key.

8. Press **ENTER** to terminate the LINE command.

The Ortho Mode

Now let's focus on an AutoCAD feature called ortho. *Ortho,* short for "orthogonal," allows you to draw horizontal or vertical lines quickly and easily. *Orthogonal,* in this context, means to draw at right angles.

Ortho is on when the ORTHO button on the status bar is depressed. You can toggle ortho on and off by clicking the ORTHO button.

1. Turn off object snap if it is currently on.

2. Click the **ORTHO** button in the status bar or press the **F8** function key. (Both actions perform the same function.)

3. Experiment by drawing lines with ortho turned on and then with ortho off. Note the difference.

HINT: ─────────────────────────────────────

Like the coordinate display feature, ortho can be toggled on and off at any time, even while you're in the middle of a command.

Chapter 10: Helpful Drawing Features

4. Attempt to draw an angular line with ortho on.

As you can see, it is not possible.

5. Clear the screen if necessary and draw the plug shown in Fig. 10-1, first with ortho off and then with ortho on. Don't worry about exact sizes and locations.

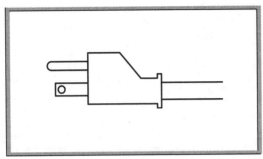

Fig. 10-1

Was it faster with ortho on?

NOTE:

In many situations, polar tracking is more convenient and easier to use than ortho. However, ortho can be very useful when polar is set to angles of other than 90° or when polar is turned off.

Tracking Time

AutoCAD keeps track of time while you work. With the TIME command, you can review this information.

1. Enter **TIME** at the keyboard or select the **Tools** pull-down menu and pick **Inquiry** and **Time**.

Text appears on the screen, providing information similar to that shown in Fig. 10-2 on the next page. Of course, the dates and times will be different.

Here's what this information means.

Current time:	Current date and time
Created:	Date and time drawing was created
Last updated:	Date and time drawing was last updated
Total editing time:	Total time spent editing the current drawing
Elapsed timer:	A timer you can reset or turn on or off
Next automatic save in:	Time before the next automatic save occurs

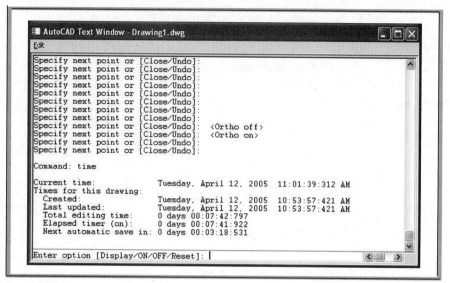

Fig. 10-2

If the current date and time are not displayed on the screen, they were not set correctly in the computer, or the computer's battery may be weak or dead.

2. If the date and time are incorrect, right-click the time located in the lower right corner of the screen, pick **Adjust Date/Time**, make the necessary changes, and pick the **OK** button.

3. Reenter the **TIME** command.

What information is different? Check the current time, total editing time, and elapsed time.

4. With AutoCAD's **TIME** command entered, type **D** for Display, and notice what displays on the screen.

This provides updated time information.

5. Enter the **OFF** option, and display the time information again.

The elapsed timer should now be off.

NOTE:

An example of when you might specify OFF is when you want to leave the computer to take a break. When you return, you turn the time back ON. This keeps an accurate record of the actual time (elapsed time) you spend working on a project.

6. Reset the timer by entering **R**, and display the time information once again.

The elapsed timer should show 0 days 00:00:00.000.

7. Last, turn on the timer and display the time information.

Notice that the elapsed timer keeps track of the time only while the timer is turned on.

Why is all of this time information important? In a work environment, the TIME command can track the amount of time spent on each project or job, making it easier to charge time to clients.

8. Press **ENTER** to terminate the TIME command, but do not close the AutoCAD Text window.

AutoCAD Text Window

As you are aware, AutoCAD displays the time information in the AutoCAD Text window. AutoCAD uses this window for various purposes, as you will see in future chapters.

Displaying the Window

1. Pick a point in the drawing area but outside the AutoCAD Text window.

This causes the graphics screen to appear in front of the AutoCAD Text window, hiding it from view.

2. Press the **F2** function key to make the AutoCAD Text window come to the front.

You can also make it reappear by picking the AutoCAD Text Window button on the Windows taskbar.

3. In the upper right corner of the AutoCAD Text window, pick the **Minimize** button (the dash).

4. Press **F2** or pick the **AutoCAD Text Window** button on the Windows taskbar to make it reappear.

 NOTE:

You can press F2 at any time to display the AutoCAD Text window. F2 serves as a toggle switch to toggle the display of the window.

5. Using the vertical scrollbar, scroll up the list of text in the AutoCAD Text window.

Observe that AutoCAD maintains a complete history of your activity. Also, notice that the AutoCAD Text window offers an Edit pull-down menu.

Copying and Pasting

1. Select the **Edit** pull-down menu from the AutoCAD Text window.

Paste to CmdLine pastes highlighted text to the Command line. This option is grayed out and not available because text has not been highlighted. The Copy item permits you to copy a selected portion of the text to the Windows Clipboard. The Clipboard is a memory space in the computer that temporarily stores information (text and graphics). After you have copied information to the Clipboard, it is easy to paste it into another software application.

Copy History copies the entire contents of the AutoCAD Text window to the Windows Clipboard. Paste enables you to paste the contents of the Clipboard to the Command line. The Options... menu item permits you to display the Options dialog box, which allows you to change various settings to meet individual needs.

2. Using the pointer, highlight the most recent time information, as shown in Fig. 10-3.

3. Pick the **Copy** item from the **Edit** pull-down menu in the AutoCAD Text window or press the **CTRL** and **C** keys to copy the text.

4. Close the AutoCAD Text window by clicking the **Close** button (the **x**) in the upper right corner of the window.

5. Minimize AutoCAD by picking the **Minimize** button (the dash) located in the upper right corner.

6. Launch a text editor such as **Notepad**. (Notepad is a standard Windows program and is part of the Accessories group of Windows utilities.)

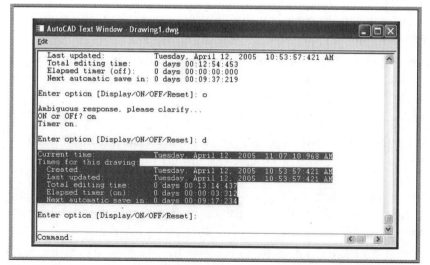

Fig. 10-3

7. In the text editor, select **Paste** from the **Edit** menu or press the **CTRL** and **V** keys.

This pastes the time information from the Windows Clipboard. At this point, you could name and save this file and/or print this information.

8. Exit the text editor without saving the file.

9. Click the **AutoCAD** button on the Windows taskbar to make the AutoCAD window reappear.

Command Line Window

Now focus your attention on the Command Line window (the window that contains the Command prompt).

1. Position the pointer so that its tip is touching any part of the Command Line window's left border.

2. With the Windows pointer, click and drag the Command Line window upward into the drawing area until a floating window forms.

You can move the window anywhere on the screen, and you can resize its width and height.

3. Right-click the left border of the Command Line window and pick the **Transparency...** option.

This displays the Transparency dialog box that was discussed in Chapter 2.

4. Move the slider to the far right and pick **OK**.

The Command Line window is now semi-transparent.

5. Right-click the left border of the Command Line window, pick the **Transparency...** option, move the slider to the far left, and pick **OK**.

6. Drag the Command Line window downward and dock it into its original position.

Copying and Pasting

It is possible to copy and paste text into the Command Line window. You may find this feature useful as you work with AutoLISP, an AutoCAD programming language.

1. At the right side of the Command Line window, scroll up using the up arrow.

2. Identify a command (any command) and highlight it by dragging the cursor across it while pressing the pick (left) button on the pointing device.

3. With the pointer inside the Command Line window, press the right button on the pointing device, causing a pop-up menu to appear.

4. Pick **Paste to CmdLine**.

AutoCAD pastes the highlighted text at the Command line.

5. With the pointer inside the Command Line window, right-click again and highlight **Recent Commands**.

6. Pick one of the commands that are listed.

AutoCAD enters the command.

7. Press the **ESC** key.

Editing Command Entries

AutoCAD allows you to correct misspellings at the Command line.

8. Type **TME**, a misspelled version of the TIME command, but do not press ENTER.

9. Using the left arrow key, back up and stop between the T and M.

10. Type the letter **I**.

 NOTE:

The Insert key serves as a toggle to turn on and off the insert mode. If the Insert function is not active, press the Insert key to turn it on.

11. Assuming that TIME is now spelled correctly, press **ENTER**.

12. Press **ESC** to cancel.

Table 10-1 briefly describes the keys that you can use to edit text on the Command line.

13. Close the AutoCAD Text window.

14. On your own, experiment with the Command line navigation options shown in Table 10-1.

Undoing Your Work

As you may recall from earlier chapters, you can undo or redo one or several actions by picking the Windows-standard Undo and Redo buttons in the docked Standard toolbar. AutoCAD also offers an UNDO command that provides further options.

Key	Action
LEFT ARROW	Moves cursor back (to the left)
RIGHT ARROW	Moves cursor forward (to the right)
UP ARROW	Displays the previous line in the command history
DOWN ARROW	Displays the next line in the command history
HOME	Places the cursor at the beginning of the line
END	Places the cursor at the end of the line
INSERT	Turns on and off the insertion mode
DEL	Deletes the character to the right of the cursor
BACKSPACE	Deletes the character to the left of the cursor
CTRL+V	Pastes text from the clipboard

Table 10-1

Reversing Multiple Operations

The basic UNDO command is similar to the Undo button on the Standard toolbar. It undoes, or reverses, one or more operations, beginning with the most recent.

1. Draw the objects shown in Fig. 10-4 at any size and location.

2. Enter the **UNDO** command at the keyboard.

3. Review the list of command options, and then enter **2** for the number of operations.

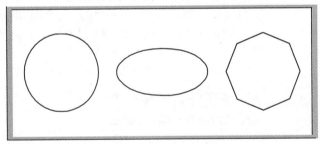

Fig. 10-4

By entering 2, you told AutoCAD to back up two steps.

NOTE: ────────

AutoCAD also includes the U command that allows you to back up one step at a time. The action of the U command can be reversed by the REDO command.

Returning to a Previous Drawing Status

Suppose you want to proceed with drawing and editing, but you would like the option of returning to this point in the session at a later time. The UNDO command's Mark and Back options allow you to do this efficiently.

Draw

1. Draw a small rectangle.

2. Enter the **UNDO** command.

3. Enter the Mark option by typing **M** and pressing **ENTER**.

AutoCAD has (internally) marked this point in the session.

4. Perform several operations, such as drawing and erasing objects and changing the buttons on the status bar.

Now suppose you decide to back up to the point where you drew the rectangle in Step 1.

5. Enter the **UNDO** command and then the **Back** option.

You have just practiced two of the most common uses of the UNDO command. Other UNDO options exist, such as BEgin and End. Refer to AutoCAD's on-line help for more information on these options.

Controlling the Undo Feature

The undo feature can use a large amount of disk space and can cause a "disk full" situation when only a small amount of disk space is available. You may want to disable the UNDO command partially or entirely by using the UNDO Control option.

Let's experiment with UNDO Control.

1. Enter the **UNDO** command and then the **Control** option.

2. Enter the **One** option.

3. Draw two arcs anywhere in the drawing area.

4. Enter the **UNDO** command.

UNDO now permits you to undo only your last operation.

5. Press **ENTER**.

6. Enter **UNDO**, **Control**, and **All**.

The undo feature is once again fully enabled.

7. Enter **UNDO** and note the complete list of options.

8. Enter **ESC** to cancel the command.

9. Exit AutoCAD without saving.

Review Questions

1. Which function key turns on the coordinate display feature?

2. Of what value is the coordinate display?

3. What is the name of the feature that forces all lines to be drawn only vertically or horizontally? What function key controls this feature?

4. Of what value is the TIME command?

5. Briefly explain each of the following components of the TIME command.
   a. Current time
   b. Created
   c. Last updated
   d. Total editing time
   e. Elapsed timer
   f. Next automatic save in

6. Which function key toggles the display of the AutoCAD Text window on and off?

7. Explain how you would copy a few lines from the AutoCAD Text window to a text editor such as Notepad.

8. Describe two methods to back up or undo your last five operations quickly.

9. Explain the use of the UNDO Mark and Back options.

Challenge Your Thinking

1. Experiment with ortho in combination with various object snap modes using the LINE command. What happens when you try to snap to a point that is not exactly horizontal or vertical to the previous point?

2. Why might you want to copy parts of the AutoCAD Text window or Command Line window to the Windows Clipboard?

Applying AutoCAD Skills

Work the following problems to practice the commands and skills you learned in this chapter.

1. Set the elapsed timer to ON. Complete the design shown in Fig. 10-5 for an over-the-hood storage area for a garage. Use ortho to keep the lines perfectly horizontal and vertical. When you have completed the drawing, set the elapsed timer OFF and write down the total time it took you to complete the drawing.

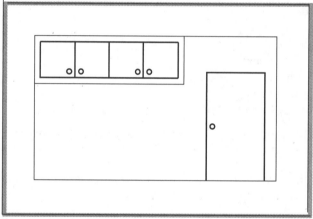

Fig. 10-5

2–4. Compare the use of ortho with object snap tracking and polar tracking by drawing the objects shown in Figs. 10-6 through 10-8. Create a new drawing to hold them. Turn on the coordinate display and note the display as you construct each of the objects.

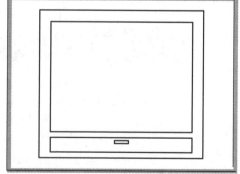

Fig. 10-6

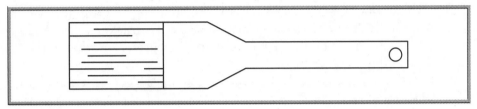

Fig. 10-7

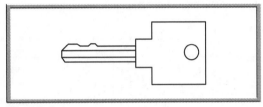

Fig. 10-8

5. Create a new drawing file and draw the lamp head shown in Fig. 10-9. Copy everything entered in the Command Line window to the Clipboard and paste it into a text editor such as Notepad. Then print it. (Read problem 6 before completing this problem.)

6. Use AutoCAD's timer to review the time you spent completing problem 5. Print this information or record it on a separate sheet of paper.

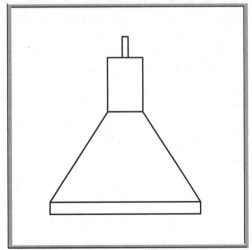

Fig. 10-9

7. Draw the simple house elevation shown in Fig. 10-10A at any size with ortho or object snap tracking and polar tracking. Prior to drawing the roof, use UNDO to mark the current location in the drawing. Then draw the roof.

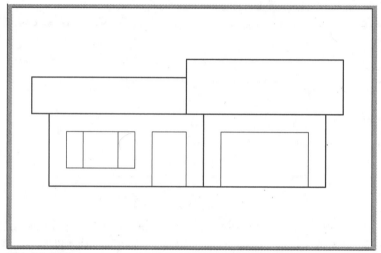

Fig. 10-10A

With UNDO Back, return to the point prior to drawing the roof. Draw the roof shown in Fig. 10-10B in place of the old roof. Use the UNDO command as necessary as you complete the drawing.

Fig. 10-10B

Using Problem-Solving Skills

Complete the following activities using problem-solving skills and your knowledge of AutoCAD.

1. As the packaging engineer for a major manufacturer, you have been asked to design a support for a new product to hold it in place within the container. The drawing, shown in Fig. 10-11, is not to scale. Create a drawing to scale according to the dimensions given. Keep track of your hours, because the time will be billed to the customer. Do not dimension it. Save the drawing and write down the total time required to complete the drawing.

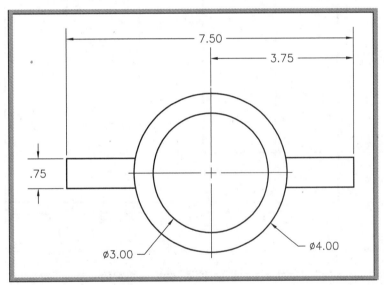

Fig. 10-11

2. The core of an electromagnet is not made up of one solid piece of metal. Rather, it is made up of a series of thin "slices," or laminations. These laminations fit together to form the core of the electromagnet. The laminations concentrate the electromagnetic field, adding strength. Draw the lamination shown in Fig. 10-12. Keep track of your time with the elapsed timer. Do not dimension, but save the drawing.

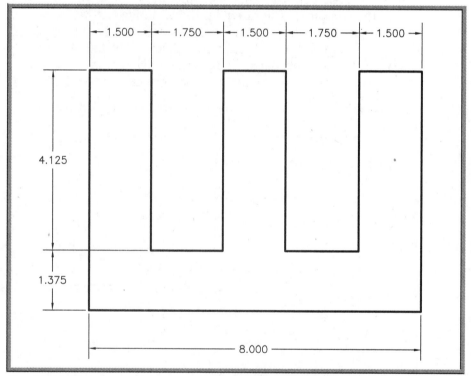

Fig. 10-12

3. Draw the front view of the tube bulkhead shown in Fig. 10-13. Before drawing the holes, use the UNDO command to mark the current location in the drawing. After completing the bulkhead, use UNDO to replace the two holes with two new ones. Draw both of the new holes with a diameter of 1.00. Change the center-to-center distance from 1.40 to 1.60.

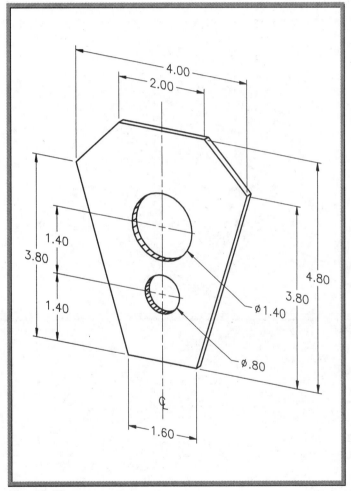

Fig. 10-13

Problem 3 courtesy of Gary J. Hordemann, Gonzaga University

4. The two identical hubs shown in Fig. 10-14 are meant to be mounted back-to-back in an assembly. Draw the front view of one hub. Before drawing the four small holes, use the UNDO command to mark the current location in the drawing. After completing the hub, use UNDO to replace the four holes with eight new ones, equally spaced circumferentially. Use a new hole diameter of .284 and a new center line diameter of 2.50.

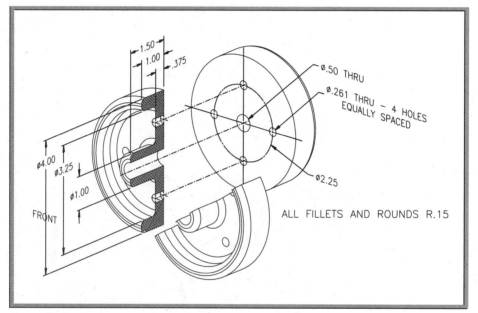

Fig. 10-14

Problem 4 courtesy of Gary J. Hordemann, Gonzaga University

Careers
Using
AutoCAD

Checker

The ability to pay attention to detail is a necessary skill for any career related to drafting. But like anyone else, even the individuals who create detailed technical drawings occasionally make mistakes. Before drawings are submitted for use in the actual building or manufacturing process, they must obtain clearance from a checker.

Checkers ensure that mathematical equations are correct, diagrams are drawn to scale, and notes and abbreviations are clear and accurate. They also ensure that the drawings contain all of the information necessary to create the part or component. Drawings that do not meet the checker's standards are returned to the drafter for revision. After the checker reviews the corrections, a supervisor, architect, or engineer reviews the corrected drawings also. Mistakes, if left unnoticed, can cost a company a great deal of money.

Communication Is the Key

Good communication skills are required of anyone who works in drafting. Teamwork is important, as checkers often work directly with both drafters and supervisors. They must be able to convey information on both verbal and visual levels.

The ability to get along with other people is also very important. A checker must be able to point out errors and suggest corrections without belittling the drafter or making the drafter feel uncomfortable.

Tim Fuller Photography

On the Job

Checkers work in a variety of drafting areas, including architecture, manufacturing, construction, transportation, and communication. Any time a drawing is created, it must be reviewed for accuracy. Checkers are not out to make drafters' lives miserable by returning drawings—which often require days, weeks, or even months of work—marked with corrections. They simply try to ensure that drawings are as complete and accurate as possible, for practical, financial, and safety reasons.

 Career Activity

- Have you ever written a paper for class and then asked a friend to proofread it for errors before turning it in? Describe how the job of a checker is similar to that of a proofreader or editor. Make a list of characteristics/skills that you think checkers and proofreaders must possess to do their job accurately. Share your list with a classmate and compare your answers.

Chapter 11 Construction Aids

Construction Aids

Objectives:

- Set and use AutoCAD's visual grid system
- Establish a snap grid that is appropriate to the current drawing task
- Use construction lines to project views in a multiview drawing

Key Terms

alignment grid   orthographic projection
aspect ratio   plane
construction lines   rays
hidden lines   snap grid
multiview drawing   xlines
offsetting

AutoCAD permits you to create a nonprinting grid system for visual layout of drawings. The snap grid complements the visual grid, assisting with the layout process. Construction lines provide a means for projecting lines from one view to another using orthographic projection. All of these construction aids make it easier to create accurate multiview drawings in the shortest amount of time possible.

Alignment Grid

The GRID command allows you to set an *alignment* grid of dots of any desired spacing, making it easier to visualize distances and drawing size. You can use the grid with both English (Imperial) and metric units. In this chapter, we will create a drawing of a plastic extrusion using metric units.

1. Start AutoCAD and set the **STARTUP** system variable to **1**.

2. Enter the **NEW** command or pick the **QNew** button.

The Create New Drawing dialog box appears.

3. Pick **Start from Scratch**, pick the **Metric** radio button, and pick the **OK** button.

4. Move the crosshairs to the upper right corner of the drawing area and read the coordinate display.

Since you are now working in the metric system, the unit of measure is millimeters (mm). Instead of a drawing area of 12 × 9, it is now approximately 420 × 297 mm, depending on the size of the window representing the drawing area on your computer.

5. Pick the **GRID** button located in the status bar.

A grid of dots appears on the screen.

6. Enter the **GRID** command at the keyboard.

Notice that the default value for the grid spacing is 10. This means that the dots are currently spaced 10 mm apart.

7. Change the spacing to 20 mm by entering **20**.

8. Reenter **GRID** and change the spacing back to **10** mm.

9. Turn off the grid by picking the **GRID** button.

10. Make the grid visible again.

NOTE:

You can also turn the grid on and off by pressing CTRL G or the F7 function key.

Snap Grid

The *snap grid* is similar to the visual grid, but it is an invisible one. You cannot see the snap feature, but you can see the effects of it as you move the crosshairs. It is like a set of invisible magnetic points. The crosshairs jump from point to point as you move the pointing device. This allows you to lay out drawings quickly, yet you have the freedom to toggle snap off at any time.

1. Pick the **SNAP** button in the status bar to turn on the snap grid.

NOTE:

The snap grid is different from object snap, which permits you to snap to points on objects.

You can also turn the snap grid on and off by pressing CTRL B or the F9 function key.

When you turn on the snap grid, it appears at first as though nothing has happened.

2. Slowly move the pointing device and watch closely the movement of the crosshairs.

The crosshairs jump (snap) from point to point.

3. Enter the **SNAP** command at the keyboard.

4. Enter **5** to specify a snap spacing of 5 mm.

5. Move the pointing device and note the crosshairs movement.

HINT:

If the grid is not on, turn it on to see better the movement of the crosshairs. Notice the spacing relationship between the snap resolution and the grid. They are independent of one another.

6. Draw a line.

Changing the Aspect Ratio

The *aspect ratio* is the ratio of height to width of a rectangular region. You can change this ratio so that the crosshairs snap one distance vertically and a different distance horizontally. This can be useful when you are laying out a drawing.

1. Enter the **SNAP** command.

Several options appear at the Command line.

2. Select the Aspect option by entering **A**.

AutoCAD asks for the horizontal spacing.

3. Enter **5**.

Next AutoCAD asks for the vertical spacing.

4. For now, enter **10**.

5. Move the crosshairs up and down and back and forth. Note the difference between the amount of vertical movement and horizontal movement.

6. Change the vertical snap back to **5**.

Chapter 11: Construction Aids

Rotating the Grid

By default, the grid appears at a 0° angle, which aligns the points in vertical and horizontal rows. However, AutoCAD allows you to change the angle of the grid. This can be useful when you want to create a drawing at an angle, while taking advantage of the snap grid.

1. Enter the **SNAP** command again, and this time enter the **Rotate** option.

2. Leave the base point at 0,0 by giving a null response. (Simply press the spacebar or **ENTER**.)

3. Enter a rotation angle of **30** (degrees).

The snap, grid, and crosshairs rotate 30° counterclockwise.

4. Draw a small object at this rotation angle near the top of the screen.

5. Return to the original snap rotation of **0** and erase all the objects.

Applying the Snap and Grid Features

1. Set snap at **5** mm both horizontally and vertically.

2. Create the drawing of a plastic extrusion shown in Fig. 11-1. Use the pointing device (not the keyboard) to specify all points, and do not include the dimensions.

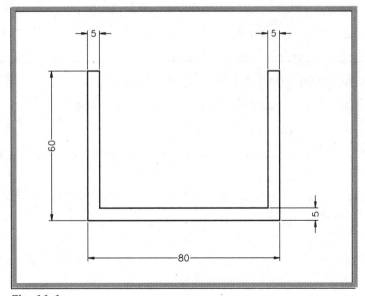

Fig. 11-1

The Drafting Settings dialog box permits you to review and make changes to the snap and grid settings.

1. From the **Tools** pull-down menu, select the **Drafting Settings...** item.

The Drafting Settings dialog box appears.

2. If the **Snap and Grid** tab is not in the foreground, select it.

As you can see, this dialog box enables you to change the *x* and *y* spacing of both the snap and grid. AutoCAD saves these settings with the drawing, so if you or someone else starts a new drawing, the old settings return.

3. Close the dialog box.

4. Save your work in a file named **snap.dwg**, but do not close the drawing or exit AutoCAD.

Construction Lines

Construction lines, also called *xlines*, are lines of infinite length. They can be used to construct objects and lay out new drawings.

Standard

Draw

1. Start a new drawing from scratch and select **Metric**.

2. Set the grid to **10** mm and the snap to **5** mm.

3. From the **Draw** toolbar, pick the **Construction Line** button.

This enters the XLINE command.

4. Pick a point anywhere on the screen.

5. Move the crosshairs and notice what happens.

A line of infinite length passes through the first point and crosshairs.

6. Pick a second point anywhere on the screen.

A line freezes into place. As you move the crosshairs, notice that a second line forms.

7. Pick another point so that the intersecting lines approximately form a right angle.

8. Press **ENTER** to terminate the XLINE command.

Unlike the snap and grid construction aids, xlines are AutoCAD objects. You can select them for use with commands, and you can snap to them using object snap.

9. Enter the **LINE** command and use the **Intersection** object snap to snap to the intersection of the two lines.

Object Snap

 HINT: ───

If Intersection is not currently set as a running object snap, open the Object Snap toolbar and pick the Intersection button.

10. Press **ESC** to terminate the LINE command.
11. Erase the xlines.

Constructing Horizontal and Vertical Xlines

The XLINE command provides several options that allow you to create specific types of construction lines.

Draw

1. Enter the **XLINE** command and enter **H** for Horizontal.
2. Pick a point anywhere on the screen; then move the crosshairs and pick a second point, and a third.

Because the Horizontal option is in effect, all of the construction lines are horizontal.

3. Press **ENTER** to terminate the XLINE command and press **ENTER** or the spacebar to reenter it.
4. Enter **V** for Vertical and pick three consecutive points anywhere on the screen.

The Vertical option forces all the lines to be exactly vertical.

Constructing Xlines at Other Angles

To create construction lines at angles other than 0 or 90 degrees, you can use the Angle option.

1. Press **ENTER** to terminate the XLINE command and reenter it.
2. Enter **A** for Angle and enter **45**.
3. Place three construction lines.
4. Terminate XLINE.
5. Erase all but one construction line.

Offsetting Construction Lines

You may find many occasions to use the Offset option of the XLINE command. *Offsetting* a line means creating the line at a specific distance from another line. For example, suppose you have created a line that represents the outer edge of a sidewalk on an architectural site drawing. If the sidewalk is to be 3 feet wide, you can create a construction line at an offset distance of 3 feet to show the width of the sidewalk.

INFOLINK

You will learn more about offsetting ordinary lines in Chapter 15.

1. Enter **XLINE** and enter **O** for Offset.

2. Enter **7** (for 7 mm), select the construction line, and pick a point on either side of it.

Creating Perpendicular Construction Lines

1. Terminate XLINE, reenter it, and select or enter the **Perpendicular** object snap.

2. Pick one of the construction lines and pick another point elsewhere on the screen.

The new construction line is perpendicular to the one you selected.

3. Terminate XLINE and erase all the objects from the screen.

Rays

Rays are a special type of construction line. They extend from a single point into infinity in a radial fashion. Rays are useful when you are laying out objects radially.

1. From the **Draw** pull-down menu, pick **Ray**.

This enters the RAY command.

2. Pick a point near the center of the screen.

3. Move the pointing device in any direction and then pick several points around the first point.

4. Terminate the RAY command.

5. Close the current drawing file and do not save your work.

Orthographic Projection

Construction lines, rays, and the grid and snap features are often used to create multiview drawings. A *multiview drawing* is one that describes a three-dimensional object completely using two or more two-dimensional views. These views commonly include a front, top, and side view, although others are often necessary. The AutoCAD features discussed in this chapter, coupled with object snap tracking, help you to produce multiview drawings using a technique known as *orthographic projection*.

As mentioned earlier in this book, *ortho* means "at right angles." Orthographic projection, therefore, is the projection of views at right angles to one another. Each adjacent view is projected at a right angle onto a plane, resulting in a two-dimensional view. A *plane* is an imaginary flat surface used to construct the two-dimensional view.

Figure 11-2 shows an example of a simple multiview drawing created using orthographic projection. Notice that the top view is located above the front view. The right view is located to the right of the front view. This positioning is standard. It is important to use the standard positions because doing so allows other people to understand the position of these views at a glance. The thin lines are temporary construction lines.

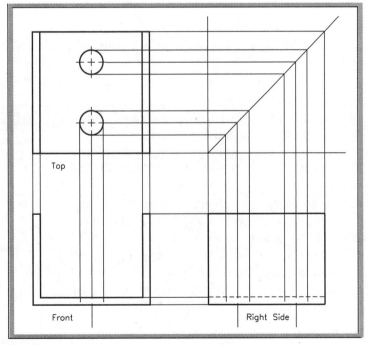

Fig. 11-2

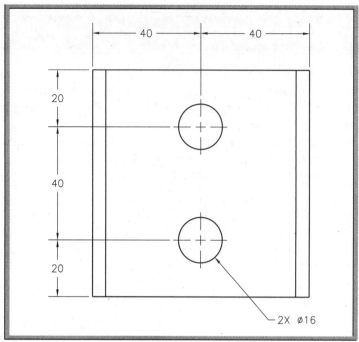

Fig. 11-3

1. Using grips, move the drawing of the plastic extrusion to the lower left area of the screen.

2. Using object snap tracking, snap, and grid, create the top view as shown in Fig. 11-2. Use the following information when constructing the top view.

   • From Fig. 11-3, obtain the length dimension of the extrusion and the location and size of the two holes.

   • The broken lines through the holes are center lines. For now, create these lines using a series of line segments. Later in this book, you will learn an easier way to draw dashed lines to create center and hidden lines.

   • You may need to turn off object snap when picking some of the points. However, object snap must be on for object snap tracking to work.

   • Using the **XLINE** command, consider drawing a vertical construction line snapped to the midpoint of one of the horizontal lines to represent the center of the holes. Then erase the xline.

3. Save your work.

INFOLINK

You will learn more about <u>linetypes</u> **in Chapter 23.**

4. After creating the top view, project the holes down to the front view. Consider the following suggestions.

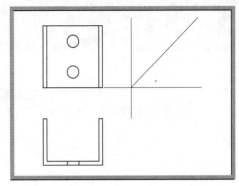

- Set **Quadrant** as a running object snap mode.
- Use object snap tracking to acquire the diameter of the holes.
- Draw hidden lines to show the holes in the front view. (*Hidden lines* are the short dashed lines in the multiview drawing that are used to indicate features that are hidden from view.) For now, create them using two short line segments.

Fig. 11-4

5. Save your work.

6. Create the right-side view by projecting lines from the top and front views. Consider the following suggestions.

- Create temporary lines as shown in Fig. 11-4. The vertical line is at an arbitrary distance from the top view.
- Draw the angled line at a 45° angle. (Use the **Angle** option of the **XLINE** command.) Use it to project the dimensions from the top view to the right-side view.
- Project lines from the top view to the 45° line. Then project vertical lines down from the intersections of the horizontally projected lines and the 45° lines to locate the right-side view.
- Use the height dimensions from the front view (Fig. 11-4) to finish the right-side view.

NOTE:

In practice, the views in a multiview drawing are laid out carefully to ensure that the distance between the front and top views is approximately equal to the space between the front and side views. Keep this in mind as you create the construction lines in Step 6.

7. Erase all temporary construction lines.

Your multiview drawing should now look similar to the multiview drawing shown in Fig. 11-2, except that your drawing will not include hidden, center, or construction lines. You will earn how to create hidden and center lines later in this book.

8. Close any open toolbars, save your work, and exit AutoCAD.

INFOLINK

You will learn more about multiview drawings in Chapter 25.

Review Questions

1. What is the purpose of the grid? How can the grid be toggled on and off quickly?

2. When would you use the snap feature? When would you toggle off the snap feature?

3. How can you set the snap feature so that the crosshairs move a different distance horizontally than vertically?

4. Explain how to rotate the grid, snap, and crosshairs 45°.

5. What is the purpose of the XLINE and RAY commands?

6. Explain the purpose of construction lines and rays. Describe a drawing situation in which you could use construction lines or rays to simplify your work.

7. What is orthographic projection? How is it used in engineering drawings?

8. What is the purpose of a multiview drawing?

Challenge Your Thinking

1. AutoCAD provides several different grid and snap systems. Write a short paper describing their differences and similarities.

2. It is possible to accomplish the same tasks using either construction lines or AutoCAD's AutoTrack™ feature. Describe a situation in which you would prefer to use one instead of the other.

Applying AutoCAD Skills

Work the following problems to practice the commands and skills you learned in this chapter. Use English (Imperial) units, not metric, for all of the problems in this section.

1–3. Draw the objects in Figs. 11-5 through 11-7 using the grid and snap settings provided beside each object.

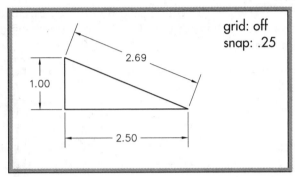

Fig. 11-5

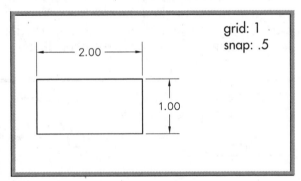

Fig. 11-6

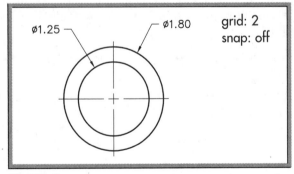

Fig. 11-7

4. Draw the two views of the rubberized hand grip for closing the safety bar on a roller coaster car (Fig. 11-8B). Begin by drawing two construction lines as shown in Fig. 11-8A.

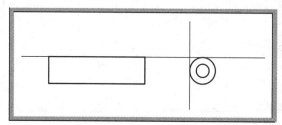

Fig. 11-8A

Draw the front view as shown. Then draw the Ø40 circle using the tangent-tangent-radius method. Locate the Ø20 circle to be concentric with the first circle, and erase the construction lines. Do not dimension it, and do not draw the center lines.

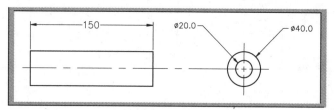

Fig. 11-8B

5. Draw three views of the slotted block shown in Fig. 11-9A. Start by drawing the front view using the LINE and CIRCLE commands. Do not worry about drawing the object exactly. Make the task easier by turning ortho on.

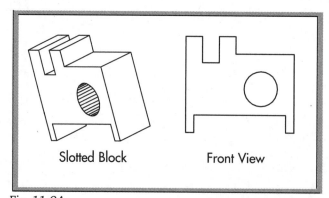

Fig. 11-9A

Use the RAY command to extend the edges in the front view into what will become the right-side view. Use the appropriate object snaps to begin the rays at the edges of the front view, as shown in Fig. 11-9B.

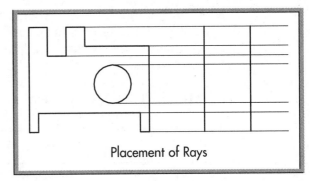

Placement of Rays

Fig. 11-9B

To begin drawing the right-side view, add two vertical lines (estimate the space between them), as shown in Fig. 11-9B. Use the construction lines (rays) to draw the visible lines. Draw the hidden lines as a series of very short lines.

When you are finished, erase the construction lines. The drawing should look like the one in Fig. 11-9C.

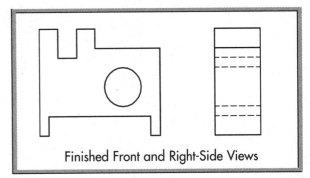

Finished Front and Right-Side Views

Fig. 11-9C

Repeat the process to draw the top view. The finished drawing should look like the one in Fig. 11-9D.

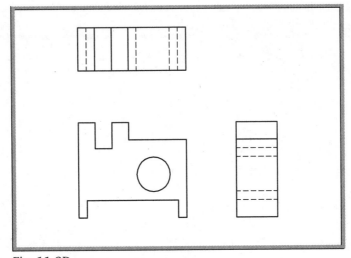

Fig. 11-9D

Problem 5 courtesy of Gary J. Hordemann, Gonzaga University

6. Draw the top, front, and right-side views of the block support shown in Fig. 11-10. Use a grid spacing of .2 and a snap setting of .1. Consider using construction lines or rays to help you position the three views. For now, draw all of the lines as continuous lines, or draw the hidden and center lines using very short lines. You may wish to change the snap spacing to .05.

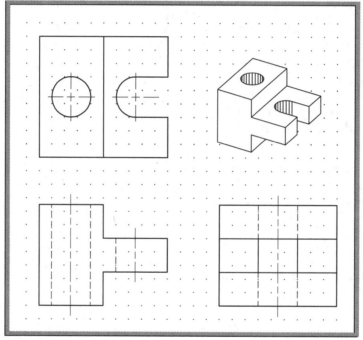

Fig. 11-10

Problem 6 courtesy of Gary J. Hordemann, Gonzaga University

Chapter 11: Construction Aids

7. Draw the three orthographic views of the bushing holder shown in Fig. 11-11. Use a grid spacing of .2 and a snap setting of .1. Consider using construction lines or rays to help you position the views. The front view is shown as a full section view; *i.e.*, it is shown as if the object has been cut in half. The diagonal lines indicate cut material; you may want to rotate the crosshairs using the SNAP command to create these lines. AutoCAD provides a way to insert such hatch patterns easily. You will learn how to do this later.

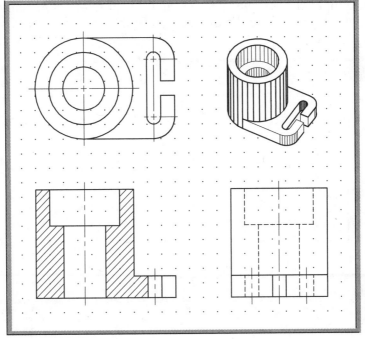

Fig. 11-11

Problem 7 courtesy of Gary J. Hordemann, Gonzaga University

Using Problem-Solving Skills

Complete the following activities using problem-solving skills and your knowledge of AutoCAD.

1–2. When the drafting department converted to CAD from board drafting, many vellum drawings had to be copied into the computer. The two drawings shown in Figs. 11-12 and 11-13 were not copied completely. Use the skills you have acquired to complete the missing views. Determine the dimensions from the .5 grid spacing. Save both drawings.

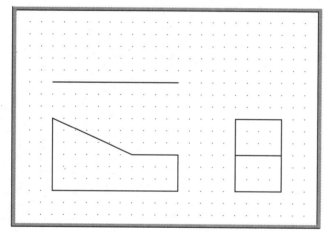

Fig. 11-12

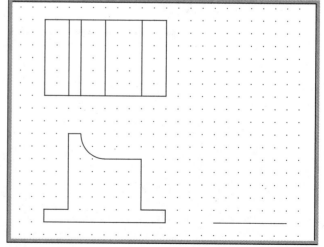

Fig. 11-13

Chapter 12 A Closer Look

Objectives:

- **Zoom in on portions of a drawing to view or add details**
- **Use the ZOOM command while another command is entered**
- **Control view resolution**

Key Terms

realtime zooming
transparent zooms
view resolution

Large drawings with detail can be difficult, and at times impossible, to view legibly on even the largest computer monitors. AutoCAD offers a zooming capability that magnifies areas of the drawing. Remarkably, this capability can magnify objects by as much as ten trillion times. While this may seem impractical for most work, scientists and astronomers produce drawings and models of incredible scale. Consider a solar system, in which distances are measured in millions or billions of miles.

Zooming

Let's apply the ZOOM command.

1. Start AutoCAD, select the **Use a Wizard** button from the Startup dialog box, pick **Quick Setup**, and pick **OK**.

HINT:

If the Startup dialog box does not appear when you start AutoCAD, enter the STARTUP system variable and enter 1. Then enter the NEW command.

2. Pick the **Architectural** radio button and **Next**.

3. Enter **20'** for the Width and **15'** for the Length of the drawing area and pick **Finish**.

HINT:

Be sure to include an apostrophe (') for the foot mark. If you don't, AutoCAD will assume that you mean inches, which would be much too small for the following office drawing. Note that when you work in inches, adding a double quote (") after the number is optional.

4. Enter **Z** for ZOOM and **A** for All.

This zooms the drawing to the 20' × 15' area.

5. Enter **GRID** and enter **1'** or **12** (inches).

6. Set the snap grid to **2"**.

7. Draw the room shown in Fig. 12-1, including the table and chair. Estimate the distance from the wall to the table.

8. Save your work in a file named **zoom.dwg**.

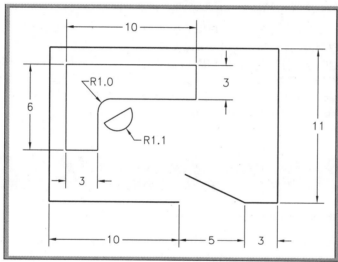

Fig. 12-1

Methods of Zooming

AutoCAD provides several options for zooming in or out on a drawing. You can enter these options by typing Z (for ZOOM) at the keyboard and then entering the first letter of the option, or you can use the many zoom buttons. These buttons are located on the Zoom toolbar and on the flyout on the Zoom Window button in the Standard toolbar. (Depress and hold with the pick button to view the flyout.)

Standard

1. Pick the **Zoom Window** button from the Standard toolbar.

This enters the ZOOM command.

2. In response to Specify first corner, place a window around the table and chair as shown in Fig. 12-2. (Pick a point for the first corner and then select a point for the opposite corner.)

The table and chair should magnify to fill most of the screen.

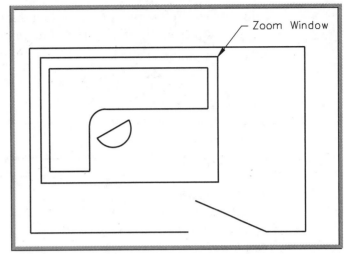

Fig. 12-2

3. Next, draw basic representations of the components that make up a CAD workstation, as shown in Fig. 12-3. Approximate their sizes, and omit the text. Save your work.

Now let's zoom in on the keyboard.

4. Pick the **Zoom Window** button again, and this time place a window around the lower half of the keyboard.

Let's zoom in further on the keyboard, this time using the Zoom In option.

5. From the Zoom toolbar or the Zoom flyout on the Standard toolbar, pick the **Zoom In** button.

Standard

6. If you cannot see the bottom of the keyboard, use the scrollbars or zoom again (using the **Zoom In** or **Zoom Window** button) so that the bottom of the keyboard fills most of the screen.

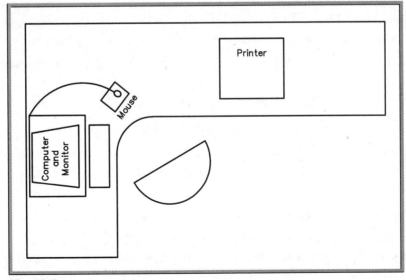

Fig. 12-3

7. In the lower left corner of the keyboard, draw a small square to represent a key, as shown in Fig. 12-4. (You may need to turn snap off first.)

8. Zoom in on the key.

9. Using the **POLYGON** command, draw a small trademark on the key as shown in Fig. 12-5.

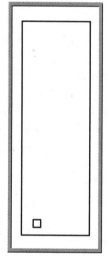

Fig. 12-4

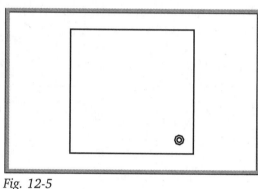

Fig. 12-5

Standard

10. Next, pick the **Zoom Previous** button located on the Standard toolbar and watch what happens.

You should now be at the previous zoom factor. The trademark should look smaller.

Zoom

11. Pick the **Zoom All** button, or enter **Z** and **A** at the keyboard.

AutoCAD zooms the drawing to its original size.

Zoom

12. Pick the **Zoom Extents** button or enter **Z** and **E** at the keyboard.

AutoCAD zooms the drawing as large as possible while still showing the entire drawing on the screen.

You can also zoom to the extent of a selected object or set of objects.

Zoom

13. Pick the **Zoom Object** button or enter **Z** and **O** at the keyboard.

14. Use a window to select the computer and keyboard and press **ENTER**.

AutoCAD zooms in, making the selected objects the extent of the drawing screen.

Zoom

15. Pick the **Zoom Extents** button or enter **Z** and **E** at the keyboard.

It is also possible to enter a specific magnification factor. AutoCAD considers the entire drawing area to be a magnification of 1. By entering a different number with the ZOOM command entered, you can change

Chapter 12: A Closer Look

the magnification relative to this. If you need to change the magnification relative to the current magnification (instead of 1), you can accomplish this by typing an x after the number.

16. Enter **Z** (or press the spacebar) and enter **2**.

This causes AutoCAD to magnify the drawing area by two times relative to the ZOOM All display (All = 1).

17. Enter **ZOOM** and **.5**.

This causes AutoCAD to shrink the drawing area by one half relative to the ZOOM All display.

Zoom

18. Pick the **Zoom In** button; pick it again.

19. Pick the **Zoom Out** button.

Zoom

20. Pick the **Zoom Extents** button again.

21. Enter **ZOOM** and **.9x**.

Zoom

Because you placed an x after the number, the zoom magnification decreases to nine tenths (90%) of its current size, not nine tenths of the ZOOM All display area.

22. Continue to practice using the **ZOOM** command by zooming in on the different CAD components of the drawing and including detail on each.

23. Save your work.

Realtime Zooming

AutoCAD's *realtime zooming* allows you to change the level of magnification by moving the cursor. This is one of the fastest ways to zoom.

Standard

1. From the Standard toolbar, pick the **Zoom Realtime** button.

This places you into realtime zoom mode. Notice that the crosshairs has changed.

2. Using the pick button, click and drag upward to zoom in on the drawing.

3. Click and drag downward to zoom out on the drawing.

4. Press **ESC** or **ENTER** to cancel realtime zooming.

NOTE:

AutoCAD offers a time-saving shortcut for using realtime zooming. At the keyboard, enter Z (for ZOOM) and then press ENTER or the spacebar. Realtime zooming becomes available.

Zooming Transparently

With AutoCAD, you can perform *transparent zooms*. This means you can use ZOOM while another command is in progress. To do this, enter 'ZOOM (notice the apostrophe) at any prompt that is not asking for a text string.

1. Enter the **LINE** command and pick a point on the screen.

2. At the prompt, enter **'Z**.

AutoCAD lists options preceded by **>>** to remind you that you're in transparent mode.

3. Zoom in on any portion of the screen.

Notice the message Resuming LINE command.

4. Pick an endpoint for the line and terminate the command.

NOTE:

Picking any of the zoom buttons also permits you to perform a transparent zoom.

Transparent zooms give you great flexibility while you are using other commands. For instance, if line endpoints require greater accuracy than the present display allows, the transparent zoom provides a solution.

5. Erase the line and enter **ZOOM All**.

Controlling View Resolution

View resolution refers to the accuracy with which curved lines appear on the screen. The VIEWRES command allows you to set the view resolution for objects in the current drawing.

1. Enter the **VIEWRES** command.

2. In reply to Do you want fast zooms? accept the Yes default value.

Notice that the circle zoom percent (screen resolution) is 1000 by default. AutoCAD allows you to enter any number from 1 to 20,000.

3. Enter **20** for the circle zoom percent.

Notice the coarse appearance of the arcs in the drawing.

4. Enter **VIEWRES** again; press **ENTER** and then enter **1000**.

Notice the smooth appearance of the arcs on the screen.

5. Save your work and exit AutoCAD.

Review Questions

1. Explain why the ZOOM command is useful.

2. Cite one example of when it would be necessary to use the ZOOM command to complete an engineering drawing and explain why.

3. Describe each of the following ZOOM options.
   a. All
   b. In
   c. Extents
   d. Previous
   e. Window

4. What is a transparent zoom? When might a transparent zoom be useful?

5. Why might you often choose to use the realtime zoom option instead of other zoom methods?

6. Explain the purpose of the VIEWRES command.

Challenge Your Thinking

1. AutoCAD drawings are stored in a vector format. Many other graphics programs store their drawings in a raster format. Find out the differences between the raster and vector formats. Can you convert a drawing in vector format to a raster format? Can you convert a raster drawing into a vector format? Explain.

2. Investigate the ZOOM Center option. What is its purpose? When might you want to use it? Explain.

Applying AutoCAD Skills

Work the following problems to practice the commands and skills you learned in this chapter.

1. As a general rule, the outside diameter of the washer face on a hex bolt head is the same as the distance across the hex flats. This dimension is approximately $1\frac{1}{2}$ times the diameter of the bolt shank. Draw the head and washer face of the $\frac{1}{4}''$ hex bolt shown in Fig. 12-6. Use the following guidelines:

   - Create a new drawing using the Quick Setup wizard, fractional units, and a drawing area of 20×15.
   - Create the outer edge of the washer face using a diameter of $\frac{7}{16}$. Place it near the center of the screen.
   - Zoom a window so that the circle almost fills the drawing area.
   - Draw a hexagon (6-sided polygon) with the same center as the circle and a distance across the flats of $\frac{7}{16}$. In other words, circumscribe the hexagon around a $\frac{7}{16}''$ circle.
   - Draw a second circle with a diameter of $\frac{1}{4}''$ to represent the shank of the bolt.
   - Enter ZOOM All and notice how small the bolt head is. Even if you had set the drawing area for the default of 12×9, it would be impossible to draw the bolt head accurately without using the ZOOM command.

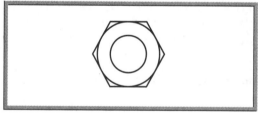

Fig. 12-6

2. Create a drawing such as an elevation plan of a building, a site plan of a land development, or a view of a mechanical part. Using the ZOOM command, zoom in on the drawing and include details. Zoom in and out on the drawing as necessary, using the different ZOOM options and transparent zooms.

3. Draw the kitchen floor plan shown in Fig. 12-7A. Then zoom in on the kitchen sink. Edit the sink to include the details shown in Fig. 12-7B. Save the file as kitchen.dwg.

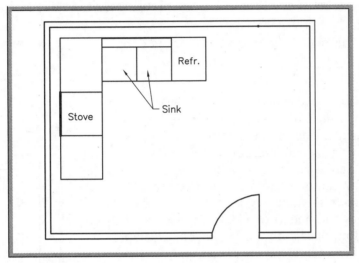

Fig. 12-7A

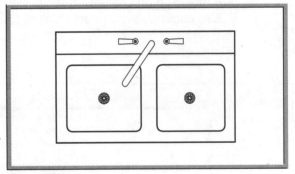

Fig. 12-7B

4. Refer to land.dwg on the Instructor Resource CD. In the middle third of the drawing, zoom in on details such as picnic tables. Add and edit objects to improve the drawing. Zoom out. Apply VIEWRES and transparent zooms.

Problem 3 courtesy of Kathleen P. King, Fordham University, Lincoln Center

Using Problem-Solving Skills

Complete the following activities using problem-solving skills and your knowledge of AutoCAD.

1. Use the ZOOM command to draw the door with a fan window shown in Fig. 12-8. Approximate all dimensions.

2. Draw the gasket shown in Fig. 12-9. Do not dimension it. Use construction lines for center lines and erase them when the drawing is complete. Are you sure the 1.50 radius is in fact a radius? Use the ZOOM command to find out. One way to create this drawing is to draw two R.8 circles and one R1.50 circle, and locate the four lines with the Tangent object snap. You can then erase the circles and use the ARC command to connect the lines. When you have completed the drawing, use the ZOOM command to ensure that all the intersections between lines and arcs are made correctly.

Fig. 12-8

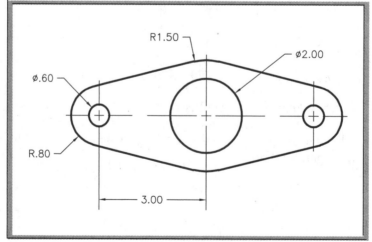

Fig. 12-9

Chapter 12: A Closer Look

Careers Using AutoCAD

Tim Fuller Photography

Cartographer

You may know people who, when they find themselves in unfamiliar territory and are in the process of becoming lost, absolutely refuse to ask for help with directions. If you are planning on taking a trip with someone like this, make sure to be prepared and bring along a map. This will help prevent extra driving time while you try to figure out where you are going. It will also save you and that friendly gas station attendant from your friend's anger when he or she is finally forced to ask for help!

Cartographers—people who draw maps—may work for cartographic service businesses. One example of such a business is Rand McNally, a premier mapmaking company based in the United States. Other cartographers work for city and county departments, colleges and universities, and the United States Department of Defense.

A Different Way To Communicate

A special language is used to create maps as symbols, scales, and legends often replace words. Cartographers must be familiar with this language in order to effectively communicate directions, distances, and landmarks to map readers.

Understanding Technology

To succeed in this field, cartographers must be well-versed in geography. The ability to draw well and print legibly is also important. The ability to draft a technical document both by hand and using a CAD system is critical.

With the rise of technology, many cartographers are now using Geographic Information Systems (GIS)—database-driven and graphic-oriented computer programs—to create maps. GIS is frequently used to produce interactive and three-dimensional maps for the Web. Because this technology is expected to become a routine tool in creating maps in the future, students interested in cartography should feel comfortable working with computers. Learning some basic graphic design techniques will also help students prepare for a career in this field.

Career Activities

- If you have access to a computer and the Internet, find a digital map service such as MapQuest or MapBlast and select a location. In what ways is this type of map different from a traditional paper map?
- Contact junior colleges and technical colleges in your area. Do they offer courses in map drafting? What prior knowledge is required?

Chapter 13 — Panning and Viewing

Objectives:

- Pan from one zoomed area to another
- Produce and use named views
- Use **Aerial View** to navigate in a complex drawing

Key Terms

panning
view box

Zooming in on a drawing solves part of the problem of working with fine detail on a drawing. A method called *panning* solves the other part. Panning permits you to move the zoom window to another part of the drawing while keeping the same zoom magnification. This helps keep you from constantly zooming in and out to examine detail.

Note the degree of detail in the architectural floor plan shown in Fig. 13-1.

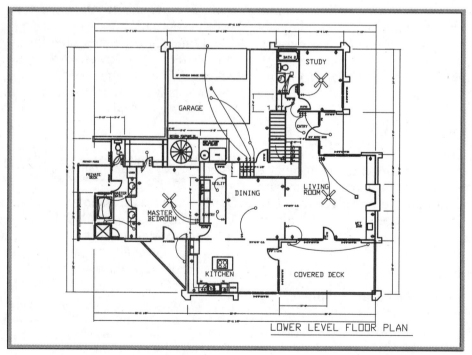

Fig. 13-1

The drafter who completed the CAD drawing in Fig. 13-1 zoomed in on portions of the floor plan in order to include detail. For example, the drafter zoomed in on the kitchen to place cabinets and appliances, as shown in Fig. 13-2.

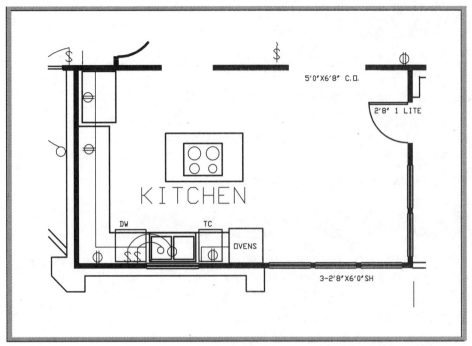

5'0"X6'8" C.D.

2'8" 1 LITE

KITCHEN

DW

TC

OVENS

3-2'8"X6'0"SH

Fig. 13-2

Suppose the drafter wants to include detail in an adjacent room but wants to maintain the present zoom magnification. In other words, the drafter wants to simply "move over" to the adjacent room. This operation can be accomplished by panning.

Panning

There are several ways to pan in AutoCAD. You can use the PAN command, various panning buttons, or the scrollbars to move around in a drawing.

1. Start AutoCAD and open the drawing named **db_samp.dwg** located in AutoCAD's Sample folder.

2. From the **File** pull-down menu, pick **Save As...**, open the folder with your name, and pick the **Save** button.

3. Zoom in on the lower right area of the drawing.

Standard

Realtime Panning

Let's pan to the upper left side of the drawing.

4. Pick the **Pan Realtime** button from the Standard toolbar, or enter **P** for PAN.

The crosshairs changes to a hand.

5. In the upper left portion of the screen, click and drag down and to the right.

The drawing window moves down and to the right.

6. Experiment further with the **PAN** command until you feel comfortable with it. Pan in different directions and at different zoom magnifications.

7. Press **ESC** or **ENTER** to exit the realtime pan mode.

Scrollbar Panning

AutoCAD's scrollbars enable you to pan horizontally and vertically. Focus your attention on the horizontal scrollbar. This is the bar at the bottom of the drawing area that contains left and right arrows at its ends.

1. Pick the left arrow once; pick it again.

This moves the viewing window to the left.

2. In the horizontal scrollbar, click and drag the movable scroll box a short distance to the right.

This moves the viewing window to the right. You can also move to the left or right by picking a point on either side of the movable scroll box.

Now focus on the vertical scrollbar. This is the bar along the right side of the drawing area. It also contains arrows at its ends.

3. Pick the up arrow a couple of times.

This moves the viewing window in the upward direction.

4. Pick the down arrow; then drag the scroll box down a short distance.

As you can see, the scrollbars offer a convenient way of panning.

Transparent Panning

You can use the PAN command transparently. This is particularly useful for reaching points that are not currently visible. To do this, enter 'PAN (notice the apostrophe) at any prompt that is not asking for a text string.

1. Enter the **LINE** command and pick a point anywhere on the screen.

2. Pick the **Pan Realtime** button or enter **'P** at the keyboard.

Standard

3. Pan to a new location, and press **ESC** or **ENTER** to exit the realtime panning mode.

4. Pick a second point to create a line segment and terminate the LINE command.

5. Erase the line.

Alternating Realtime Pans and Zooms

AutoCAD provides a fast method of alternating between realtime pans and zooms.

Standard

1. Pick the **Zoom Realtime** button from the Standard toolbar.

2. Zoom in on the drawing.

3. Right-click the pointing device.

This displays a shortcut menu.

4. Select **Pan** from the menu and pan to a new location.

5. Right-click the pointing device to display the menu.

6. Pick one of the **Zoom** options.

7. When you are finished, pick **Exit** from the shortcut menu or press **ESC** or **ENTER** to exit the realtime zoom and pan mode.

Capturing Views

Imagine that you are working on an architectural floor plan. You've zoomed in on the kitchen to include details such as the appliances, and now you're ready to pan over to the master bedroom. Before leaving the kitchen, you foresee a need to return to the kitchen for final touches. However, by the time you're ready to do this final work on the kitchen, you may be at a different zoom magnification or at the other end of the drawing. The VIEW command solves the problem.

1. Using realtime panning and zooming, find and zoom in on the four offices in the lower right area of the drawing.

2. From the **View** pull-down menu, select **Named Views...**, or enter **VIEW** at the AutoCAD prompt.

This enters the VIEW command and displays the View dialog box, as shown in Fig. 13-3 on page 168.

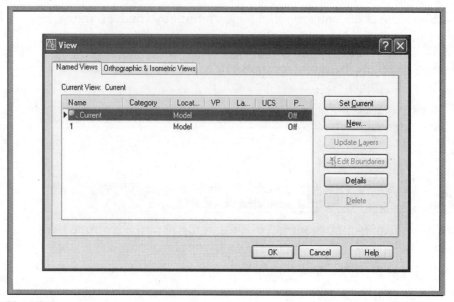

Fig. 13-3

3. Pick the **New...** button.

4. For View name, enter **Executives** and pick **OK**.

AutoCAD adds the named view to the list.

5. Pick the **OK** button.

6. Pan to a new location on the drawing.

7. Enter the **VIEW** command, pick the **Executives** view from the list, pick the **Set Current** button in the dialog box, and pick **OK**.

AutoCAD restores the named view.

8. Enter **VIEW** again, pick the **Executives** view, and pick the **Details** button.

This displays information about the named view.

9. Browse the information and then pick **OK**.

NOTE:

The Orthographic & Isometric Views tab does not apply to two-dimensional drawings. It enables you to restore orthographic or isometric views of three-dimensional drawings.

10. Pick **OK** to exit the View dialog box.

Aerial View

Aerial View is a fast and easy-to-use navigational tool for zooming and panning.

1. From the **View** pull-down menu, pick **Aerial View**.

This displays the Aerial View window with db_samp.dwg inside, as shown in Fig. 13-4.

The buttons in the Aerial View window should look familiar. The white rectangle represents the zoomed area.

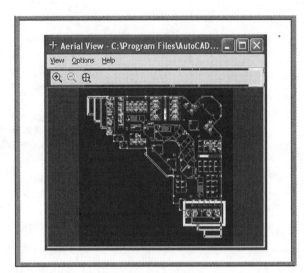

2. Move the crosshairs into the Aerial View window and click once.

A new crosshairs appears.

3. Move the window (called a *view box*) around.

AutoCAD zooms in on the area defined by the view box. Meanwhile, Aerial View maintains a view of the entire drawing. The view box, depicted by a white rectangle, shows the size and location of the current view in the graphics screen.

Fig. 13-4

4. Right-click to freeze the view box.

Aerial View shows the location of the current zoom window.

5. Press the pick button to activate the view box.

6. Press the pick button again to change the size of the view box and then pick another point to lock in a new size.

7. Move the view box to a new location and right-click.

8. On your own, experiment with the buttons and menu items in the Aerial View window.

9. Close the Aerial View window.

10. **ZOOM All** and exit AutoCAD without saving.

Chapter 13
Review & Activities

Review Questions

1. Explain why panning is useful.

2. Is it possible to pan diagonally using the scrollbars? Explain.

3. Explain why named views are useful.

4. Explain the method of alternating between realtime pans and zooms.

5. What is the primary benefit of using Aerial View?

Challenge Your Thinking

1. Many zoom and pan variations were introduced in the last two chapters. Describe how and when you might use a combination of them, but explain also why it may be impractical to use all of them. Which methods of zoom and pan are the fastest, and which seem to be the most practical for most applications?

2. Is Aerial View an alternative to using the PAN command? Explain when you would use each most effectively.

Applying AutoCAD Skills

Work the following problems to practice the commands and skills you learned in this chapter.

1. Open the Welding Fixture-1.dwg file located in the Sample folder. Use AutoCAD's zoom feature to magnify the parts list located in the lower right area of the drawing. Use Realtime Zoom and Realtime Pan to study all parts of the list. What is the description of item 10? What is the required quantity of this part?

2. Draw each of the views shown in Fig. 13-5 using the dimensions shown. However, do not include the dimensions on the drawing. Zoom in on one of the views and store it as a named view. Then pan to each of the other views and store each as a named view. Restore each named view and alter each shape. Be as creative as you wish, but be certain that all three of the views are consistent with one another.

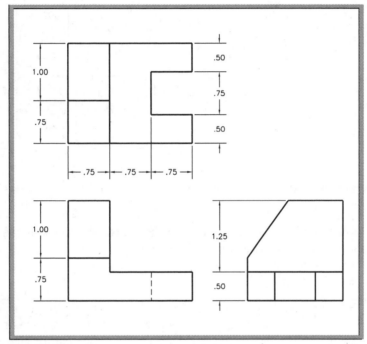

Fig. 13-5

3. Zoom in on one of the three views from the drawing in problem 2. Use the scrollbars to pan to the adjacent views.

4. Perform dynamic pans and zooms on the drawing in problem 2.

5. Open the kitchen.dwg file you created in problem 3 of Chapter 12. Add the details for the refrigerator and the stove as shown in Fig. 13-6. Open Aerial View and use it to position the zoom window as you work. Save the file as kitchen2.dwg.

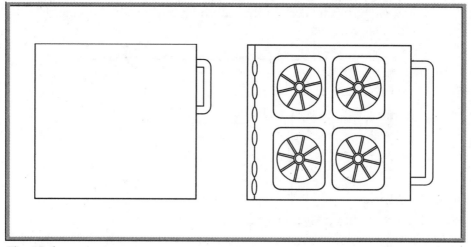

Fig. 13-6

6. Using the VIEW command, save the following views of the kitchen:
   Sink: closeup of the sink
   Fridge: closeup of the refrigerator
   Stove: closeup of the stove

7. Refer to land.dwg on the Instructor Resource CD. Zoom in on a small section of the drawing. Use the PAN command to move around the drawing. Use the VIEW command to save and restore views. Apply transparent zooms and pans. Perform dynamic and realtime zooms and pans.

Using Problem-Solving Skills

Complete the following activities using problem-solving skills and your knowledge of AutoCAD.

1. As the person responsible for computer allocation, you need to know how many offices are without computers. Open the AutoCAD file db_samp.dwg located in AutoCAD's Sample folder (Fig. 13-7). Using dynamic zoom and pan as necessary, make a list (by office number) of offices that do not have computers. Do not save any changes you may have made to the drawing.

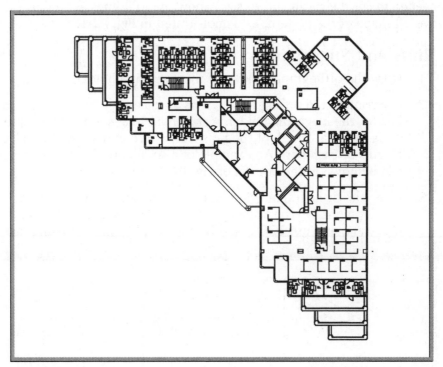

Fig. 13-7

2. Reopen db_samp.dwg and use Aerial View to make a list of rooms with designations of COPY, I.S., STOR., or COFFEE. Which of these rooms should have computers? Do you think office number 6001 should have a computer? Why or why not? Do not save your work.

Part 2 Project

Patterns and Developments

Many manufactured products begin as flat pieces of metal, plastic, or other material. Examples include cereal boxes, heating and cooling ducts, and even items of clothing. Manufacturers use flat *patterns* to cut material to the proper size and shape. The construction of these patterns is known as a *development*. Fig. P2-1 shows a box constructed for a company that distributes breakfast cereals, as well as an example of a pattern.

Description

Study the magazine holder shown in Fig. P2-2. It is to be manufactured from cardboard, and you are responsible for developing the pattern needed to cut the cardboard to size. Show the fold lines as well as the overall shape of the pattern, as demonstrated in Fig. P2-1.

Hints and Suggestions

1. Draw the orthographic views of the magazine holder.

2. Referring to the orthographic views as necessary, create a stretchout line for the magazine holder. A *stretchout line* is a line that represents the total length of the finished pattern and shows the relative lengths of each part. An example of a stretchout line for the cereal box is shown at the bottom of Fig. P2-2.

3. Using the orthographic views and the stretchout line, create the development. Use object snaps and other construction aids as necessary, and use ZOOM and PAN to simplify your drawing task.

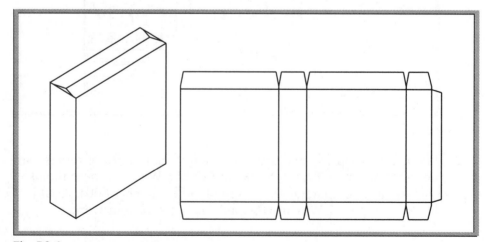

Fig. P2-1

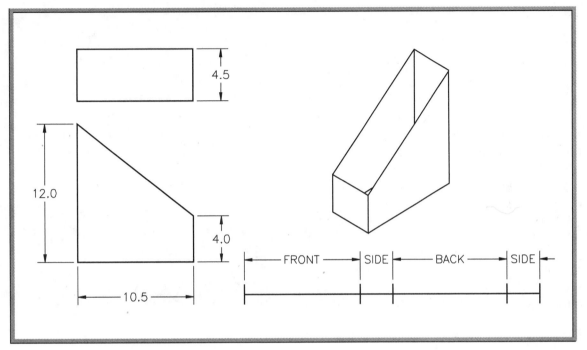

Fig. P2-2

Summary Questions

Your instructor may direct you to answer these questions orally or in writing. You may wish to compare and exchange ideas with other students to help increase efficiency and productivity.

1. Pinpoint any difficulties you encountered with this project and explain how you solved them.

2. Which construction aids did you use to create the orthographic views of the magazine holder? The development? Explain how these aids helped.

3. How would you create a pattern for a soup can? In what ways would the development be different from that for the magazine holder? Explain.

Careers

Using AutoCAD

Architectural Drafter

What type of house do you live in? Perhaps it's an older home, where floorboard creaks and groans are part of its charm. Or maybe it is tall, narrow, and practically bumps shoulders with your neighbor's house, like many homes located in the city. If you live in the country, your home may sprawl across your property, surrounded by plenty of grass and shade trees.

Although many individuals are responsible for the actual construction of homes, as well as factories, shopping centers, skyscrapers, schools, and other buildings, architectural drafters play an important role in the planning stages of a building's design.

Taxi/Jean Louis Batt

Climbing the Architectural Ladder

Architectural drafters are employed by research and engineering firms, corporate development departments, and architectural and construction companies. Apprenticeships are sometimes offered for students interested in entering this field, and taking classes in high school or at a junior college or vocational school will also help provide basic drafting principles. Architectural drafters may begin as tracers or junior drafters. After gaining experience, they often advance to senior drafting or checking positions.

Paying Attention to Detail

Architectural drafters must be able to visualize objects in two and three dimensions. They must also exhibit good eye-hand coordination, which is a skill necessary to create fine, detailed drawings. The ability to pay attention to detail is a critical characteristic

for an architectural drafter to possess; just one minor design flaw can result in a major problem, such as a lopsided doorway or an imperfect foundation. For information on student involvement in architecture, visit the American Institute of Architecture Students (AIAS) Web site.

Career Activities

- What types of detail do you think architectural drafters must be aware of when creating drawings? Think of some common features many buildings have on both the outside and the inside. Compare your list with a classmate's.
- Has employment in this field been steady over the past few years? What are the forecasts for future employment?

Chapter 14 Solid and Curved Objects

Objectives:

- **Produce solid-filled objects**
- **Control the appearance of solid-filled objects**
- **Create and edit polylines**
- **Create and edit splines**

Key Terms

B-splines	splines
control points	system variable
polyline	traces
screen regeneration	vertex

Many technical drawings contain thick lines and solid objects. Drawings for printed circuit boards, for instance, include relatively thick lines called *traces*. The trace lines represent the metal conductor material that is etched onto the PC board. Drawings and models that use curved lines often take advantage of polylines and spline curves. Examples are consumer products and business machines ranging from phones, pagers, and cosmetic products to printers, copiers, and fax machines. Many of the plastic injection-molded parts on these products are designed using polylines and spline curves. You will learn more about polylines and spline curves later in this chapter.

Note the heavy lines in the drawing shown in Fig. 14-1. They were used to make the drawing descriptive and visually appealing.

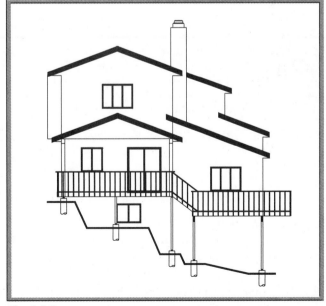

Fig. 14-1

Drawing Solid Shapes

The SOLID command produces solid-filled objects.

1. Start AutoCAD, select the **Start from Scratch** button from the Startup dialog box, pick **Imperial** units, and pick **OK**.

HINT:

If the Startup dialog box does not appear when you start AutoCAD, enter the STARTUP system variable and enter 1. Then enter the NEW command.

2. From the **Draw** pull-down menu, pick **Surfaces** and **2D Solid**.

This enters the SOLID command.

3. Produce a solid-filled object similar to the one in Fig. 14-2. Pick the points in the exact order shown.

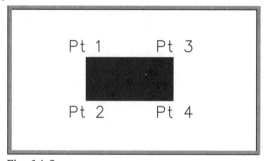

Fig. 14-2

HINT:

Consider using snap and grid.

4. Pick a fifth and sixth point.

AutoCAD adds a new piece to the solid object.

5. Experiment with the **SOLID** command. (If you pick the points in a different order, AutoCAD creates an hourglass-shaped object.)

6. Press **ENTER** to terminate the SOLID command.

Keep the objects on the screen because you will need them for the following section.

Controlling Object Fill

With the FILL command, you can control the appearance of solids, wide polylines, multilines, and hatches.

FILL is either on or off. When FILL is off, only the outline of a solid is represented on the screen. This saves time when the screen is regenerated.

INFOLINK

You will learn more about multilines in Chapter 15 and hatches in Chapter 18.

1. Enter the **FILL** command.

2. Enter **OFF**.

To see the effect of turning FILL on or off, you must force AutoCAD to recalculate all of the vectors on the screen. This is known as *screen regeneration*. To force the screen to regenerate, AutoCAD provides the REGEN command.

3. Enter the **REGEN** command to force a screen regeneration.

The objects are no longer solid-filled.

4. Reenter the **FILL** command and turn it **ON**.

5. Enter **REGEN** to force another screen regeneration.

6. Erase all objects on the screen.

Polylines

A *polyline* is a connected sequence of line and arc segments that is treated by AutoCAD as a single object. Polylines are often used in lieu of conventional lines and arcs because they are more versatile.

Drawing a Thick Polyline

Let's use the PLINE command to create the electronic symbol of a resistor, as shown in Fig. 14-3.

1. Set snap at **.5**.

2. Pick the **Polyline** button from the Draw toolbar, or enter **PL** (the PLINE command alias).

This enters the PLINE command.

3. Pick a point in the left portion of the screen.

The PLINE options appear.

Draw

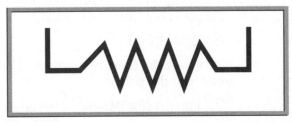

Fig. 14-3

4. Enter **W** (for Width) and enter a starting and ending width of **.15** unit. (Notice that the ending width value defaults to the starting width value.)

5. Draw the object by approximating the location of the endpoints. If you make a mistake, undo the segment. Press **ENTER** when you have finished the object.

6. Select anywhere on the polyline to display its grips.

Notice that the entire polyline is treated as a single object.

7. Save your work in a file named **poly.dwg**.

Editing a Polyline

AutoCAD supplies a special editing command called PEDIT to edit polylines. PEDIT allows you to perform many advanced functions, such as joining polylines, fitting them to curve algorithms, and changing their width. Let's edit the polyline using the PEDIT command.

1. Display the Modify II toolbar.

2. Pick the **Edit Polyline** button from the Modify II toolbar.

This enters the PEDIT command.

3. Pick the polyline.

The PEDIT options appear.

4. Enter **W** and specify a new width of **.2** unit.

The width of the polyline changes. As you can see, using polylines can be an advantage when there is a possibility that you will need to change the width of several connected line segments at a later time.

Let's close the polyline, as shown in Fig. 14-4.

5. Enter **C** for Close.

The PEDIT Fit option creates a smooth curve based on the locations of the vertices of the polyline. In this context, a *vertex* (plural *vertices*) is any endpoint of an individual line or arc segment in a polyline. Let's do a simple curve-fitting operation.

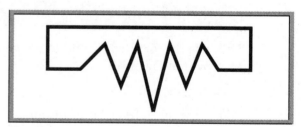

Fig. 14-4

6. Enter **F** for Fit.

Notice how the drawing changed. The curve passes through all of the vertices of the polyline.

7. Enter **D** (for Decurve) to return it to its previous form.

Modify II

Chapter 14: Solid and Curved Objects

Next, let's move one of the vertices as shown in Fig. 14-4.

8. Enter **E** for Edit Vertex.

Notice that a new set of choices becomes available. Also notice the x at one of the vertices of the polyline.

9. Move the x to the vertex you want to change by pressing **ENTER** several times.

10. Enter **M** for Move and pick a new point for the vertex.

11. To exit the PEDIT command, enter **X** (for eXit) twice.

12. Save your work.

PEDIT contains many more editing features. Experiment further with them on your own.

Exploding Polylines

The EXPLODE command gives you the ability to break up a polyline into individual line and arc segments.

1. Enter the **UNDO** command and select the **Mark** option. (We will return to this point at a later time.)

2. Pick the **EXPLODE** button from the Modify toolbar.

3. Pick the polyline and press **ENTER**.

Modify

Notice the message: Exploding this polyline has lost width information. The UNDO command will restore it.

This is the result of applying EXPLODE to a polyline that contains a width other than the default. You now have an object that contains many line objects for easier editing, but you have lost the line width.

4. To illustrate that the object is now made up of several objects, pick one of them.

Drawing Curved Polylines

In some drafting applications, there is a need to draw a series of continuous arcs to represent, for example, a river on a map. If the line requires thickness, the ARC Continue option will not work, but the PLINE Arc option can handle this task.

1. Enter the **PLINE** command and pick a point anywhere on the screen.

2. Enter **A** for Arc.

Draw

A list of arc-related options appears.

3. Move the crosshairs and notice that an arc begins to develop.

4. Enter the **Width** option and enter a starting and ending width of **.1** unit.

5. Pick a point a short distance from the first point.

6. Pick a second point, and a third.

7. Press **ENTER** when you are finished.

8. Enter **UNDO** and **Back**.

Spline Curves

Splines are curves that use sampling points to approximate mathematical functions that define a complex curve. AutoCAD allows you to create splines, also referred to as *B-splines*, using two different methods. You can use the PEDIT command to transform a polyline into a spline curve, or you can use the SPLINE command to generate the spline directly.

Transforming a Polyline into a Spline

The PEDIT Spline option uses the vertices of the selected polyline as the control points of the curve. *Control points* are points that exert a "pull" on a curve, influencing its shape. The curve passes through the first and last control points and is pulled toward the other points, but it does not necessarily pass through them. The more control points you specify in an area, the more pull they exert on the curve.

Modify II

1. Pick the **Edit Polyline** button from the Modify II toolbar and select the polyline.

2. Enter **S** for Spline.

Do you see the difference between the Spline and Fit options? The Spline option uses the vertices of the selected polyline as control points, whereas the Fit option produces a curve that passes through all vertices of the polyline.

3. Enter **D** to decurve the object.

4. Enter **X** to exit the PEDIT command.

The type of spline created by the Spline option depends on the setting of the SPLFRAME system variable. (A *system variable* in AutoCAD is similar to a command, except that it holds temporary settings and values instead of performing a specific function.) The following note provides more information about the types of splines AutoCAD provides through the PEDIT command.

Chapter 14: Solid and Curved Objects

NOTE: ──

In connection with PEDIT, AutoCAD offers two spline options: quadratic B-splines and cubic B-splines. An example of each is shown below.

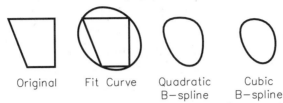

Original   Fit Curve   Quadratic B-spline   Cubic B-spline

The system variable SPLINETYPE controls the type of spline curve to be generated. Set the value of SPLINETYPE at 5 to generate quadratic B-splines. Set its value at 6 to generate cubic B-splines.

The PEDIT Decurve option enables you to turn a spline back into its frame, as illustrated in the previous Step 3. You can view the spline curve and its frame simultaneously by setting the SPLFRAME system variable to 1.

Producing Splines Directly

With the SPLINE command, you can create spline curves using a single command. Spline curves are used extensively in industry to produce free-form shapes for automobile body panels and the exteriors of aircraft and consumer products. The spline curve itself is the building block for producing interesting and sometimes extremely complex designs.

1. From the **File** pull-down menu, pick **Save As...**, and enter **splines** for the file name.

2. Erase all the objects in the drawing.

3. From the Draw toolbar, pick the **Spline** button.

Draw

This enters the SPLINE command.

4. Pick a point and then a second point.

5. Slowly move the end of the spline back and forth and notice how the spline behaves.

6. Continue to pick a series of points.

7. Press **ENTER** three times to complete the spline and terminate the command.

Draw

8. Enter **SPLINE** again and pick a series of points.

9. Enter **C** for Close.

10. Press **ENTER** to terminate the command.

NOTE:

The Fit Tolerance option changes the tolerance for fitting the spline curve to the control points. A setting of 0 causes the spline to pass through the control points. By changing the fit tolerance to 1, you allow the spline to pass through points within 1 unit of the actual control points.

You can easily convert a spline-fit polyline into a spline curve. Enter the SPLINE command, enter O for Object, and pick the polyline.

Editing Splines

Using grips, you can interactively edit a spline entity. The SPLINEDIT command provides many additional editing options.

1. Without any command entered, select one of the spline curves to display the grips.

2. Lock onto one of the grips and move it to a new location.

This is the fastest way to edit a spline.

3. Press **ESC** to clear the grips.

4. From the Modify II toolbar, pick the **Edit Spline** button.

Modify II

This enters the SPLINEDIT command.

5. Select a spline.

The editing options appear. The Close option closes the spline curve. The Move vertex option enables you to edit a spline by moving its vertices. The Refine option permits you to enhance the spline data.

6. Enter **R** for Refine to display the refinement options.

7. Enter **A** for Add control point.

8. Pick several points on the spline to add new control points, and then press **ENTER**. (Turn off object snap before adding new control points.)

9. Press **ENTER** twice.

The rEverse option reverses the order of the control points in AutoCAD's database. The Undo option undoes the last edit operation. The Fit Data option permits you to edit fit data using several options.

Modify II

10. Pick the **Edit Spline** button and pick a spline.

11. Enter **F** for Fit Data and pick the **Help** button to read about the fit options.

12. Experiment with the **Fit Data** options on your own.

13. Close the Modify II toolbar, save your work, and exit AutoCAD.

Chapter 14: Solid and Curved Objects

Review Questions

1. How would you draw a solid-filled triangle using the SOLID command?

2. What is the purpose of FILL, and how is it used?

3. What object types does FILL affect?

4. What is a polyline?

5. In what situations might you decide to use the PLINE command instead of the LINE and ARC commands?

6. Briefly describe each of the following PLINE options.
   a. Arc
   b. Close
   c. Halfwidth
   d. Length
   e. Undo
   f. Width

7. Briefly describe each of the following PEDIT command options.
   a. Close
   b. Join
   c. Width
   d. Edit vertex
   e. Fit
   f. Spline
   g. Decurve
   h. Ltype gen
   i. Undo

8. Explain a situation in which you might need to explode a polyline.

9. Describe the SPLINETYPE system variable.

10. When you create a spline curve using the SPLINE command, what does a fit tolerance setting of 0 indicate?

11. What is the fastest method of editing a spline?

12. What editing function can you perform by entering SPLINEDIT, Refine, and Add control point?

Challenge Your Thinking

1. Discuss possible uses for the SOLID and FILL commands. Under what circumstances might you use them? Give specific examples.

2. Discuss everyday applications of spline curves.

Applying AutoCAD Skills

Work the following problems to practice the commands and skills you learned in this chapter.

1. Construct the building in Fig. 14-5 using the PLINE and SOLID commands. Specify a line width of .05 unit. Estimate the size and shape of the roof.

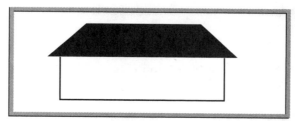

Fig. 14-5

2. After you have completed problem 1, place the solid shapes as indicated in Fig. 14-6. Estimate their sizes and locations.

Fig. 14-6

3. Use the PLINE command to draw the road sign indicating a sharp left corner (Fig. 14-7).

Fig. 14-7

4. The SI symbol shown in Fig. 14-8 is used on metric drawings that conform to the SI system and use what is called *third-angle projection*. This projection system, which places the top view of a multiview drawing above the front view, is the normal way of showing views in the United States. Draw the symbol, including the outlines of the letters S and I; then use the SOLID command to fill in the letters.

Fig. 14-8

Problem 4 courtesy of Gary J. Hordemann, Gonzaga University

5. Create the approximate shape of the racetrack shown in Fig. 14-9 using PLINE. Specify .4 unit for both the starting and ending widths. Select the Arc option to draw the figure.

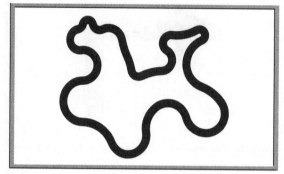

Fig. 14-9

6. Draw the symbols in Fig. 14-10 using the PLINE and PEDIT commands.

Fig. 14-10

7. Draw the car in Fig. 14-11 using the PLINE and PEDIT commands.

Fig. 14-11

8. The artwork for one side of a small printed circuit board is shown in Fig. 14-12. Reproduce the drawing using donuts and wide polylines. Use the grid to estimate the widths of the polylines and the sizes of the donuts.

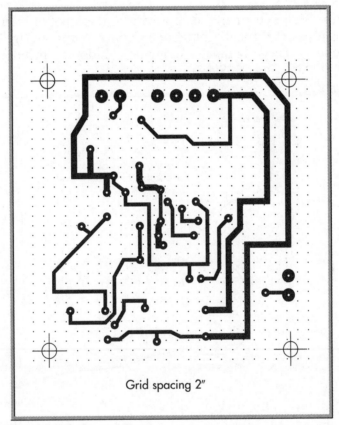

Grid spacing 2″

Fig. 14-12

Problem 8 courtesy of Gary J. Hordemann, Gonzaga University

? Using Problem-Solving Skills

Complete the following activities using problem-solving skills and your knowledge of AutoCAD.

1. Use the SPLINE command to approximate the route of Interstate 89 as it passes through Chittenden County, Vermont (Fig. 14-13). Include the Lake Champlain shoreline, but do not include the text. Save the drawing as 89.dwg.

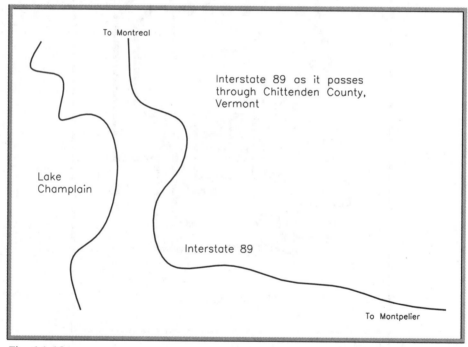

To Montreal

Interstate 89 as it passes through Chittenden County, Vermont

Lake Champlain

Interstate 89

To Montpelier

Fig. 14-13

2. Model railroads can be built to many different scales. One of the more popular scales is HO, in which ⅛″ = 1′. With the PLINE command, design an HO scale railroad track layout in a figure-eight pattern. Make the center intersection at 90 degrees. At a convenient point in your layout, add a spur to terminate in a railroad yard with three parallel tracks for storing the rolling stock. Refer to books in your local library or search the Internet for any additional information you may need to complete this problem.

3. Construct the drawing shown in Fig. 14-1.

Chapter 14: Solid and Curved Objects

Chapter 15 — Adding and Altering Objects

Objectives:

- **Create chamfered corners**
- **Break pieces out of lines, circles, and arcs**
- **Produce fillets and rounds**
- **Offset lines and circles**
- **Create and edit multilines**

Key Terms

chamfers	image tiles	multiline
fillets	mline	rounds

Most consumer products contain rounded or beveled edges for improved aesthetics and functionality. AutoCAD offers quick ways to create rounded inside corners, called *fillets*, and outside corners, called *rounds*, as well as beveled edges, called *chamfers*. Other useful tools include breaking lines and offsetting them to create parallel lines. AutoCAD also gives you the flexibility to produce multiple lines at one time. This is especially useful for architectural drawing, as are the other capabilities mentioned here.

Creating Chamfers

The CHAMFER command enables you to place a chamfer at the corner formed by two lines.

1. Start AutoCAD and open the drawing named **gasket.dwg**.

2. Pick the **Chamfer** button from the Modify toolbar.

3. Enter **D** for Distance.

Modify

The distance you specify is the distance from the intersection of the two lines (the corner) to the start of the bevel, or chamfer. You can set the chamfer distance for the two lines independently.

4. Specify a chamfer distance of **.25** unit for both the first and second distances.

5. Place a chamfer at each of the corners of the gasket by picking the two lines that make up each corner.

Modify

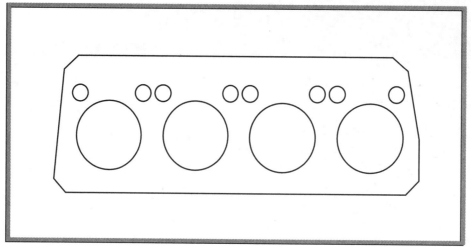

Fig. 15-1

When you're finished, the drawing should look similar to the one in Fig. 15-1, with the possible exception of the holes.

6. Obtain AutoCAD on-line help to learn about the other options offered by the CHAMFER command.

 HINT:

Enter the CHAMFER command and pick the Help button or press F1.

Breaking Objects

Let's remove (break out) sections of the gasket so that it looks like the drawing in Fig. 15-2 on the next page.

As you know, the bottom edge of the gasket was drawn as a single, continuous line. Therefore, if you were to use the ERASE command, it would erase the entire line. The BREAK command, however, allows you to "break" certain objects such as lines, arcs, and circles.

1. From the Modify toolbar, pick the **Break** button.

This enters the BREAK command.

Modify

2. Turn off object snap if it is on.

3. Pick a point on the line where you'd like the break to begin. Since the locations of the breaks in Fig. 15-2 are not dimensioned, you may approximate the location of each start point.

Chapter 15: Adding and Altering Objects

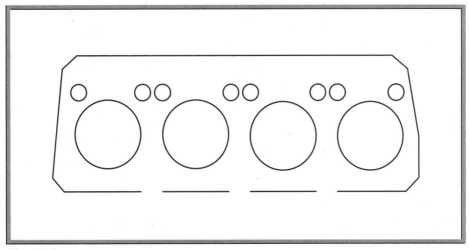

Fig. 15-2

4. Pick the point where you'd like the break to end.

The piece of the line between the first and second points disappears. Let's break out two more sections of approximately equal size as shown in Fig. 15-2.

5. Repeat Steps 1 through 3 to create each break.

6. Experiment with the other BREAK options on your own. Refer to AutoCAD's on-line help for more information about these options.

7. Insert arcs along the broken edge of the gasket as shown in Fig. 15-3.

8. Save your work.

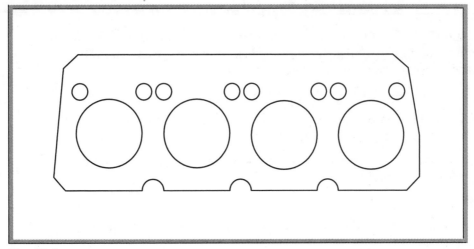

Fig. 15-3

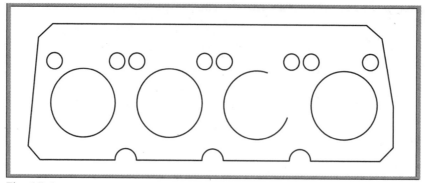

Fig. 15-4

Let's break out a section of one of the holes in the gasket as shown in Fig. 15-4.

Modify

9. Enter the **BREAK** command.

10. Pick a point anywhere on the circle. (You may need to turn off object snap.)

11. Instead of picking the second point, enter **F** for first point.

12. Pick the first (lowest) point on the circle.

13. Working counterclockwise, pick the second point.

A piece of the circle disappears.

Standard

14. Pick the **Undo** button on the Standard toolbar or enter **U**.

Creating Fillets and Rounds

In AutoCAD, the FILLET command creates both fillets and rounds on any combination of two lines, arcs, or circles. Let's change the chamfered corners on the gasket to rounded corners.

1. Erase each of the four chamfered corners.

Modify

2. From the Modify toolbar, pick the **Fillet** button, enter **R** for Radius, and enter a new radius of **.3**.

3. Produce fillets at each of the four corners of the gasket by picking each pair of lines.

4. Save your work.

The gasket drawing should now look similar to the one in Fig. 15-5.

Chapter 15: Adding and Altering Objects

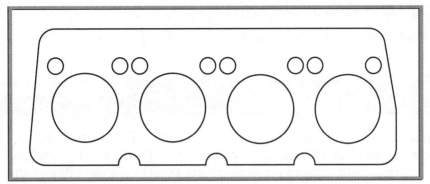

Fig. 15-5

Extending to Form a Corner

Let's move away from the gasket and try something new.

1. Above the gasket, draw lines similar to the ones in Fig. 15-6. Omit the numbers.

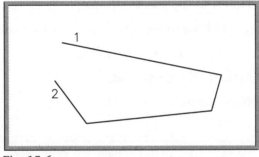

Fig. 15-6

Modify

2. Enter the **FILLET** command and set the fillet radius at **0**.

3. Select lines 1 and 2.

This technique works with the CHAMFER command, too.

Trim Option

The Trim option is useful for making slotted holes.

1. Above the gasket drawing, draw two relatively short lines that are parallel.

Modify

2. Enter the **FILLET** command, enter **T** for Trim, and press **ENTER** to accept the Trim default value.

3. In response to Select first object, pick the end of one of the two lines.

4. In response to Select second object, pick the same end of the other line.

AutoCAD fits an arc to the two lines.

5. Repeat Steps 2 through 4 for the other ends of the two lines.

Offsetting Objects

Offsetting a line results in another line at a specific distance, or offset, from the original line. Offsetting a circle or arc results in another circle or arc that is concentric (shares the same center point) with the original one.

Offsetting Circles

Modify

1. From the Modify toolbar, pick the **Offset** button.

This enters the OFFSET command.

2. For the offset distance, enter **.2**.

3. Select one of the holes in the gasket drawing.

4. Pick a point inside the hole in reply to Specify point on side to offset.

5. Select another hole and pick a side to offset.

6. Press **ENTER** to terminate the command.

7. Delete the two offset circles.

Offsetting Through a Specified Point

Modify

1. Using the **LINE** command, draw a triangle of any size and shape.

2. Pick the **Offset** button and enter **T** (for Through).

3. Pick any one of the three lines that make up the triangle. (You may need to turn off object snap.)

4. Pick a point a short distance from the line and outside the triangle.

The offset line appears. Notice that the line runs through the point you picked.

5. Do the same with the remaining two lines in the triangle so that the object looks similar to the one in Fig. 15-7. Press **ENTER** when you're finished.

6. Enter **CHAMFER** and set the first and second chamfer distances at **0**.

7. Pick two of the new offset lines.

Fig. 15-7

Chapter 15: Adding and Altering Objects

8. Do this again at the remaining two corners to complete the second triangle.

9. Clean up the drawing so that only the gasket remains.

10. Save your work and close the drawing file.

Drawing Multilines

The OFFSET command is useful for adding offset lines to existing lines, arcs, and circles. If you want to produce up to 16 parallel lines, you can do so easily—and simultaneously—with the MLINE command. This command produces parallel lines that exist as a single object in AutoCAD. The object is called a *multiline* or *mline* object.

1. Begin a new drawing from scratch and use **Imperial** units.

2. From the **Draw** pull-down menu, pick **Multiline**.

This enters the MLINE command.

3. Create a large triangle and enter **C** to close it.

Notice that the corners of the triangle meet perfectly. The triangle is a single object—an mline object.

4. Erase the triangle.

Creating Multiline Styles

The Multiline Style dialog box permits you to change the appearance of multilines and save custom multiline styles. Let's use this dialog box to produce the drawing of a room, as shown in Fig. 15-8.

Fig. 15-8

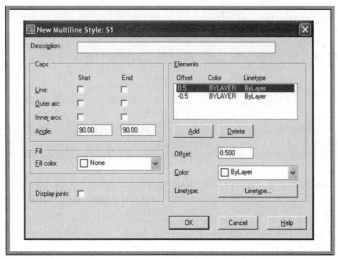

Fig. 15-9

1. Set snap at **.25** unit and turn on either polar tracking or ortho.

2. From the **Format** pull-down menu, pick the **Multiline Style...** item.

The Multiline Style dialog box appears.

3. Pick the **New...** button.

4. Type **S1** and pick the **Continue** button.

The New Multiline Style dialog box appears as shown in Fig. 15-9.

5. In the box located to the right of Description, type **This style has end caps.**

NOTE: ────────────────────────────

AutoCAD does not require you to enter a description for multiline styles you create. However, it is good practice to document your styles so that you can see at a glance which style is appropriate if you need to change or update the drawing later.

6. On the left side under Caps, check the **Line Start** and **Line End** boxes.

NOTE: ────────────────────────────

Many other options are available in the New Multiline Style dialog box. You can control the number of line elements and the color, linetype, and lineweight of each element. You can also set the distance between lines by changing the Offset values.

Chapter 15: Adding and Altering Objects

7. Pick the **OK** button.

8. Pick the **Save...** button.

9. In the Save Multiline Style dialog box, pick the **Save** button to store the S1 style in the file named acad.min.

Notice that the current style is still STANDARD. Before you can use a new style, you must make it the current style.

10. Select **S1** and pick the **Set Current** button.

Look in the Preview area in the Multiline Style dialog box. Notice that both ends of the S1 multiline are now "capped" with short connecting lines.

11. Pick the **OK** button.

When you create new multilines, the S1 style will apply.

Scaling Multilines

You can set the distance between lines in a multiline using the Scale option of the MLINE command.

1. Enter **MLINE** at the keyboard.

2. Enter **S** for Scale and enter **.25**.

3. With **MLINE** entered, create the drawing shown in Fig. 15-8, approximating its size.

HINT:

Use the Undo button if you need to back up one line segment.

4. Press **ENTER** to terminate MLINE.

5. Save your work in a file named **multi.dwg**.

Let's add the interior walls shown in Fig. 15-10.

6. Display the Multiline Style dialog box.

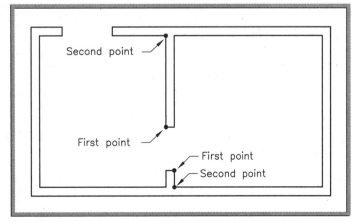

Fig. 15-10

7. Create a new multiline style named **S2** with the following description: **This style has a start end cap**.

8. Uncheck the End check box located under Caps to the right of Line and pick **OK**.

9. Save S2 and make it the current style.

10. Pick **OK** to exit the Multiline Style dialog box.

11. Using **MLINE**, draw the interior walls as shown in Fig. 15-10. Press **ENTER** to complete each wall. (You may need to turn object snap off before performing this step.)

You will edit the wall intersections in the following section.

Editing Intersections

Using the Multiline Edit Tools dialog box, it is possible to edit the intersection of multilines.

1. From the **Modify** pull-down menu, pick **Object** and then the **Multiline...** item.

This enters the MLEDIT command and displays the Multilines Edit Tools dialog box, as shown in Fig. 15-11. The twelve squares in the dialog box are called *image tiles*.

2. Pick the **Open Tee** image tile in the second row and second column.

Focus your attention on either of the two interior walls.

3. In reply to Select first mline, pick one of the two interior walls.

4. In reply to Select second mline, pick the adjacent exterior wall.

This causes AutoCAD to break the exterior wall.

5. Edit the second interior wall by repeating Steps 4 and 5.

6. Press **ENTER** to terminate the MLEDIT command.

7. On your own, explore the remaining options in the Multiline Style and Multilines Edit Tools dialog boxes.

8. Save your work and exit AutoCAD.

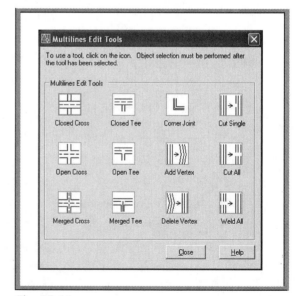

Fig. 15-11

Review Questions

1. What is the function of the CHAMFER command?

2. How is using the BREAK command different from using the ERASE command?

3. If you want to break a circle or arc, in which direction do you move when specifying points: clockwise or counterclockwise?

4. In what toolbar is the Fillet button found?

5. How do you set the fillet radius?

6. What can you accomplish by setting either FILLET or CHAMFER to 0?

7. Explain the purpose of the OFFSET command.

8. How might multilines be useful? Give at least one example.

9. Explain the purpose of the Multiline Style and Multilines Edit Tools dialog boxes.

Challenge Your Thinking

1. Explore the Angle option of the CHAMFER command. How is it different from the Distance option? Discuss situations in which each option (Angle and Distance) may be useful.

2. To help potential clients understand floor plans, the architectural firm for which you work draws its floor plans showing the walls as solid gray lines the thickness of the wall. Exterior walls are 6" thick, and interior walls (those that make up the room divisions) are 5" thick. Explain how you could achieve these walls using MLINE and the Multiline Style dialog box.

3. AutoCAD allows you to save multiline styles and then reload them later. Explore this feature and write a short paragraph explaining how to save and restore multilines.

Applying AutoCAD Skills

Work the following problems to practice the commands and skills you learned in this chapter.

1. Create the first drawing on the left in Fig. 15-12. Don't worry about exact sizes and locations, but do use snap and ortho. Then use FILLET to change it to the second drawing. Set the fillet radius at .2 unit.

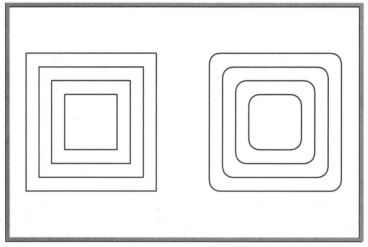

Fig. 15-12

2. Create the triangle shown in Fig. 15-13. Then use CHAMFER to change it into a hexagon. Set the chamfer distances at .66 unit.

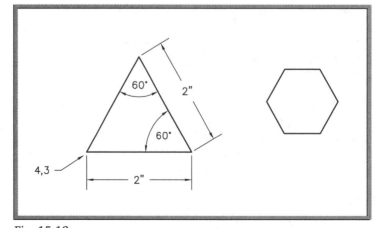

Fig. 15-13

3. Create the roller and cradle for a conveyor. Begin by drawing a circle and three lines as shown in Fig. 15-14A. Specify a fillet radius of .25 and create the fillets on the left side as shown in Fig. 15-14B. Then fillet the right side. The finished roller and cradle should look like the one in Fig. 15-14C.

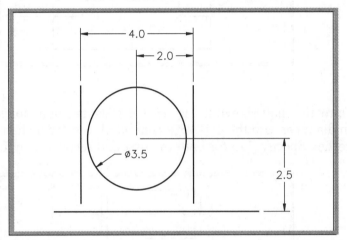

Fig. 15-14A

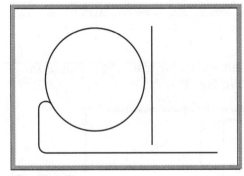

Fig. 15-14B

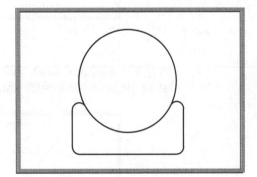

Fig. 15-14C

4. Draw the steel base shown in Fig. 15-15. Use the FILLET command to place fillets in the corners as indicated.

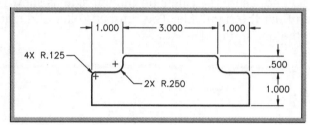

Fig. 15-15

5. Draw the shaft shown in Fig. 15-16. Make it 2 units long by 1 unit in diameter. Use the CHAMFER command to place a chamfer at each corner. Specify .125 for both the first and second chamfer distances.

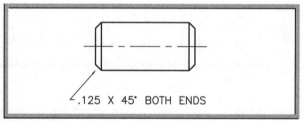

Fig. 15-16

6. Using the ARC and CHAMFER commands, construct the detail of the flat-blade screwdriver tip shown in Fig. 15-17.

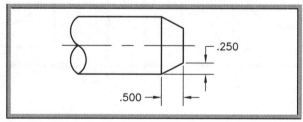

Fig. 15-17

7. Create the picture frame shown in Fig. 15-18. First create the drawing on the left using the LINE and OFFSET commands. Then modify it to look like the drawing on the right using the CHAMFER command.

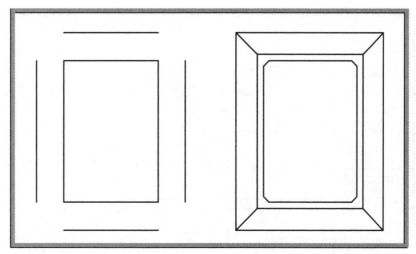

Fig. 15-18

8. Create the A.C. plug shown in Fig. 15-19 using the MLINE, MLSTYLE, CHAMFER, and FILLET commands. Measure an electrical plug to obtain real sizes.

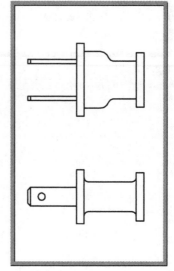

Fig. 15-19

Problems 7 and 8 courtesy of Dr. Joseph K. Yabu, San Jose State University

9. Draw the front view of the rod guide end cap shown in Fig. 15-20. Use the OFFSET command to create the outer circular shapes and the inner horizontal lines. Use the BREAK and FILLET commands to create the inner shape with four fillets. Do not dimension the drawing.

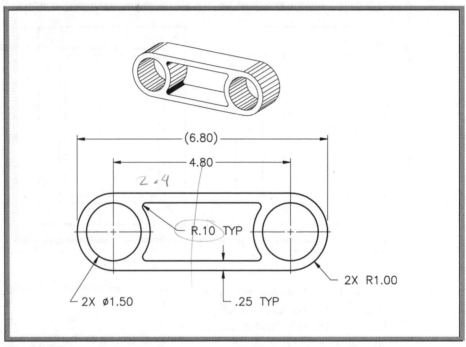

Fig. 15-20

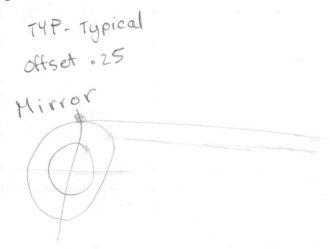

Problem 9 courtesy of Gary J. Hordemann, Gonzaga University

Chapter 15: Adding and Altering Objects

10. Draw the objects and rooms shown in Fig. 15-21. Use multilines to produce the walls. Copy and move the office furniture into the rooms. Omit the text in the drawing.

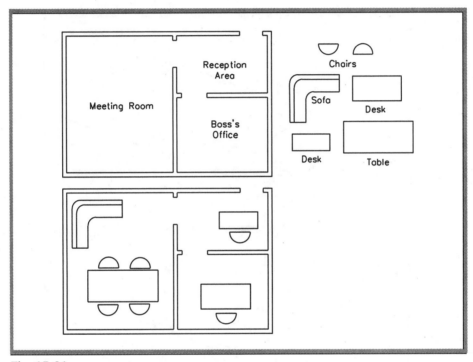

Fig. 15-21

11. Draw the front view of the ring separator shown in Fig. 15-22. Start by using the CIRCLE and BREAK commands to create a pair of arcs. Then use OFFSET to create the rest of the arcs. Do not dimension the drawing.

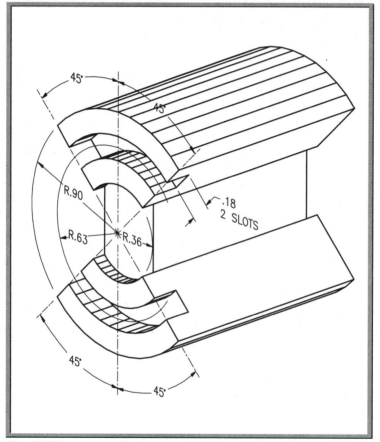

Fig. 15-22

Problem 11 courtesy of Gary J. Hordemann, Gonzaga University

Chapter 15: Adding and Altering Objects

Using Problem-Solving Skills

Complete the following activities using problem-solving skills and your knowledge of AutoCAD.

1. Kitchens are arranged around their three most important elements: the sink, the range, and the refrigerator. The traffic pattern is arranged in a triangle, with a vertex at each of the three items. As a result, kitchens are usually arranged in one of three patterns: U-shaped, L-shaped, and parallel. Draw an example of each type of kitchen using the MLINE command with capped ends. (You may want to find examples of each type of kitchen layout in architectural references before you begin drawing to familiarize yourself with the options.) Do not label the appliances, but draw the range using four circles to represent the burners, the refrigerator as a rectangle, and the sink as a smaller rectangle with the corners rounded, as shown in Fig. 15-23. Show the triangular traffic pattern for each kitchen you design. Which seems most convenient to you? Why?

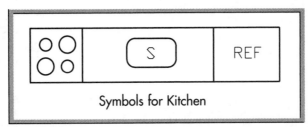
Symbols for Kitchen

Fig. 15-23

2. Select one of the kitchen arrangements you created in problem 1, and design a kitchen around it. Include the walls, door and window placements, and any other appliances or furniture that you think a kitchen should include.

Careers Using AutoCAD

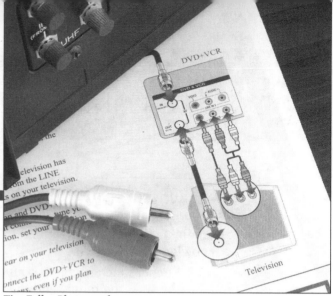

Technical Illustrator

You have just purchased a new stereo and you are eager to hear how your favorite CD sounds through the system's powerful speakers. In order to figure out how to connect each component properly, you consult the instruction sheet:

Connect wire A from component 1 to component 2. Connect wire B from component 1 to component 3. Connect wires C and D from component 1 to external speakers.

How do you know which is wire A? How is it different from wire D? How can you be certain not to confuse component 1 with component 2? Fortunately, the instructions to your new stereo also include a detailed parts list as well as an illustrated example of the setup procedure.

Assisting the User

Technical illustrators produce drawings for instruction manuals and appliance or equipment operating documents. These illustrations are helpful when paired with a written set of instructions; users are then able to identify specific parts and verify that they have followed the directions properly. Technical illustrators may consult directly with the person writing the instructions or with the manufacturer of the product in order to ensure that their understanding of the item, its parts, and the intended procedure is accurate.

Attention to Detail

Like many other types of drafting careers, technical illustrators must pay close attention to detail. If an unusual piece of equipment is

Tim Fuller Photography

mentioned in writing but a visual example is not provided, the person following the directions may become confused. A combination of illustrations and text offers multiple opportunities to understand exactly how to build or operate an item, which can prevent improper assembly or misuse.

Many drafters are drawn to the field of technical illustration, probably because the two have many requirements in common. Even though technical illustrators rarely create dimensioned technical drawings, their work must be accurate and complete. In fact, many technical illustrators have a strong drafting background. In addition, technical illustrators must have a good knowledge of the products they are illustrating.

Career Activities

- Have you ever used instructions that were text-based only, with no accompanying illustrations? What about instructions that provided drawings only, but no text? Which type of instructions do you find easiest to follow? Why?
- Investigate technical illustrator careers using the Internet. Has employment in this field been steady over the past few years? What are the forecasts for future employment?

Chapter 16 Moving and Duplicating Objects

Objectives:

- **Change an object's properties**
- **Move and copy objects**
- **Mirror objects and parts around an axis**
- **Produce rectangular and polar arrays**

Key Terms

array   polar array
circular array   rectangular array

Earlier in this book, you learned how to move and copy objects using the grips method. Now you will learn how to perform these and other operations using commands. Why would you want to? The commands give you additional options. Also, if you choose to write scripts or programs that automate certain AutoCAD operations, you could embed these commands into them, so knowing how they work is important.

You will also learn methods of arraying objects, both in rectangular and circular fashion. The arraying features are useful for many purposes, including architectural drafting, facilities planning, auditorium and stadium design, and countless machine design applications. The problems at the end of this chapter provide examples.

Changing Object Properties

The Properties window provides a method of changing several of an object's properties.

1. Start AutoCAD and open the drawing named **gasket.dwg**.

2. Select one of the holes and right-click.

3. Pick the **Properties** item from the shortcut menu.

This displays the Properties window, as shown in Fig. 16-1 on page 212. Review the list of object properties, including radius and diameter. Note that the properties listed in the dialog box depend on the object selected. In this case, the options presented are specific to the circle you selected in Step 2.

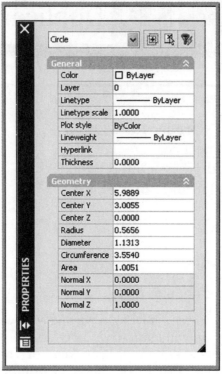

Fig. 16-1

4. Change the diameter of the circle by entering a new number into the Diameter text box and press **ENTER**.

The diameter of the circle changes.

You will use this dialog box many times throughout this book.

5. Change the diameter back to its original value, and close the dialog box.

6. Press the **ESC** key to remove the object selection.

Moving Objects

Let's move the gasket to the top of the screen using the MOVE command.

Modify

1. From the Modify toolbar, pick the **Move** button, or enter **MOVE** at the keyboard.

HINT:

M is the command alias for MOVE.

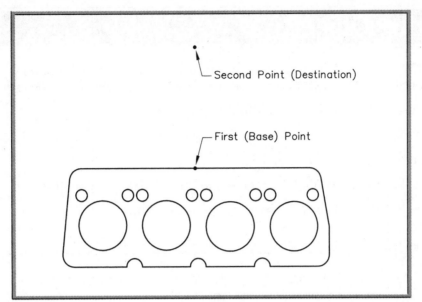

Second Point (Destination)

First (Base) Point

Fig. 16-2

AutoCAD displays the Select objects prompt.

2. Place a window around the entire gasket drawing and press **ENTER**.

3. In reply to Specify base point or displacement, place a base point somewhere on or near the gasket drawing as shown in Fig. 16-2.

4. Move the pointing device in the direction of the second point (destination).

5. Pick the second point.

The gasket should move accordingly.

6. Practice using the **MOVE** command by moving the drawing to the bottom of the screen.

7. Reenter the **MOVE** command.

Modify

8. In response to the Select objects prompt, pick two of the four large holes and press **ENTER** (or right-click).

9. Specify the first point (base point). Place the point anywhere on or near either of the two holes.

10. In reply to Specify second point or displacement, move the pointing device upward. (Polar tracking should be on.)

11. Enter **3** to move the holes vertically 3 units.

12. Undo the move.

Copying Objects

The COPY command is almost identical to the MOVE command. The only difference is that the COPY command does not move the object; it copies it.

1. Erase all of the large holes in the gasket except for one.

2. From the Modify toolbar, pick the **Copy** button.

This enters the COPY command.

Modify

3. Select the remaining large hole and press **ENTER** or right-click.

4. Select the center of the hole for the base point.

![lightbulb icon] **HINT:**

Use the Center object snap. Also, turn on ortho or polar tracking before completing the following step.

5. Move the crosshairs and place the hole in the proper location.

6. Repeat Step 5 until all four large holes are in place; then press **ENTER**.

7. Practice using the **COPY** command by erasing and copying the small holes.

8. Save your work.

Mirroring Objects

There are times when it is necessary to produce a mirror image of a drawing, detail, or part. A simple copy of the object is not adequate because the object being copied must be reversed, as was done with the butterfly in Fig. 16-3. One side of the butterfly was drawn and then mirrored to produce the other side. The mirroring feature can save you a large amount of time when you are drawing a symmetrical object or part.

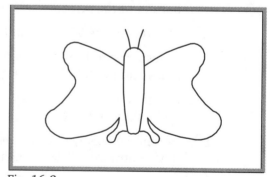

Fig. 16-3

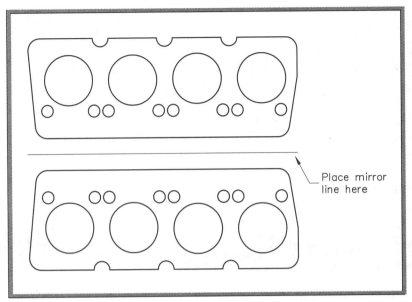

Fig. 16-4

Mirroring Vertically

The same is true if the engine head gasket we developed is to be reproduced to represent the opposite side of an eight-cylinder engine. In this case, we will mirror the gasket vertically (around a horizontal plane).

1. If necessary, move the gasket to the bottom of the screen to allow space for another gasket of the same size.

2. From the Modify toolbar, pick the **Mirror** button.

This enters the MIRROR command.

Modify

3. Select the gasket by placing a window around it, and press **ENTER**.

4. Create a horizontal mirror line near the gasket by selecting two points on a horizontal plane as shown in Fig. 16-4.

AutoCAD asks if you want to delete the source objects.

5. Enter **N** for No, or press **ENTER** since the default is No.

AutoCAD mirrors the gasket as shown in Fig. 16-4.

Mirroring at Other Angles

You can also create mirror images around an axis (mirror line) other than horizontal.

Modify

1. Draw a small triangle and mirror it with an angular (*e.g.,* 45°) mirror line.

NOTE:

You can also mirror objects using grips. Without a command entered, pick an object, pick (activate) one of the grip handles, and then right-click. From the shortcut menu, pick Mirror. Notice that if you mirror using this method, the selected grip will lie on the mirror line. To create a different mirror line, you must first enter B (for Base point).

2. Erase the triangles, save your work, and close the drawing.

Producing Arrays

An *array* is an orderly grouping or arrangement of objects. AutoCAD's arraying capability can save a large amount of time when you create a drawing that includes many copies of an object arranged in a rectangular or circular fashion.

1. Begin a new drawing.

2. Select **Use a Wizard** and **Quick Setup** and pick **OK**.

3. Pick **Architectural** for the units and pick the **Next** button.

4. Enter a width of **24′** and a length of **18′**, and pick the **Finish** button.

5. Set the grid at **1′** and the snap at **1″**; **ZOOM All**.

6. Zoom in on the lower left quarter of the drawing area and create the chair shown in Fig. 16-5. Use the following information.

   - Create the 1′-3″ radius arc first. Use the snap grid to snap to the start and end points of the arc.
   - Produce the 9″ arc by offsetting by 6″ from the 1′-3″ arc.
   - Create the first 3″ arc using the endpoints of the larger arcs; then create the second using the **MIRROR** command.
   - The 10″ arc passes through the centers of the 3″ arcs.

7. **ZOOM All** and save your work in a file named **chair.dwg**.

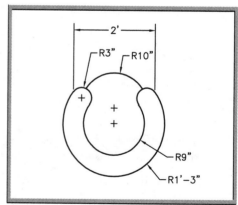

Fig. 16-5

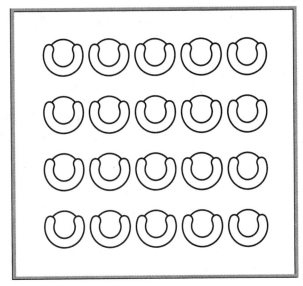

Fig. 16-6

Creating Rectangular Arrays

A *rectangular array* is one in which the objects are arranged in rows and columns. An example is shown in Fig. 16-6.

Modify

1. Pick the **Array** button from the docked Modify toolbar.

This displays the Array dialog box. Notice that Rectangular Array is selected.

2. Enter **4** for Rows and **5** for Columns.

3. Enter **3'** for Row offset and **34"** for Column offset.

These offsets define the distance between the rows and columns. The distance is measured from the center of one chair to the center of the next.

4. Pick the **Select objects** button located in the upper right area of the dialog box.

5. Select the chair, press **ENTER**, and pick the **Preview** button.

AutoCAD displays an array of 20 chairs, as shown in Fig. 16-6, and a window to accept, modify, or cancel the array you've created.

6. Pick **Accept**.

7. Save your work and close the drawing.

NOTE:

You can produce rectangular arrays at any angle by changing Angle of array in the Array dialog box.

Creating Circular Arrays

Next, we're going to produce the cycle wheel shown in Fig. 16-7. Notice the arrangement of the spokes. We can create them quickly by drawing only two lines and then using a circular array to create the remaining lines. A *circular array* (also called a *polar array*) is one in which the objects are arranged radially around a center point.

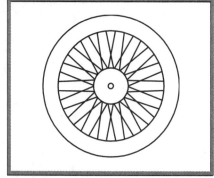

Fig. 16-7

1. Begin a new drawing from scratch.

2. Draw a wheel similar to the one in Fig. 16-8. Make the wheel large enough to fill most of the screen. Use the **Center** object snap to make the circles concentric.

3. Draw two crossing lines similar to the ones in Fig. 16-9. Use the **Nearest** object snap to begin and end the lines precisely on the appropriate circles.

4. Save your work in a file named **wheel.dwg**.

5. Pick the **Array** button or enter **AR**, the command's alias.

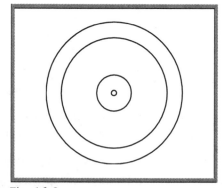

Fig. 16-8

6. Use the following information to complete the options in the Array dialog box.

   • Select **Polar Array**.
   • Select the button to the right of Center point and pick the wheel's center.
   • Enter **18** for Total number of items.
   • Do not change Angle to fill because you want to array the two spokes 360 degrees.
   • Pick the **Select objects** button, select the two spokes, press **ENTER**, and pick **Preview**.

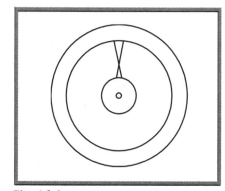

Fig. 16-9

The wheel should look similar to one in Fig. 16-7.

7. Pick **Accept** to finish the array and end the command.

8. Practice creating polar arrays using other objects. At least once, specify less than 360 degrees for Angle to fill.

9. Save your work and exit AutoCAD.

Chapter 16: Moving and Duplicating Objects

Review Questions

1. What is the purpose of the Properties window?

2. In what toolbar is the Move button located?

3. Explain how the MOVE command is different from the COPY command.

4. Describe a situation in which the MIRROR command would be useful.

5. During a MIRROR operation, can the mirror line be specified at any angle? Explain.

6. Name the two types of arrays.

7. State one practical application for each type of array.

8. What is the effect on a polar array if you specify an angle to fill of less than 360 degrees? Explain.

9. Explain how a rectangular array can be created at any angle.

Challenge Your Thinking

1. Experienced AutoCAD users take the time necessary to analyze the object to be drawn before beginning the drawing. Discuss the advantages of doing this. Keep in mind what you know about the AutoCAD commands presented in this chapter.

2. You have been asked to complete a drawing of the dialpad for a pushbutton telephone. You have decided to create one of the pushbuttons and use the ARRAY command to create the rest. Each pushbutton is a 13 mm × 8 mm rectangle, and the buttons are arranged as shown in Fig. 16-10. What should you enter for the space between rows and space between columns?

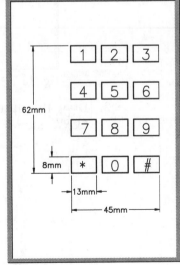

Fig. 16-10

Applying AutoCAD Skills

Work the following problems to practice the commands and skills you learned in this chapter.

1. Create the shim shown in Fig. 16-11. Begin by drawing the left half of the shim according to the dimensions shown, using construction lines for center lines. Then mirror the objects to create the right half of the shim. Erase any unneeded lines. Save the drawing as shim.dwg.

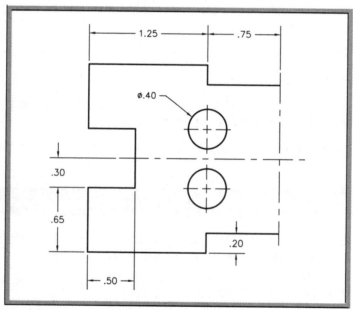

Fig. 16-11

2. Draw the chisel shown in Fig. 16-12 using the ARC, FILLET, LINE, COPY, and MIRROR commands.

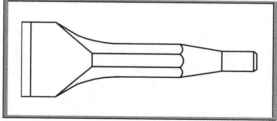

Fig. 16-12

Problem 2 courtesy of Joseph K. Yabu, Ph.D., San Jose State University

3. Draw the key shown in Fig. 16-13 using the ARC, FILLET, LINE or PLINE, COPY, and MIRROR commands.

4. Apply the commands introduced in this chapter to create the piping drawing shown in Fig. 16-14.

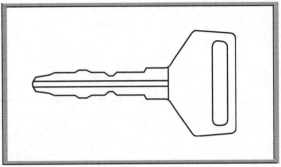

Fig. 16-13

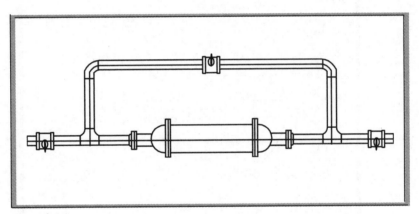

Fig. 16-14

5. Develop a plan for an auditorium with rows and columns of seats. Design the room any way you'd like. Save your work as auditorium.dwg.

6. Create the cooling fan shown in Fig. 16-15.

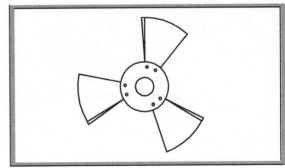

Fig. 16-15

Problems 3 and 6 courtesy of Joseph K. Yabu, Ph.D., San Jose State University
Problem 4 courtesy of Dr. Kathleen P. King, Fordham University, Lincoln Center

7. Draw the front view of the bar separator shown in Fig. 16-16. Draw one-fourth of the bar separator using the LINE, ARC, and FILLET commands with a grid setting of .1 and a snap of .05. Use the MIRROR command to make half of the object; then use it again to create the other half.

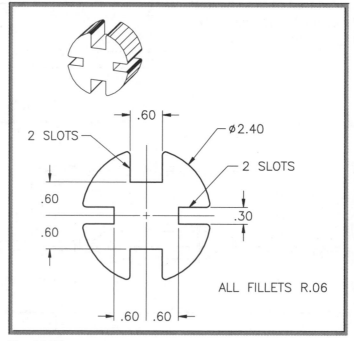

Fig. 16-16

8. Draw the air vent shown in Fig. 16-17 using the ARRAY command. Other commands to consider are POLYGON and COPY.

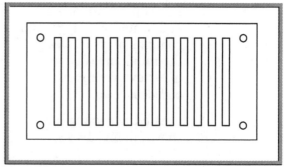

Fig. 16-17

Problem 7 courtesy of Gary J. Hordemann, Gonzaga University
Problem 8 courtesy of Joseph K. Yabu, Ph.D., San Jose State University

9. Draw the front view of the wheel shown in Fig. 16-18.

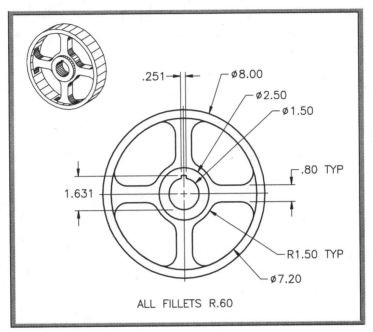

Fig. 16-18

Begin by drawing one-fourth of the object. Draw the three arcs and four lines as shown on the left in Fig. 16-19. Center the arcs on a known point. Next, use the FILLET command to create the cavity shown on the right. Erase the outside lines and the outside arc. Then use the MIRROR command to make three more copies of the cavity. Complete the drawing by adding arcs and circles. Try constructing the keyway by first drawing half of the shape, then using MIRROR to generate the other half. For more practice, try drawing with one-eighth of a wheel.

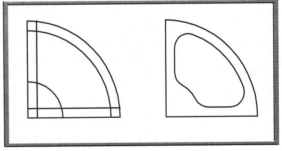

Fig. 16-19

Problem 9 courtesy of Gary J. Hordemann, Gonzaga University

10. Load the drawing named gasket.dwg. Erase each of the holes. Replace the large ones according to the locations shown in Fig. 16-20, this time using the ARRAY command. Reproduce the small holes using the ARRAY command, but don't worry about their exact locations. The diameter of the large holes is 1.25; the small holes have a diameter of .30.

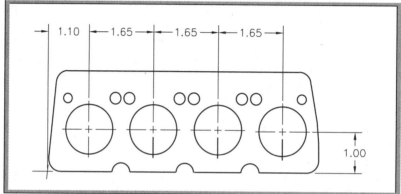

Fig. 16-20

11. Draw the snowflakes in Fig. 16-21 using the ARRAY command. Since snowflakes are six-sided, you may want to begin by drawing a set of concentric circles. Then use the ARRAY command to insert 6, 12, or more construction lines. The example in Fig. 16-22 shows the beginning of one of the snowflakes. Note that you only need to array three lines at 15° intervals for a total of 45° (two lines if you use the MIRROR command).

Fig. 16-21

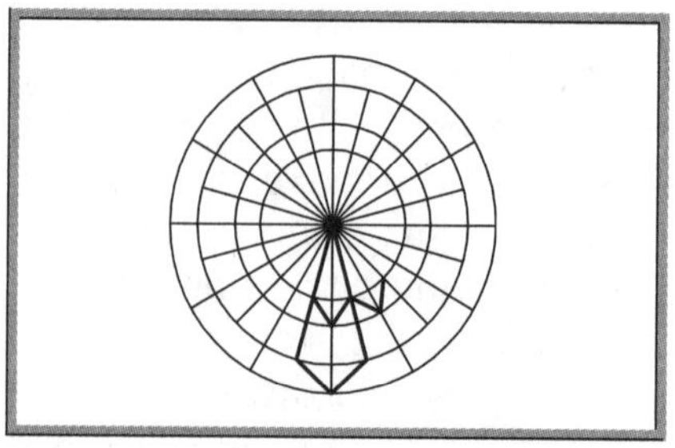

Fig. 16-22

Problem 11 courtesy of Gary J. Hordemann, Gonzaga University

12. Draw the top view of the power saw motor flywheel shown in Fig. 16-23. Use the ARRAY command to insert the 24 fins and arcs. You may find the OFFSET command useful in creating the first fin.

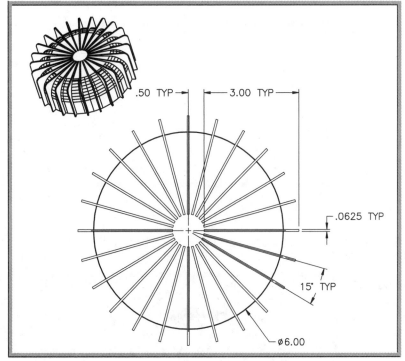

Fig. 16-23

Using Problem-Solving Skills

Complete the following activities using problem-solving skills and your knowledge of AutoCAD.

1. You are creating problems for an AutoCAD textbook and need to simulate the grid displayed with a 0.5 spacing. With the POINT command on the Draw toolbar, create a rectangular array within a drawing space of 12,9 so that the grid completely fills the space. If you find the point is too small to show on the screen, obtain help to find out how to make it larger. When you have completed the grid, turn AutoCAD's grid on and off to see how your grid corresponds to the real grid. Save the drawing as grid.dwg.

Problem 12 courtesy of Gary J. Hordemann, Gonzaga University

You have received a design change for the shim you created in Fig. 16-11. Open shim.dwg and use the MOVE and COPY commands to arrange the holes in a pattern approximately as shown in Fig. 16-24. Save the drawing as shimrev.dwg.

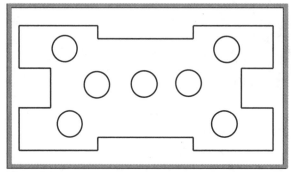

Fig. 16-24

3. Draw the wheel shown in Fig. 16-25 (you decide how).

Fig. 16-25

Problem 2 adapted from a drawing by Bob Weiland

Chapter 16: Moving and Duplicating Objects

Careers Using AutoCAD

Tim Fuller Photography

HVAC Designer

Have your grandparents ever told stories about how they didn't have air conditioning when they were growing up? Or have you read stories about the days before electricity, when a single fireplace was used to warm an entire household during the coldest months of winter? Fortunately, we live during a time when heating and air-conditioning systems are fairly commonplace. Certainly you have a furnace in your home, and almost any store you visit in summertime will have its air conditioning going full blast. HVAC (heating, ventilation, and air conditioning) designers are the people we have to thank for designing the heating and air-conditioning units that make our lives more comfortable.

Understanding Efficiency

HVAC designers create plans for heating and cooling systems for residential, commercial, industrial, and institutional uses. The most efficiently designed systems use ventilation techniques to help improve the quality of air indoors. They can also reduce the amount of energy used to heat or cool a building. This can help businesses and homeowners save money. Energy-efficient systems are also beneficial to the environment.

Combining Skills

Individuals working with HVAC systems should be familiar with standard codes for heating and cooling systems. Math skills are important in order to determine heat loss and air flow calculations. After gaining some drafting experience, HVAC designers may move into positions focusing on heating and drafting analysis and procedures. Computer skills are helpful for all individuals involved in this field, including knowledge of CAD-related programs and word processing programs.

Career Activities

- Interview someone who can tell you about employment of HVAC designers in your area. What education is required? Are apprenticeships available?
- Has employment in this field been steady over the past few years? What are the forecasts for future employment?

Chapter 17 Modifying and Maneuvering

Objectives:

- Stretch objects to change their overall shape
- Scale objects using a scale factor or reference length
- Rotate objects to exact angles
- Trim and extend lines to specific boundaries
- Trim and extend multilines
- Join individual lines to specific boundaries

Key Terms

colinear
rotating
scale

The more you create and edit drawings with AutoCAD, the more you will discover a need to edit objects in many ways. You will find several uses for trimming, extending, joining, and lengthening lines as you experiment with new ideas. One of the benefits of using AutoCAD is that you are not penalized for drawing something at the wrong size or in the wrong place. You can easily correct it using one of the commands covered in this or another chapter.

Stretching Objects

1. Start AutoCAD and choose the **Quick Setup** wizard.

2. For the units, use **Architectural**, and for the area, enter **120'** and **90'**.

Be sure to include the foot symbols (') so that AutoCAD knows that these are feet, not inches.

3. Enter **ZOOM All**.

4. Draw the site plan shown in Fig. 17-1 according to the dimensions shown. With snap set at **2'**, place the lower left corner of the property line at absolute point **10,10**. Omit dimensions. All points in the drawing should fall on the snap grid.

5. Save your work in a file named **site.dwg**.

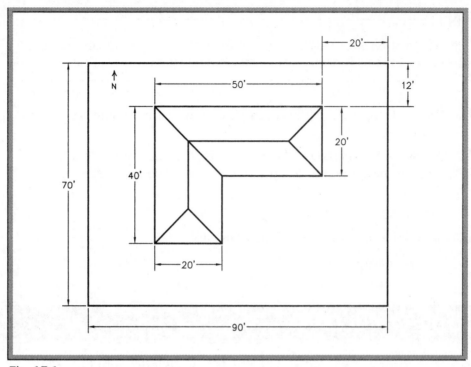

Fig. 17-1

Modify

6. Pick the **Stretch** button from the Modify toolbar or enter **S**, the STRETCH command's alias.

This enters the STRETCH command.

7. Select the east end of the house as shown in Fig. 17-2 and press **ENTER**. Use the crossing object selection procedure. (See the hint on page 230.)

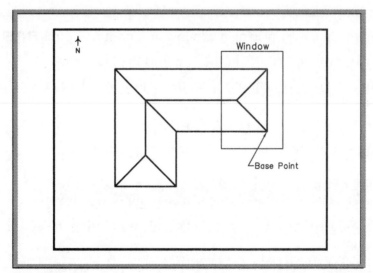

Fig. 17-2

HINT:

In reply to Select objects, pick a point to the right of the house and move the cursor to the left to create a green crossing window (Fig. 17-2).

8. Pick the lower right corner of the house for the base point as shown in Fig. 17-2.

9. In reply to Second point of displacement, stretch the house **10′** to the right and pick a point. (Snap should be on.) The house stretches dynamically.

The house should now be longer.

10. Stretch the south portion of the house **10′**.

11. Save your work.

Scaling Objects

Since the house is now slightly too large, let's scale it down to fit the lot. When you *scale* an object, you increase or decrease its overall size without changing the proportions of the parts to each other. You can use the SCALE command to scale by a specific scale factor, or you can use the existing size of the object as a reference length for the new size.

Using a Scale Factor

Modify

1. Pick the **Scale** button from the Modify toolbar or enter **SC**, the command's alias.

2. Select the house by placing a window around it and press **ENTER**.

3. Pick the lower left corner of the house as the base point.

At this point, you can dynamically drag the house into place or enter a scale factor.

4. Enter **.5** in reply to Specify scale factor.

Entering .5 reduces the house to half (.5) of its previous size.

Scaling by Reference

Suppose the house is now too small. Let's scale it up using the SCALE Reference option.

Modify

1. Enter **SCALE**, select the house as before, and press **ENTER**.

2. Pick the lower left corner of the house again for the base point.

3. Enter **R** for Reference.

4. For the reference length, pick the lower left corner of the house, and then pick the corner just to the right of it as the second point.

This establishes the length of the south wall (10′) as the reference length.

5. In response to Specify new length, move the crosshairs 4 feet to the right of the second point (snap should be on) and pick a point.

The south wall becomes 14′ long, and the rest of the house is scaled by reference to match.

Rotating Objects

Let's rotate the house on the site. Unlike scaling, *rotating* doesn't change the actual size of the house. It merely places it at a different angle. Using the ROTATE command, you can rotate objects by a specific number of degrees. You can also drag the objects to a new angle if the precise angle is not important.

Rotating by Dragging

Let's rotate the house by dragging it into place.

Modify

1. Enter **ROTATE**, select the house, and press **ENTER** after you have made the selection.

2. Pick the lower left corner of the house for the base point.

3. Drag the house in a clockwise direction a few degrees and pick a point.

4. Pick the **Undo** button to undo the rotation.

Specifying an Angle of Rotation

AutoCAD also allows you to specify the angle of rotation. In practice, this is a more useful option than dragging because you can control the placement of objects more precisely.

Modify

1. Pick the **Rotate** button from the Modify toolbar, select the entire house, and press **ENTER**.

2. Pick the lower left corner of the house for the base point.

3. Turn off ortho and snap. As you move the crosshairs, notice that the object rotates dynamically.

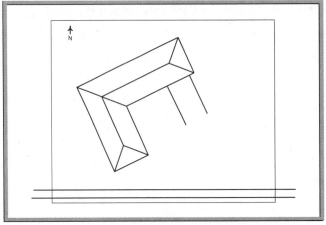

Fig. 17-3

4. For the rotation angle, enter **25** (degrees).

The house rotates 25° counterclockwise. Your drawing should now look similar to the one in Fig. 17-3 without the additional lines.

Trimming Lines

The TRIM command allows you to trim lines that extend too far. You can use TRIM to clean up intersections and to convert construction lines and rays into useful parts of a drawing.

Draw

1. Draw two horizontal xlines 3′ apart to represent a sidewalk, as shown in Fig. 17-3. Make the south edge of the sidewalk 2′ from the south property line.

2. With ortho off, draw a partial driveway as shown in Fig. 17-3. Make the driveway 13′ wide.

 HINT: ─────────────────────────

Copy the line that makes up the east edge of the house for the first line of the driveway. Use the OFFSET command to create the second line of the driveway exactly 13′ from the first line.

Modify

3. Pick the **Trim** button from the Modify toolbar or enter **TR**, the command's alias.

4. Select the east property line as the cutting edge and press **ENTER**.

5. Select the sidewalk lines on the right side of the property line.

Chapter 17: Modifying and Maneuvering

The sidewalk should now end at the property line.

6. Press **ENTER** to end the command.

7. Trim the west end of the sidewalk to meet the west property line using the **TRIM** command.

8. Obtain on-line help to learn about the Project and Edge options.

Extending Lines

The most accurate way to extend the driveway lines so that they meet the south property line is to use the EXTEND command.

Modify

1. Pick the **Extend** button from the Modify toolbar.

2. Select the south property line as the boundary edge and press **ENTER**.

3. In reply to Select object to extend, pick the ends of the two lines that make up the driveway and press **ENTER**.

The driveway extends to the south property line.

Modify

4. Using the **TRIM** command, remove the short intersecting lines so that the sidewalk and driveway look like those in Fig. 17-4.

 HINT:

Use a crossing window to select all four objects, press ENTER, and then select the parts you want to remove.

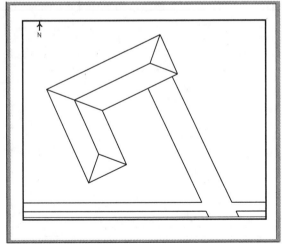

Fig. 17-4

5. Save your work.

Working with Multilines

The TRIM and EXTEND commands work with multilines as well.

1. Open the drawing named **multi.dwg**.

2. Open the **Multiline Style** dialog box, set the style **S1** as the current style, and pick **OK**.

3. Create a new multiline as shown in Fig. 17-5.

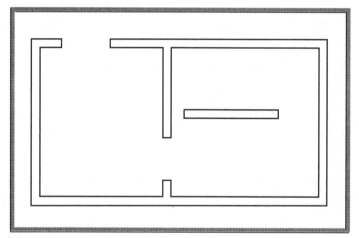

Fig. 17-5

4. Use the **EXTEND** command to extend the new wall to the interior wall. In reply to Enter mline junction option, type **O** for Open.

Your drawing should look like the one shown in Fig. 17-6.

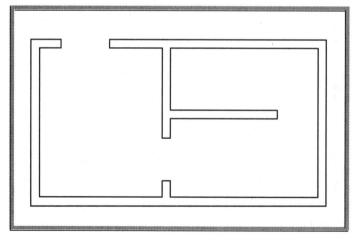

Fig. 17-6

5. Reenter the **MLINE** command and create another wall as shown in Fig. 17-7.

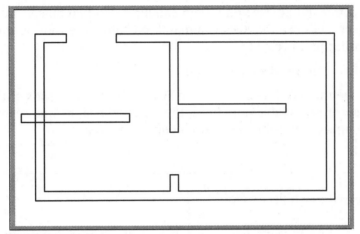

Fig. 17-7

6. Use the **TRIM** command to trim away the part of the wall that extends beyond the exterior wall of the building. This time, select the **Merge** option at the Enter mline junction option prompt.

The drawing should now look like the one in Fig. 17-8.

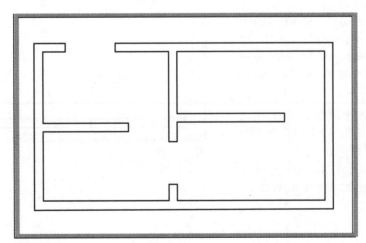

Fig. 17-8

7. Save your work and close multi.dwg.

Joining Lines

The JOIN command allows you to join similar individual objects into a single object. Let's use the JOIN command to change the roof on the house created earlier in this chapter.

1. If you closed **site.dwg**, reopen it.

2. Change the roof line as shown in Fig. 17-9. On both ends of the house, erase the diagonal lines, then draw the short line segments to connect the roof ridge to each end of the house.

HINT: ——————————————————————

Use the Midpoint object snap.

3. Pick the **Join** button from the **Modify** toolbar or enter **J**, the command's alias.

4. Select the first line segment in reply to Select source object, select the second line segment in response to Select lines to join to source, and press **ENTER**.

5. Repeat Steps 3 and 4 for the other end of the house.

The roof ridge line on each side of the house is now a single line.

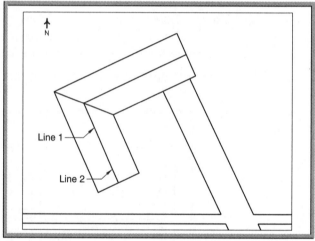

Fig. 17-9

NOTE: ——————————————————————

The JOIN command only works for two lines that are *colinear* (lie on the same infinite line).

6. Save your work and exit AutoCAD.

Chapter 17: Modifying and Maneuvering

Review Questions

1. Explain the purpose of the STRETCH command.

2. Using the SCALE command, what number would you enter to enlarge an object by 50%? to enlarge it to 3 times its present size? to reduce it to ½ its present size?

3. Explain how you would dynamically scale an object up or down.

4. Can you rotate an object dynamically? Explain.

5. How would you specify a 90° clockwise rotation of an object accurately?

6. Explain the purpose of the TRIM command.

7. Describe a situation in which the EXTEND command would be useful.

8. Explain the similarities between the EXTEND and JOIN commands.

Challenge Your Thinking

1. Refer to site.dwg, which you completed in this chapter. If you were doing actual architectural work, you would need to place the features on the drawing much more precisely than you did in this drawing. Describe a way to place a 14'-wide driveway exactly 2 feet from the corner of the house. The driveway should be perpendicular to the house.

2. Both the ZOOM command and the SCALE command make objects appear larger and smaller on the screen. Write a paragraph explaining the differences between the two commands. Explain the circumstances under which each command should be used.

Applying AutoCAD Skills

Work the following problems to practice the commands and skills you learned in this chapter.

1. Create a new drawing and draw the kitchen range symbol shown in Fig. 17-10A. The square is 5 units on each side, the large circles are R.8, and the small circles are R.6. Locate the circles approximately as shown: equidistant from the adjacent sides, with the large circles in the upper left and lower right corners.

    Notice that this is an incorrect symbol of a kitchen range. The large and small circles should be reversed. You can accomplish this by rotating the circles 90°. To make sure the circles stay in their same relative locations, you will rotate them around the center of the square. To do this, first draw two diagonal lines as shown in Fig. 17-10B. Then enter the ROTATE command and select the four circles and two diagonal lines. For the base point of the rotation, snap to the intersection of the two lines. After you have completed the rotation, erase both of the diagonal lines. Then scale the range to 25% of its current size. Save the drawing as range.dwg.

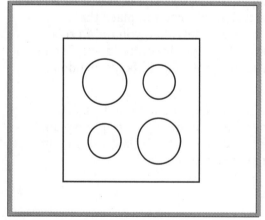

Fig. 17-10A

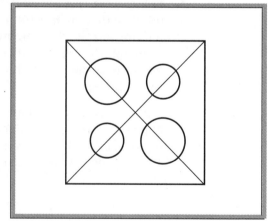

Fig. 17-10B

2. Load site.dwg, the drawing you created in this chapter. Perform each of the following operations on the drawing.

- Change the roof line by adding angled (hipped) ridges on both ends of the house.
- Stretch the driveway by placing a (crossing) window around the house and across the driveway. Stretch the driveway to the north so that the house is sitting farther to the rear of the lot.
- Add a sidewalk parallel to the east property line. Use the TRIM and BREAK commands to clean up the sidewalk corner and the north end of the new sidewalk.
- Reduce the entire site plan by 20% using the SCALE command.
- Stretch the right side of the site plan to the east 10′.
- Rotate the entire site plan 10° in a counterclockwise direction.
- Place trees and shrubs to complete the site plan drawing. Use the ARRAY command to create a tree or shrub as shown in Fig. 17-11. Duplicate and scale the tree or shrub using the COPY and SCALE commands.

When you finish, your drawing should look similar to the one in Fig. 17-11. Save your work.

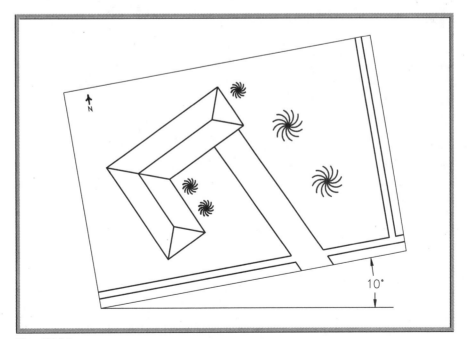

Fig. 17-11

3. Marketing wants to use the original design of your TV remote control device, but you must change it to meet competitive standards. Create the first drawing of the device (shown on the left in Fig. 17-12). Then copy the drawing to make the following changes using the commands you learned in this chapter.

- Scale the device to 90% of its original size.
- Change the two buttons at the bottom into one large button.
- Reduce the overall length.

The final device should look similar to the drawing on the right. Save the drawing as remote.dwg.

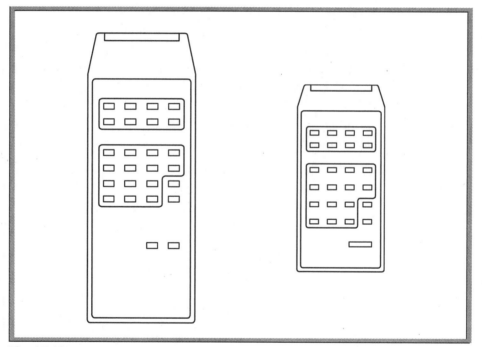

Fig. 17-12

4. Draw the front view of the tube bundle support shown in Fig. 17-13. Proceed as follows:

- Draw the outside circle and one of the holes.
- Offset both circles to create the width of the material.
- Draw a line connecting the center of the outside circle and the center of the hole.
- Offset the line to create the internal support structure that connects the hole with the center of the support.
- Draw another line from the center of the large circle at an angle of 30° to the first line.
- Use TRIM to create half of one cavity.
- Use MIRROR to create the other half of the cavity.
- Fillet the three corners.
- Use ARRAY to insert five more copies of the hole and cavity.

When you have completed the drawing, use the SCALE command to shrink it by a factor of .25 using the obvious base point. Save the drawing as tubebundle.dwg.

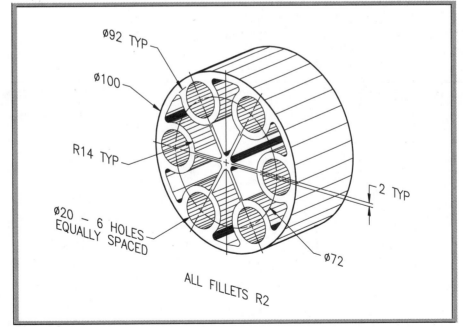

Fig. 17-13

Problem 4 courtesy of Gary J. Hordemann, Gonzaga University

Chapter 17: Modifying and Maneuvering

Chapter 17
Review & Activities continued

Using Problem-Solving Skills

Complete the following activities using problem-solving skills and your knowledge of AutoCAD.

1. You have a full single garage on your lot. The garage measures 16' × 25' and has a 9' door opening. Draw a floor plan of the garage with 6"-thick walls. Show the opening, but do not include the door. Your spouse's car necessitates a larger garage. Use the EXTEND command to convert the full single garage to a full double garage measuring 25' × 25'. Include two 9' door openings separated by a 1' center support. Trim where necessary. Save the drawing as garage.dwg.

2. Local zoning laws will not permit you to have a full double garage on your narrow lot. Use the TRIM command to convert the proposed full double garage to a small double garage measuring 20' × 20' with one large 16' door opening. Maintain the 6"-thick walls. Save the drawing as garage2.dwg.

3. Create the house shown in Fig. 17-14 on a 100' × 35' lot.
   • Make the drawing limits the size of the lot.
   • Create a new multiline style named FRAME with two elements a total of 4" apart to represent the walls.
   • Estimate the size of the house and of each room.
   • Use MLEDIT, TRIM, and EXTEND to create the wall intersections.
   • Save the file as frame.dwg.

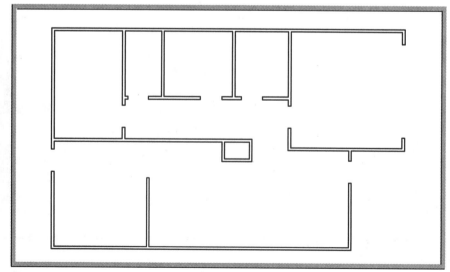

Fig. 17-14

242   *Chapter 17: Modifying and Maneuvering*

4. Create the two armchairs shown in Fig. 17-15A. Approximate their sizes, but align them as shown in the illustration so that the backs of the chairs are colinear. Use the chairs to make a matching sofa. To do this, make a copy of both chairs, placing the copy at any empty space on the screen. Then remove the inside arms of both chairs and extend the horizontal lines to meet, as shown in Fig. 17-15B. Finally, use the JOIN command to join the lines that represent the back of the sofa so that the back of the sofa consists of a single line. Join the other horizontal lines as well. Save the file as furniture.dwg.

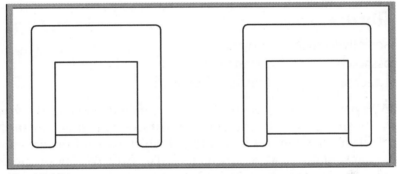

Fig. 17-15A

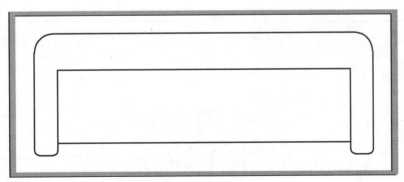

Fig. 17-15B

Chapter 18

Hatching and Sketching

Objectives:

- Hatch objects and parts according to industry standards to improve the readability of drawings
- Edit a hatch by changing, adding, and removing boundaries
- Change the characteristics of an existing hatch
- Use the **Tool Palettes** window to apply hatching
- Use sketching options to create objects with irregular lines

Key Terms

associative hatch   hatch
crosshatch   island
full section   section drawings

AutoCAD's hatching and sketching features can help enhance the visual appearance and readability of a drawing. For example, on the map shown in Fig. 18-1, hatching makes it easy to tell which parts of the area are national forest and which are urban areas. The irregular lines representing the roads were created using the SKETCH command.

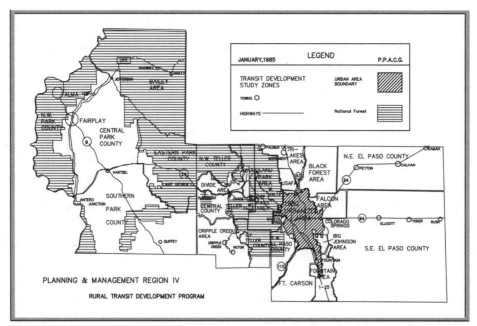

Fig. 18-1   Courtesy of David Salamon, Pikes Peak Area Council of Governments

Hatching and sketching can enhance the visual appearance and readability of a drawing. They also increase the size of the drawing file, so use them wisely.

Hatching

A *hatch* or *crosshatch* is a repetitive pattern of lines or symbols that shows a related area of a drawing. Hatching is used extensively in engineering drawing and production drafting to show cut surfaces. For example, when you draw a section of a mechanical part, you use hatching to reflect the cut, or internal, surface. *Section drawings* are used to show the interior detail of a part. Figure 18-2 shows an example of a *full section* (one that extends all the way through the part). The drawing on the left is the front view, and the drawing on the right is the section view.

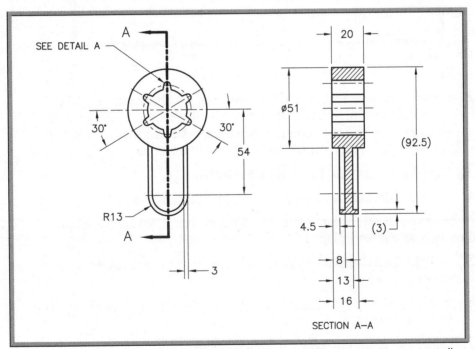

Fig. 18-2   Courtesy of Julie H. Wickert, Austin Community College

Creating a Boundary Hatch

The BHATCH command creates associative hatches. An *associative hatch* is one that updates automatically when its boundaries are changed.

1. Start AutoCAD and begin a new drawing from scratch.

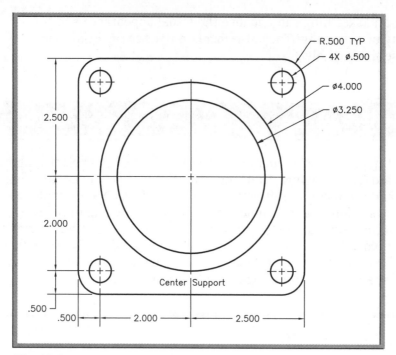

Fig. 18-3

2. Create a sectional view of the center support shown in Fig. 18-3. (You will add the hatch later in this exercise.) Use the dimensions shown in the figure to create it accurately. Do not include the text (Center Support), the center lines, or the dimensions.

3. Save your work in a file named **hatch.dwg**.

Draw

4. From the docked Draw toolbar, pick the **Hatch** button.

This enters the BHATCH command and displays the Hatch and Gradient dialog box, as shown in Fig. 18-4.

5. Pick the **More Options** arrow (**>**) located in the lower right corner of the dialog box.

6. In the Island display style area of the dialog box, pick **Normal**, unless it is already selected.

7. Under the Hatch tab, pick the Pattern down arrow.

A list of hatch patterns appears. ANSI31 is the standard crosshatch pattern.

8. Pick the **...** button located to the right of the down arrow you just picked.

The Hatch Pattern Palette appears with ANSI hatch patterns. ANSI stands for the American National Standards Institute, an industry standards organization in the United States.

Fig. 18-4

9. Pick the **ISO** tab.

ISO hatch patterns appear. ISO stands for the International Standards Organization, which also produces industry standards.

10. Pick the **Other Predefined** tab.

AutoCAD displays other predefined hatch patterns that are available to you.

11. Pick the **ANSI** tab, pick **ANSI31**, and pick **OK**.

Notice the Angle and scale area in the dialog box. Angle permits you to rotate the hatch pattern, while Scale enables you to scale it. The scale value should be set at the reciprocal of the plot scale so that the hatch pattern size corresponds to the drawing scale. This information will be useful when you begin to print or plot your drawings.

12. Under Boundaries, pick the **Add: Select objects** button, select the entire drawing, and press **ENTER**.

The Hatch and Gradient dialog box reappears.

13. Pick the **Preview** button located in the lower left area of the dialog box to preview the hatching.

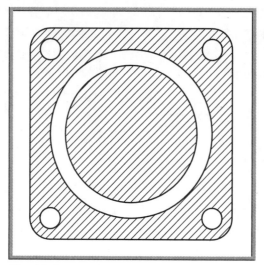

Fig. 18-5

14. As instructed by AutoCAD, right-click to accept the hatch.

Your drawing should now look similar to the one in Fig. 18-5.

15. Save your work.

Editing a Hatch

By default, hatches created with the BHATCH command are associative. As a reminder, an associative hatch is one that updates automatically when you change the size or shape of the hatched object. Associative hatches also update when you add or remove islands inside the hatched area.

Changing the Boundary

1. Select the inner cylindrical part of the center support. Be sure to select the circle and not the hatching.

2. Right-click to display the shortcut menu and pick **Properties**.

3. In the Properties dialog box, change Diameter to **2.75** and press **ENTER**.

The diameter changes, along with the associative hatch.

4. Change the diameter of the lower right hole to **.375** and close the Properties dialog box.

Once again, AutoCAD updates the hatch pattern according to the new hatch boundary.

5. Undo the two changes you just made.

Adding and Removing Islands

The boundaries of a hatch can also be edited by adding and removing islands from the hatch. An *island* is an area within an object, such as a hole, that is not hatched.

The HATCHEDIT command permits you to change the boundaries of an associative hatch. To enter this command, you can enter HATCHEDIT at the keyboard, right-click and use the shortcut menu, or pick the Edit Hatch button from the Modify II toolbar.

Modify II

1. Display the Modify II toolbar and pick the **Edit Hatch** button.

This enters the HATCHEDIT command.

2. Pick the hatch in the drawing.

This causes the Hatch Edit dialog box to display. As you can see, it is very similar to the Hatch and Gradient dialog box.

3. Under Boundaries, pick the **Remove boundaries** button.

4. Select the inner cylindrical part of the center support and press **ENTER**.

5. Pick the **Preview** button. Then right-click to accept the edit.

AutoCAD computes the area of a hatch and updates it automatically when you edit the hatch.

Standard

6. Select the hatch and pick the **Properties** button on the Standard toolbar.

Notice that the hatch area is 11.4336 square inches.

Modify II

7. Reopen the **Hatch Edit** dialog box.

8. Pick the **Add: Select objects** button and add the center cylinder back into the hatched area. Pick **OK** when you are finished.

9. Select the hatch and check the area displayed in the Properties dialog box.

The hatch area is automatically updated to 19.7294 square inches.

10. Close the Properties dialog box and the Modify II toolbar and save your work.

Changing Other Hatch Characteristics

The Hatch Edit dialog box also allows you to change other characteristics of a hatch. For example, you can choose a different hatch pattern or change the angle or scale of the current hatch pattern.

Modify II

1. Open the **Hatch Edit** dialog box and select the hatch.

2. Make the following changes:
Pattern: **ESCHER**
Angle: **30**
Scale: **1.3**

3. Pick the **OK** button.

AutoCAD applies the changes to the drawing.

4. Experiment further with creating and changing associative hatch objects on your own.

5. Using the **Edit Hatch** button, change the hatch pattern to **ANSI31**, the angle to **0**, and the scale to **1**, and pick **OK**.

6. Save your work.

Tool Palettes Window

The Tool Palettes window provides another method of inserting hatch patterns.

1. Move the center support to the right area of the drawing area, produce a copy of it in the left area, and remove the hatch pattern from the copy.

2. From the Tools pull-down menu, pick Tool Palettes Window.

This displays the Tool Palettes window.

3. Review the contents of the Tool Palettes window by clicking each of its seven tabs.

4. Pick the **Hatches** tab and pick the hatch pattern for steel.

 HINT:

Rest the pointer on each of the patterns to see their names.

The steel hatch pattern locks onto the crosshairs.

5. Click inside the center support's cylindrical feature.

AutoCAD hatches it only.

6. Undo the hatching.

7. Select the steel hatch pattern again from the Tool Palettes window and pick a point just inside the outer boundary of the center support. Do not select the cylindrical area.

AutoCAD hatches the entire center support.

8. Erase the second center support and center the first one.

9. Close the Tool Palettes window and save your work.

You will learn more about the customizable Tool Palettes window later in the book.

INFOLINK

See Chapter 32 for more about the Tool Palettes window.

Sketching

The SKETCH command is rarely used to create "sketches" because AutoCAD offers many other commands that make sketching (in its traditional sense) a faster, more accurate process. However, the SKETCH command is used for other purposes, such as to show irregular lines on maps like the one shown in Fig. 18-1.

Let's try some freehand sketching.

1. Using **Save As...** from the **File** pull-down menu, create a new drawing named **sketch.dwg**, and erase the current drawing.

2. Enter **SKPOLY** at the keyboard and specify a value of **1**.

SKPOLY is a system variable that controls whether the SKETCH command creates lines (0) or polylines (1). Setting SKPOLY to 1 allows you to smooth sketched curves using the PEDIT command.

3. Turn off snap and enter the **SKETCH** command.

4. Specify **.1** unit for the Record increment. (This should be the default value.)

The sketching options appear at the Command line. Table 18-1 provides a brief description of each of the SKETCH options.

Option	Function
Pen	Raises/lowers pen
eXit	Records all temporary lines and exits
Quit	Discards all temporary lines and exits
Record	Records all temporary lines
Erase	Selectively erases temporary lines
Connect	Connects to an existing line endpoint
. (period)	Line to point

Table 18-1

5. To begin sketching, pick a point where you'd like the sketch to begin. The pick button toggles the pen down.

6. Move the pointing device to sketch a short line.

7. Pick a second point (to toggle the pen up), and type **X** to exit.

8. Move to an open area on the screen and sketch the golf course sand trap shown in Fig. 18-6. Set the **Record increment** at **.2**.

It's okay if your sketch doesn't look exactly like the one in Fig. 18-6.

9. When you're finished sketching the sand trap, type **R** for Record or **X** for eXit.

10. Practice sketching by using the remaining SKETCH options. Draw anything you'd like.

11. When you're finished, save your work and exit AutoCAD.

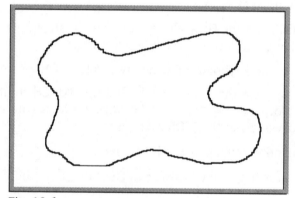

Fig. 18-6

Review Questions

1. Explain why hatch patterns are useful.

2. With which command can you change a hatch pattern after the hatch has been applied to an object?

3. The BHATCH command permits you to hatch drawings. Describe a second method.

4. Briefly describe the purpose of each of the following options of the SKETCH command.
   Pen
   eXit
   Quit
   Record
   Erase
   Connect
   . (period)

5. SKETCH requires a Record increment. What does this increment determine? (If you're not sure, specify a coarse increment such as .5 or 1 and notice the appearance of the sketch lines.)

6. What is the purpose of the SKPOLY system variable?

Challenge Your Thinking

1. Find out what the Inherit Properties button on the Hatch and Gradient dialog box does. Explain how it can save you time when you create a complex drawing that has several hatched areas.

2. Investigate the differences and similarities between ANSI and ISO standards. Which (if either) do companies in your area prefer?

3. Investigate the Gradient tab of the Hatch and Gradient dialog box. What does this tab allow you to do? When might you want to use this feature?

Applying AutoCAD Skills

Work the following problems to practice the commands and skills you learned in this chapter.

1. Draw the wall shown in Fig. 18-7. Replace the angled break line on the right with an irregular line drawn with the SKETCH command to indicate a continuing edge. Hatch the wall in the brick pattern. Save the drawing as brickwall.dwg.

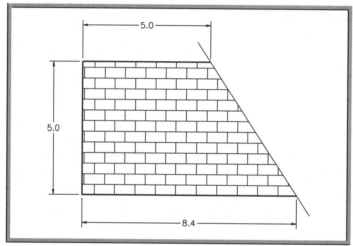

Fig. 18-7

2. Create the simplified house elevation shown in Fig. 18-8. Use the SKETCH command to define temporary boundaries for the hatch patterns on the roof.

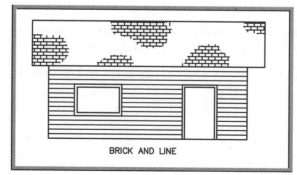

Fig. 18-8

3. Use the SKETCH command to create a map of the United States. Use Fig. 18-9 as a guide.

4. Use the SKETCH command to draw a logo like the one shown in Fig. 18-10. Before sketching the lines, be sure to set SKPOLY to 1. Use the BHATCH Solid pattern to create the filled areas.

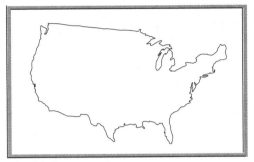

Fig. 18-9

5. Draw the right-side view of the slider shown in Fig. 18-11 as a full section view. Assume the material to be aluminum. See Appendix D for a table of dimensioning symbols. For additional practice, draw the top view as a full section view.

Fig. 18-10

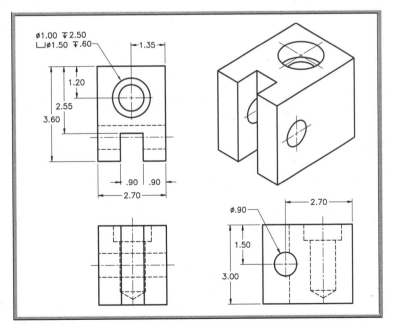

Fig. 18-11

Problems 4 and 5 courtesy of Gary J. Hordemann, Gonzaga University

Using Problem-Solving Skills

Complete the following activities using problem-solving skills and your knowledge of AutoCAD.

1. Your company is trying some new bushings for the swings in the child's playground set it manufactures. Draw the front view and full section of the bushing shown in Fig. 18-12. Hatch the section as appropriate. Save the drawing as bushing.dwg.

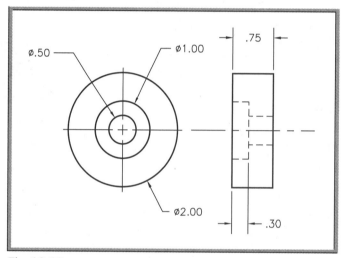

Fig. 18-12

2. An alternate design of the bushing described in problem 1 provides a hard insert to act as a wear surface within the bushing (Fig. 18-13). Make the change in the sectional view and hatch accordingly. Save the drawing as altbushing.dwg.

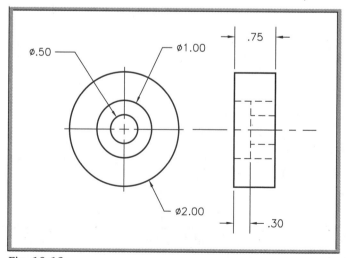

Fig. 18-13

3. The drawing shown in Fig. 18-14 was used with a computer-driven, high-pressure water cutter to cut a shield out of a piece of ¼" brass. Use the SKETCH command with SKPOLY set to 1 to draw a similar shield.

Fig. 18-14

Part 3 Project

Signage

Throughout history, people have used visual cues in the form of signs to communicate information. Signs are typically easy to interpret, quick to read (if indeed they contain any actual text at all) and unobjectionable. Signs without text, by the way, are often considered universal in nature: they are capable of being understood in different cultures regardless of the language spoken. The purpose of a sign can be varied—advertising, public service, directions, warnings, identification—almost any message that needs to be transmitted. A very common use of signs is to regulate vehicular traffic. They identify routes taken, geographic direction of the vehicle, speed limits, expressway exit/entrance numbers, nearby hospitals, and sources of food and fuel for the traveler, among other uses.

Description

In AutoCAD, create the traffic signs assigned by your instructor or improve upon some of the signs as they currently exist. For example, you might create a more effective graphic. Actual sizes and legal descriptions of many signs are available on the Internet.

Alternatively, your instructor may ask you to design your own original signs. Suggestions include:

elephant crossing	no junk food
cash only	idea factory
no fish here	no fishing
no tire-changing	barbecues available
wet concrete	drive safely or don't drive

Use your imagination or seek advice from the instructor for further ideas.

Hints and Suggestions

1. Consider using DRAW > SURFACE > 2D SOLID to create the sign as a solid color; alternatively, you may want to use POLYGON with the Solid hatch pattern to color in the background.

2. Consider using gradients from the Gradient tab of the Hatch and Gradient dialog box. Would a gradient be effective on a highway sign?

3. Polylines allow a choice of thickness as well as color for lines, arrows, and other images that might be appropriate for your sign's communication.

4. If you use polylines, filleting closed polygons can be accomplished in a single command.

5. Utilize the various modifying commands to reuse pieces of one sign that might lend itself to another section or sign.

Summary Questions

Your instructor may direct you to answer these questions orally or in writing. You may wish to compare and exchange ideas with other students to help increase efficiency and productivity.

1. Why is technical drawing sometimes referred to as a "universal language"?

2. What might be a reason to keep all signs in a particular category, such as "informational" or "regulatory," based on a single color scheme or shape?

3. What characteristics make traffic signs easy to read as drivers go by?

4. How is it that so many countries can use similar traffic signs, yet not even have a shared language or culture? Explain your reasoning.

5. Do you think there is a relationship between the size of a sign and its importance to the driver? Explain.

Careers
Using
AutoCAD

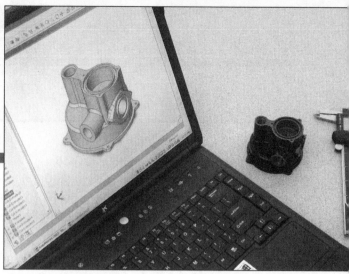

Tim Fuller Photography

Engineer

Imagine performing the same tasks, day after day, during the course of your entire career. Without the opportunity to learn new things and encounter challenging work, the workday would become very dull. Most jobs, however, provide opportunities for workers to fine-tune their skills. Young employees often start out in entry-level positions and, by showing their employers that they are hard-working, loyal, and motivated individuals, they are able to advance their status. Some workers move around within one company while others may work for several employers, holding jobs at different levels. Both of these career paths provide a learning platform for employees interested in advancing their status and earnings.

Growing on the Job

Drafters have a variety of opportunities to move upward in the workplace. Those who develop a particular area of interest, such as mechanical, electrical, architectural, or aeronautical, may eventually wish to work as an engineer. Enrolling in college programs and receiving a degree in a concentrated area is a requirement for any drafter interested in advancing to an engineering position.

Thinking About It

Engineers use their expertise to assist drafters, builders, and maintenance crews. They provide detailed working instructions and drawings for others involved in the produc-

tion process, and they work closely with drafters to design systems and products. Engineers are problem-solvers; they are able to think clearly and logically in order to derive possible solutions to both existing and potential problems. The ability to work with others is also an important quality. Engineers often find themselves acting as project managers. Being able to prioritize and organize tasks helps them maintain an efficient working team.

Career Activities

- Write a report about the various types of engineers and their typical duties. Explain why engineering is a logical path of advancement for drafters.
- Are employment forecasts encouraging for engineers?
- Do you think it would be better to gain experience as a drafter before pursuing an engineering degree, or to obtain a degree before working in the field? Explain your answer.

Chapter 19 Notes and Specifications

Objectives:

- Create and edit text dynamically
- Specify the position and orientation of text
- Create and use new text styles
- Review options for fonts and adjust the quality of TrueType fonts
- Produce multiple-line text using AutoCAD's text editor
- Import text from an external word processor
- Format text to create lists

Key Terms

font	notes
justify	specifications
mtext	title block

Notes and specifications are a critical part of most engineering drawings. Although the two terms are often used interchangeably, notes and specifications are technically two different types of text. *Notes* generally refer to the entire drawing, rather than to any one specific feature. For example, a note might describe the thickness of a part, if the thickness is uniform throughout the part. *Specifications*, on the other hand, provide information about size, shape, and surface finishes that apply to specific portions of an object or part.

Without text, the drawings would be incomplete. The drawing in Fig. 19-1 on page 262 shows the amount of text that is typical in many drawings. Some drawings contain even more text.

As you can see, the text is an important component in describing the drawing. With traditional drafting, the text is placed by hand, consuming hours of tedious work. With CAD, you can place the words on the screen almost as fast as you can type them.

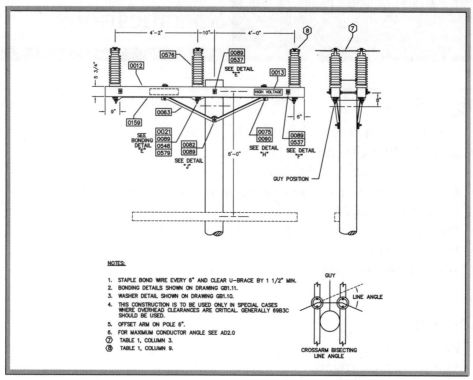

Fig. 19-1

Types of Text

AutoCAD supplies two different types of text objects. The one you use should depend on the type and amount of text you are inserting. The DTEXT command inserts one or more lines of text as single-line objects. The MTEXT command, on the other hand, treats one or more paragraphs of text as a single object. This chapter presents both types of text.

Dynamic Text

The DTEXT command enables you to display AutoCAD text dynamically in the drawing as you type it. It also allows you to *justify*, or align, the text in several ways, including left, right, and center justification.

1. Start AutoCAD and start a new drawing from scratch.

2. Set snap at **.25**.

3. From the **Draw** pull-down menu, select **Text** and then **Single Line Text**, or enter **DT** (the DTEXT command alias).

This enters the DTEXT command.

4. In response to Specify start point of text, place a point on the left side of the screen. The text will be left-justified beginning at this point.

5. Reply to the Specify height prompt by moving the crosshairs up **.25** unit from the starting point and picking a point.

6. Enter **0** (degrees) in reply to Specify rotation angle of text.

A blinking cursor appears inside a narrow text box.

7. At the Enter text prompt, type your name using both upper- and lowercase letters and press **ENTER**.

You should again see the Enter text prompt.

8. Type your P.O. box or street address and press **ENTER**.

Notice where the text appears in relation to the previous line.

9. Type your city, state, and zip code and press **ENTER**.

10. Press **ENTER** again to terminate the DTEXT command.

Let's enter the same information again, but this time in a different format.

11. Reenter the **DTEXT** command (press the space bar), and **J** for the Justify option.

The justification options appear.

12. Enter the **C** (Center) option.

13. Place the center point near the top center of the screen and set the text height by entering **.2** at the keyboard. Do not insert the text at an angle.

14. Repeat Steps 7 through 10. Be sure to press **ENTER** twice at the end.

When you're finished, your text should be centered like the example in Fig. 19-2. If it isn't, try again.

15. Save your work in a file named **text.dwg**.

```
┌─────────────────────────────────────┐
│           Ms. Jane Doe               │
│      507 East 104th Avenue           │
│        Cadsville, CA 30160           │
└─────────────────────────────────────┘
```

Fig. 19-2

Multiple-Line Text

The MTEXT command creates a multiple-line text object called *mtext*. AutoCAD uses a text editor to create mtext objects.

1. Erase the text from the upper half of the screen.

2. From the Draw toolbar, pick the **Multiline Text** button, or enter **T** at the keyboard.

This enters the MTEXT command and displays the text style and height at the Command line.

3. Pick a point in the upper left area of the screen.

Notice the new list of options.

HINT:

Recall that you can access the list of options at the dynamic input cursor by pressing the down arrow key on the keyboard.

4. Enter **W** for Width.

5. Enter **4.5** in reply to Specify width or, with ortho and snap on, pick a point about **4.5** units to the right of the first point.

AutoCAD displays the multiline text editor with a ruler and the Text Formatting toolbar just below the docked Layers toolbar. Notice that the font is Txt and the default text height is .2000 unit.

NOTE:

A *font* is a set of characters, including letters, numbers, punctuation marks, and symbols, in a particular style. Arial and Romans are examples of fonts used in AutoCAD.

6. From the drop-down list, change the font from Txt to **Arial**.

7. Type the following sentence.

   I'm using AutoCAD's text editor to write these words.

8. Highlight the word **using** by double-clicking and then pick the **B** (bold) button.

9. Highlight the word **words** and pick the **U** (underline) button.

The sentence should now look like the one in Fig. 19-3.

10. Pick the **OK** button.

11. Save your work.

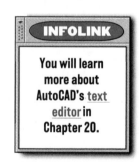

INFOLINK

You will learn more about AutoCAD's text editor **in Chapter 20.**

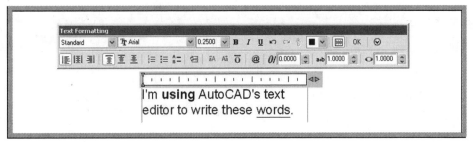

Fig. 19-3

Importing Text

Instead of typing text within AutoCAD, you can create it using a standard word processing program and then import it. Many people find it easier to create lengthy text passages using word processors with which they are familiar.

1. Minimize AutoCAD.

2. From the Windows **Start** menu, pick **Programs** (or **All Programs**), **Accessories**, and **Notepad**.

3. Enter the following text:

   AutoCAD's text editor permits you to import text from a file.
   This text will appear in AutoCAD shortly.

4. From Notepad's **File** pull-down menu, pick **Save As...**, select the folder with your name, and name the file **import.txt**.

5. Exit Notepad and maximize AutoCAD.

6. From the Draw toolbar, pick the **Multiline Text** button.

7. Near the top center of the screen, pick the first corner.

8. Produce a rectangle that measures about **1** unit tall by **3** units wide.

The multiline text editor appears.

9. Right-click inside the text box to display the shortcut menu.

10. Select **Import Text...** from the menu.

11. Locate and select the new file named **import.txt**.

This imports the text into the multiline text editor.

12. Highlight the text and pick the **Arial** font from the drop-down list.

The text changes to the Arial font.

13. Pick the **OK** button.

14. Save your work.

Draw

Formatting Text

Multiple line text in AutoCAD offers many of the same formatting capabilities as standard word processing programs. For example, you can format text into bulleted, numbered, or lettered lists. You can also create paragraph indents and tabs.

Draw

A

1. From the Draw toolbar, pick the **Multiline Text** button, or enter **T** at the keyboard.

2. Pick a point in an open area of the screen, enter **W** for Width, and **6** for the width.

3. Type the following lines. Be sure to press **ENTER** at the end of each line.
   Notes:
   All dimensions in inches.
   Material: Aluminum 6061-T6 bar stock.
   All fillets and rounds R.125 unless otherwise specified.
   Break all sharp corners.

4. Highlight all of the typed text, change the font to **Arial**, and make the text height **.2000** units.

5. Highlight the first line and underline it.

6. Highlight every line except the first line and pick the **Numbering** button from the **Text Formatting** toolbar.

7. If there is too much space between the numbers and text, pick the small rectangle in the ruler bar and slide it to the left to align the text closer to the numbers.

The list should now look like the one in Fig. 19-4.

8. Pick the **OK** button.

9. Save your work.

Notes:
1. All dimensions in inches.
2. Material: Aluminum 6061-T6 bar stock.
3. All fillets and rounds R.125 unless otherwise specified.
4. Break all sharp corners.

Fig. 19-4

Text Styles and Fonts

It's possible to create new text styles using the STYLE command. During their creation, you can expand, condense, slant, and even draw the characters upside-down and backwards.

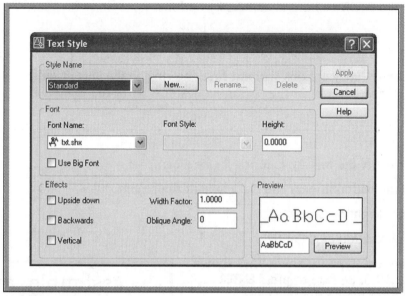

Fig. 19-5

Creating a New Text Style

1. From the **Format** pull-down menu, select **Text Style**, or enter **ST** at the keyboard for the STYLE command.

The Text Style dialog box appears as shown in Fig. 19-5.

 HINT: ───────────────────────────────

You can also open the Text Style dialog box by picking the Text Style Manager button from the docked Styles toolbar. It is located at the right of the docked Standard toolbar.

Styles

2. Pick the **New...** button, enter **comp1** for the new text style name, and pick the **OK** button.

3. Under Font Name, display the list of fonts by picking the down arrow.

4. Using the scrollbar, find the font file named **complex.shx** and select it.

Notice that the text sample in the Preview box changes to show the appearance of the complex.shx font.

5. Study the other parts of the dialog box and then pick the **Apply** and **Close** buttons.

You are ready to use the new comp1 text style. Notice that the comp1 name appears in the Styles toolbar.

6. Enter the **DTEXT** command and **Justify** option.

HINT:

You can enter Justify options without first entering Justify. Just enter the capitalized letter(s) in the option names directly at the Command line.

7. Right-justify the text by entering **R** (for Right).

8. Place the endpoint near the right side of the screen.

9. Set the height at **.3** unit.

10. Set the rotation angle at **0**.

11. For the text, type the three lines shown in Fig. 19-6. Be sure to press **ENTER** twice after typing the third line.

With the STYLE command, you can develop an infinite number of text styles. Try creating other styles of your own design. The romans.shx font is recommended for most engineering drawings.

> Assemble and
> ream for No. 0
> taper pin.

Fig. 19-6

As you create more text styles within a drawing file, you may occasionally want to check their names.

12. Select the down arrow in the Styles toolbar to display the list of defined text styles.

If you want to learn more about the text styles, you will need to display the Text Style dialog box.

Styles

13. Select the **Text Style Manager** button from the Styles toolbar.

14. In the Text Style dialog box, under Style Name, pick the down arrow and select **Standard**.

The font name (txt.shx) and other characteristics of the style appear. This is the default text style.

15. Pick the **Close** button.

Text Fonts

AutoCAD supports TrueType fonts and AutoCAD compiled shape (SHX) fonts.

1. Reopen the **Text Style** dialog box and pick the **New** button.

2. Type **tt** for the new text style and pick the **OK** button or press **ENTER**.

Chapter 19: Notes and Specifications

3. Under **Font Name**, find and select **Swis721 BT**.

The Swis721 BT font displays in the preview area. Notice the overlapping Ts located at the left of the font name. This indicates that it is a TrueType font. Notice also that Roman displays under Font Style.

4. Under **Font Style**, display the list of options, pick **Italic**, and notice how the font changes in the Preview area.

Note the 0.0000 value in the text box under Height. This 0 value indicates that the text is not fixed at a specific height, giving you the option of setting the text height when you enter the DTEXT command.

5. Pick the **Apply** and **Close** buttons.

6. Enter the **DTEXT** command and pick a point anywhere in the drawing area.

7. Enter **.2** for the height and **0** for the rotation angle.

8. Type **This is TrueType.** and press **ENTER** twice.

This is TrueType.

Fig. 19-7

The TrueType text should look like the text in Fig. 19-7.

Text Quality

The TEXTQLTY system variable sets the resolution of text created with TrueType fonts. A value of 0 represents no effort to refine the smoothness of the text. A value of 100 represents a maximum effort to smooth the text. Lower values decrease resolution and increase plotting speed. Higher values increase resolution and decrease plotting speed.

INFOLINK

You will learn more about plotting in Chapter 24.

1. Enter the **TEXTQLTY** system variable at the keyboard.

2. Press the **ESC** key.

3. Experiment with other TrueType fonts on your own.

4. Save your work.

Setting the Current Style

Let's set a new current text style.

1. Select the down arrow in the docked Styles toolbar and pick the **comp1** text style.

The comp1 style is now the current text style.

2. Place some new text on the screen and then terminate the DTEXT command.

3. Change back to the **tt** text style.

4. Save your work.

Applying Text in Drawings

Now that you have learned many ways of creating text, let's apply that knowledge to the creation of a title block. A *title block* is a portion of a drawing that is set aside to give important information about the drawing, the drafter, the company, and so on.

1. Begin a new drawing from scratch and save it as **title.dwg**.

2. Set the grid at **.25** and snap at **.0625**.

3. Create the title block shown in Fig. 19-8 using the following information:

   - For Dynamic Design, Inc., create a new style named **Swiss** using the **Swis721 Ex BT** font. Make the text **.25** tall.

   - Create a new text style named **Roms** using the **romans.shx** font, and use it to produce the small text, which is **.06** tall.

   - Use the **Justify** options of the **DTEXT** command to position the text accurately.

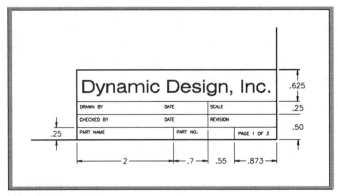

Fig. 19-8

4. Save your work and exit AutoCAD.

NOTE:

Many companies use their own customized title blocks. Autodesk also supplies several variations of title blocks in the Templates folder. If you need a ready-to-use title block for a drawing, you may want to use the Select File dialog box to find one that suits your needs.

Chapter 19: Notes and Specifications

Review Questions

1. What is the most efficient way to enter a single line of text? What command should you use?

2. What might be a benefit of using the MTEXT command?

3. What is the purpose of the DTEXT Style option?

4. Name at least six fonts provided by AutoCAD.

5. What command do you enter to create a new text style?

6. Briefly describe how you would create a tall, thin text style.

7. Explain how to import text from a word processor into AutoCAD. Why might you choose to do this?

8. What is the purpose of a title block?

9. Explain the differences between the DTEXT command and the MTEXT command. Describe a situation in which each command would clearly be a better choice.

Challenge Your Thinking

1. Obtain on-screen help and then experiment with the COMPILE command. What are the advantages and disadvantages of compiling fonts? When might you want to use this option?

2. In addition to standard AutoCAD and TrueType fonts, AutoCAD can also work with PostScript fonts. Find out more about PostScript fonts. Write a paragraph comparing and contrasting TrueType and PostScript fonts. Are there any situations in which one type might be preferable to the other? Explain.

3. Which word processor do you use on your computer? Make a list of all the word processing programs you have used. Most likely, any of these can be used to import text into AutoCAD. If possible, try importing text from several word processors.

Applying AutoCAD Skills

Work the following problems to practice the commands and skills you learned in this chapter.

1. Type your name and today's date in Windows Notepad. Import this text into the drawings you created in Chapter 17. Save your work.

2. Create a new text style using the following information:
   Style name: cityblueprint
   Font file: CityBlueprint
   Height: .25 (fixed)
   Width factor: 1
   Oblique angle: 15

3. Create a new text style using the following information:
   Style name: ital
   Font file: italic.shx
   Height: 0 (not fixed)
   Width factor: .75
   Oblique angle: 0

4. Use the DTEXT command to place the text shown in Fig. 19-9. Use the cityblueprint text style you created in problem 2. Right-justify the text. Do not rotate the text.

> Apply a light coat
> of primer after
> sand-blasting the surface.

Fig. 19-9

5. Use the MTEXT command and the ital text style you created in problem 3 to create the text shown in Fig. 19-10. Set the text height at .3 unit. Rotate the text 90°. Using mtext, set the width of the text line to 4 units.

> You may need to rotate the page to read this.

Fig. 19-10

6. Open the remote.dwg file you created in Chapter 17, "Applying Auto-CAD Skills" problem 3. If you have not yet created the drawing, do so now. Then add the text as shown in Fig. 19-11. Use all of the commands from Chapters 1 through 19 to your advantage.

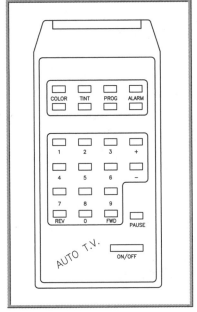

Fig. 19-11

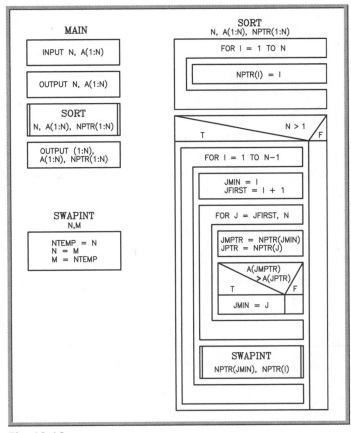

Fig. 19-12

7. Shown in Fig. 19-12 is a block diagram algorithm for a program that sorts numbers into ascending order by a method known as *sorting by pointers*. Use the LINE and OFFSET commands to draw the boxes. Then insert the text using the romans.shx font. For the words MAIN, SORT, and SWAPINT, use the romanc.shx font and make the text larger. For the symbol >, use the symath.shx font, character N.

Problem 7 courtesy of Gary J. Hordemann, Gonzaga University

Using Problem-Solving Skills

Complete the following activities using problem-solving skills and your knowledge of AutoCAD.

1. Most companies have standard operating procedures for creating engineering drawings. Your company is new, and the procedures have not yet been formalized; however, it has become standard practice to place a rectangular border around drawings and enclose the drawing identification in a title block. Open shaftbearing.dwg, which you created in Chapter 9 ("Applying AutoCAD Skills" problem 2). Draw a border around the periphery of the drawing. Across the bottom, draw a title block .75 unit high and divide it into three equal parts. In the first part, place your name; in the second part, place the name of the drawing; and in the third part, place today's date. Save the drawing.

2. Open the shim.dwg or shimrev.dwg drawing from Chapter 16. Make a border and title block and complete the information as in problem 1, above. Add the following information to the drawing in the form of drawing notes. Use the text style and font of your choice, or as specified by your company's procedures. Save the drawing.

   Notes:
   1. Material: SAE 1010 Carbon Steel
   2. Break sharp edges
   3. All dimensions $+/-0.01$ unless otherwise specified

Careers
Using
AutoCAD

Tim Fuller Photography

Entrepreneur

One of the exciting things about our country is that we have the freedom to choose our own career paths. If you are interested in health care, you might decide to pursue a career in medicine or nutrition. If you wish to provide a specific type of product or service that few other companies offer, you may consider starting your own business.

Entrepreneurs—people who create and run their own businesses—are found in nearly every field. In 2000, about 10,000 of the noted 213,000 drafting jobs were held by self-employed individuals.

Working on Your Own

Experienced, independent individuals work in many areas of drafting, but few have entered the profession on their own without first obtaining specialized training. Most have worked initially for a drafting company to learn about drafting techniques. After accumulating years of experience, these individuals decided to obtain their own clients and work for themselves. Some drafting entrepreneurs use their business as their primary means of income, while others meet with clients in their spare time, in addition to working a full-time job.

Wanted: Motivated, Reliable Professionals

A successful entrepreneur must be self-motivated, patient, and persistent. Building a new business takes time, and when you're the boss, your company's fate is entirely in your hands. Good time management and decision-making skills are helpful, too. Entrepreneurs must also be able to communicate with others; good relationships with both clients and employees are necessary in order to obtain business and to keep it. One of the most important characteristics, however, is reliability. Clients and employees must know that they can take you at your word.

Career Activities

- Have you ever tried to run your own business? Maybe you've sold lemonade to your friends and neighbors on hot summer days. Or maybe you've decided to earn some extra money by mowing lawns, or by cleaning homes. What skills did you use when trying to obtain business? What did you do to keep customers coming back for your services?

- Interview an entrepreneur in your area. How did the person become interested in becoming an entrepreneur? What was the most difficult problem the person faced? How was the problem solved?

Chapter 20

Text Editing and Spell Checking

Objectives:

- Edit text using AutoCAD's text editor
- Create special text characters
- Find and replace text
- Check the spelling of text in a drawing using AutoCAD's spell checker

Key Terms

in-place text editor
special characters

AutoCAD offers several commands and features for editing and enhancing text and mtext objects. Its text editor and spell checker help you produce error-free notes and specifications. With AutoCAD, it is possible to find and replace text using a feature similar to that in word processors such as Microsoft® Word.

Editing Text

AutoCAD provides two commands to change the contents of text. DDEDIT edits text created with the DTEXT command. MTEDIT edits multiple-line text created with the MTEXT command. In each case, AutoCAD displays an edit box, or *in-place text editor*, that allows you to edit the selected text.

1. Start AutoCAD and open the file named **text.dwg**.

2. Create a new text style named **roms** using the **romans.shx** text font. Apply all of the default values and close the dialog box.

3. Using the **DTEXT** command, enter the following text at a height of **.15**.

   NOTE: ALL ROUNDS ARE .125

Editing Dynamic Text

You can edit text that was created with the DTEXT command right on the screen.

1. Double-click the text you just created.

The text becomes highlighted against a blue background.

2. Move the pointer to the area containing the text and pick a point between the words ALL and ROUNDS.

3. Enter **FILLETS AND** so that the line now reads NOTE: ALL FILLETS AND ROUNDS ARE .125.

4. Press **ENTER**.

5. Press **ENTER** again to terminate the command.

Editing Mtext Objects

You can use a similar method to edit mtext.

1. Double-click the mtext object located in the upper left area of the screen.

This displays an in-place text editor and the Text Formatting toolbar.

2. With the pointing device, position the cursor in the text and change a couple of words in the same way as you would in most word processors or text editors.

3. Change the style to **roms** and the height to **.125** and pick **OK**.

HINT:

To change the text height, you must first highlight the text.

The mtext changes according to the changes you made in the dialog box.

Justifying Mtext

1. Repeat Step 1 from the previous section.

2. Change the text style to **tt**.

3. Highlight the text and right-click.

4. From the pop-up menu, rest the pointer on **Justification**.

AutoCAD offers nine different ways to justify mtext objects. Top Left is the default setting. Text can be center-, left-, or right-justified with respect to the left and right text boundaries. Also, text can be middle-, top-, or bottom-aligned with respect to the top and bottom text boundaries.

5. Select **Top Right** and pick **OK** on the Text Formatting toolbar.

AutoCAD right-justifies the text along the top and right boundaries.

Creating Special Characters

The in-place text editor permits you to insert special characters such as the degree (°) and plus/minus (±) symbols. *Special characters* are those that are not normally available on the keyboard or in a basic text font.

Draw

1. Enter the **MTEXT** command.

2. In an empty area of the screen, create a rectangle measuring about **1** unit tall by **2** units wide.

3. Enter the following specification, but omit the special characters.

   Drill a ø.5 hole at a 5° angle using a tolerance of ±.010.

4. Pick a point prior to .5 and right-click.

5. From the shortcut menu, select **Symbol** and **Diameter**.

AutoCAD inserts special codes that represent the diameter symbol. They will become the diameter symbol after you close the text editor.

NOTE:

When using DTEXT, you can enter these codes, although it is usually faster to add them later. To see a list of the codes, enter the HELP command and search the index for "special characters."

6. Pick a point after 5, right-click, and select **Symbol** and **Degrees**.

The actual degree symbol appears.

7. Pick a point prior to .010, right-click, and select **Symbol** and **Plus/Minus**. The actual plus/minus symbol appears.

8. Pick the **OK** button on the Text Formatting toolbar.

9. Double-click the text to redisplay the in-place text editor.

10. Move the pointer inside the text box, right-click, and rest the pointer on **Symbol**.

The menu includes several commonly used symbols such as angle, centerline, delta, ohm, and not equal.

11. Select **Other...** from the menu.

This displays a character map, giving you access to many additional characters.

12. Close the character map and pick the **OK** button to close the text editor.

Finding and Replacing Text

Suppose you have lengthy paragraphs of text and you want to find and replace certain words.

1. Rest the crosshairs on top of the mtext object located in the upper left area of the screen.

2. Right-click and pick the **Find...** item.

This displays the Find and Replace dialog box, as shown in Fig. 20-1.

3. In the Find text string text box, enter **write**, and in the Replace with text box, enter **draft**.

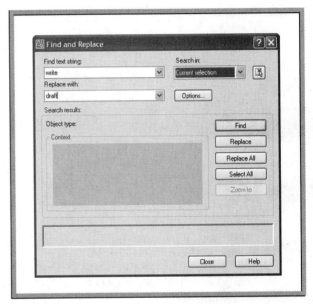

4. Pick the **Find** button.

AutoCAD displays the text in the dialog box and finds and highlights the word write.

5. Pick the **Replace** button.

AutoCAD replaces the word.

6. Pick the **Options...** button.

This displays the Find and Replace Options dialog box.

Fig. 20-1

If you check the Match case check box, AutoCAD finds the text only if the case of all characters in the text match the text in the Find text box. If you check the Find whole words only check box, AutoCAD matches the text in the Find text box only if it is a single word. If the text is part of another text string, AutoCAD ignores it.

The Include area of the dialog box specifies the type of objects you want to include in the search. By default, all options are selected.

7. Pick the **Cancel** button.

8. In the upper right area of the dialog box, pick the down arrow under Search in.

As you can see, you can search in the current selection or in the entire drawing.

9. Pick the **Close** button.

Using AutoCAD's Spell Checker

AutoCAD's spell checker examines text for misspelled words.

Fig. 20-2

1. Enter **sp** to enter the SPELL command.

2. In reply to Select objects, pick the text **NOTE: ALL FILLETS AND ROUNDS ARE .125** and press **ENTER**.

If you spelled the words correctly, AutoCAD displays the message box shown in Fig. 20-2.

3. Pick **OK**.

NOTE:

If AutoCAD displays the Check Spelling dialog box, pick the Cancel button.

4. Using the **DTEXT** command, add the term **Computer-aided** to the screen.

5. Enter the **SPELL** command, enter **All**, and press **ENTER**.

AutoCAD displays the Check Spelling dialog box because it found what could be a misspelled word. "Computer-aided" is not included in the spell checker's dictionary. As you can see, it has suggested "Computer" to replace "Computer-aided."

6. Pick the **Lookup** button to review alternative suggestions. Then pick the **Ignore** button to leave Computer-aided as it is.

The Context area of the dialog box displays the phrase in which AutoCAD located the current word. The functions of the remaining buttons in this dialog box are described in Table 20-1.

7. If AutoCAD finds other words that might be misspelled, pick the **Ignore All** button repeatedly, if necessary.

8. Pick the **x** in the upper right corner to close the dialog box.

Button	Function
Ignore All	Skips all remaining words that match the current word.
Change	Replaces the current word with the word highlighted in the Suggestions box.
Change All	Replaces the current word in all selected text objects.
Add	Adds the current word to the current custom dictionary. (The maximum word length is 63 characters.)
Change Dictionaries	Displays the Change Dictionaries dialog box, which permits you to change the dictionary against which AutoCAD checks spelling.

Table 20-1

Other Options

AutoCAD provides other options for changing the appearance of mtext.

1. Double-click the mtext in the top right corner of the drawing area.

2. Move the pointer inside the text box and right-click.

3. Select the **Indents and Tabs** item.

This displays the Indents and Tabs dialog box. It permits you to adjust the indentation and tabs for mtext objects.

 NOTE:

You can also display the Indents and Tabs dialog box by right-clicking the ruler above the mtext.

4. Pick the **Cancel** button and right-click inside the text box.

Notice that Find and Replace is available in this shortcut menu.

5. Rest the pointer on **Change Case**.

This allows you to change the mtext to all upper- or lowercase letters. Below Change Case is AutoCAPS. It converts all newly typed and imported text to uppercase, but it does not affect existing text.

6. Pick the **OK** button, save your work, and exit AutoCAD.

Review Questions

1. How does selecting and right-clicking text permit you to edit dynamic text? mtext?

2. Why might the character map be useful?

3. How would you add a diameter symbol to mtext?

4. Explain a situation in which AutoCAD's spell checker may find a word that is spelled correctly. What might cause this?

5. Suppose you have paragraphs of text that contain several words or phrases that need to be replaced with new text. What is the fastest way of replacing the text?

Challenge Your Thinking

1. Look again at the mtext object you edited in this chapter. If you changed the word text to test and the word write to rite, what misspellings do you think the spell checker would find? Try it and see. Then write a short paragraph explaining the proper use of a spell checker.

2. As you read in this chapter, AutoCAD allows you to change the dictionary against which it checks the spelling of words. AutoCAD also allows you to create one or more custom dictionaries. Under what circumstances might you want to use a custom dictionary? Might you ever need more than one? Explain.

Applying AutoCAD Skills

Work the following problems to practice the commands and skills you learned in this chapter.

1. Open shaftbearing.dwg, which you updated in the previous chapter to include text. Change the date to be the date you actually created the original drawing. Add your middle initial to your name. If you already have your middle initial, remove it. Spell-check the title block. Did your name appear for correction? If it did, add it to the dictionary. Save the drawing as shaft-rev1.dwg.

2. Edit the drawing notes in shimrev.dwg. Remove note number 2 (Break sharp edges). Renumber note 3, and add periods at the end of the notes. Spell-check the notes and the title block. Your name should not appear for correction. Save the drawing.

3. Refer again to shimrev.dwg. Your method of entering +/− for a plus/minus value in the original note 3 is not acceptable to your supervisor. You are asked to change to the standard format of ±. Erase your notes, reenter the text using the MTEXT command, and insert the proper symbol. Save the drawing.

Using Problem-Solving Skills

Complete the following activities using problem-solving skills and your knowledge of AutoCAD.

1. Open truetype.dwg located in AutoCAD's Sample folder and save it in your named folder as truetype2.dwg. Zoom in so that you can read the individual lines and notice that the file lists the alphabet in each of the TrueType fonts offered within AutoCAD. Search and replace each occurrence of the alphabetic listing with the phrase ALL DIMENSIONS ARE IN INCHES UNLESS OTHERWISE SPECIFIED. This will show you how the phrase would look in each of the available TrueType fonts. Plot the drawing or zoom around the drawing as necessary to determine which fonts are best for notes and specifications on drawings. Select the five most appropriate fonts. Then create a new drawing, create a new text style for each of the fonts, and create example text in each style. Save the drawing as techtext.dwg and plot it for future reference.

2. The border and title block shown in Fig. 20-3 are suitable for a paper size of 8.5″ × 11″. The border allows .75″ white space on all four sides.

Draw the border and title block using the dimensions in the detail drawing shown in Fig. 20-4. The three sizes of lettering in the title block are .0625″, .125″, and .25″. Use the romans.shx font. Replace "Bulldog Engineering" with your own name and logo. Position the text precisely and neatly. Use the snap and grid as necessary. Omit words, such as BY in DRAWN BY, and misspell a few words. Then, using AutoCAD's editing and spell-checking capabilities, fix the text.

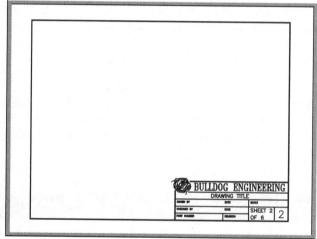

Fig. 20-3

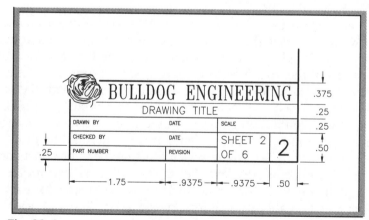

Fig. 20-4

Problem 2 courtesy of Gary J. Hordemann, Gonzaga University

Chapter 20: Text Editing and Spell Checking

Chapter 21 Tables

Objectives:

- **Create a table and populate it with text**
- **Apply styles to tables**
- **Edit tables**
- **Use formulas in tables to perform calculations**

Key Terms

cells
formula
table

Engineering drawings often contain a parts list or bill of materials presented in tabular form. These tables list the requirements for the raw materials used to manufacture parts, all the individual detail parts that make up an assembly, and other critical information. AutoCAD allows you to create tables in your drawings and provides easy editing capabilities. You can even use tables to perform calculations as you would in a spreadsheet program.

Creating Tables

A *table* is a set of rows and columns that includes text. The basic units of a table—the individual boxes within the table's grid—are called *cells*. Creating a table in AutoCAD is similar to creating one in Microsoft Word®. After creating a table, you are permitted to populate it with text.

1. Start AutoCAD, create a new drawing, and save it as **table.dwg**.

2. From the docked Draw toolbar, pick the **Table** button or enter the **TABLE** command.

Draw

This displays the Insert Table dialog box, as shown in Fig. 21-1 on page 286.

3. In the Insertion Behavior area, pick **Specify insertion point**.

4. In the Column & Row Settings area, enter **3** for the number of columns and **5** for the number of rows, and pick the **OK** button.

A table appears that you can drag on the screen.

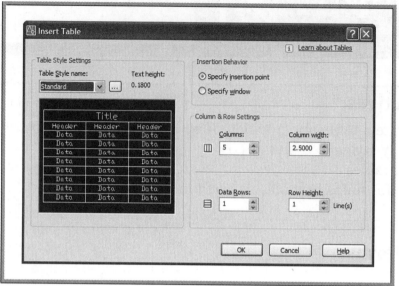

Fig. 21-1

5. Insert the table in the center of the lower part of the screen.

The Text Formatting toolbar appears.

6. Select **tt** for the text style and enter **.18** for the text height.

7. Type **Fasteners** and then press the tab key.

8. Select **tt** for the text style, enter **BOLT**, and press the tab key.

9. Select **tt** for the text style, enter **.3125 x 16 x .75** and press the tab key.

10. Select **tt** for the text style, enter **10**, and press the tab key.

The process repeats.

11. Pick the **OK** button to terminate the process.

Table Styles

In most cases, the same or similar text styles are used in all of a table's cells. AutoCAD allows you to create a table style to streamline the process of entering text.

1. Undo the insertion of the table.

Draw

2. Pick the **Table** button from the Draw toolbar.

3. In the Table Style Settings area of the Insert Table dialog box, pick the button located to the right of the down arrow (the one containing the three dots).

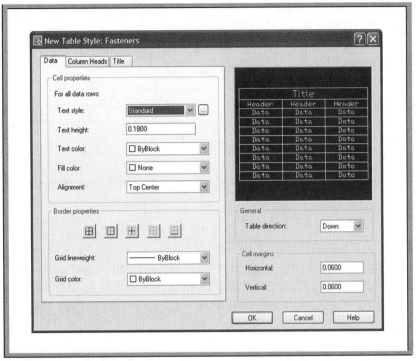

Fig. 21-2

This displays the Table Style dialog box.

4. Pick the **New** button, enter **Fasteners** for the new style name, and pick the **Continue** button.

The New Table Style dialog box appears, as shown in Fig. 21-2.

5. With the Data tab selected, select **tt** for the text style and enter **.125** for the text height.

The Border properties area of the dialog box controls the appearance of the border lines. The General area allows you to change a table so that it reads from bottom to top instead of from top to bottom. The Cell Margins area controls the spacing between the border of the cell and the cell contents.

6. Pick the **Column Heads** tab, enter **tt** for the text style, and enter **.15** for the text height.

7. Pick the **Title** tab, enter **tt** for the text style, and enter **.18** for the text height.

8. Pick the **OK** button; pick the **Close** button in the Table Style dialog box.

9. In the Insert Table dialog box, select **Fasteners** for the table style and pick the **OK** button.

10. Insert the table in the lower half of the screen.

FASTENERS		
Type	Size	Quantity
HEX BOLT	.375 x 16 x 1	10
HEX NUT	.375 - 16	10
PLAIN WASHER	.5 x 1 x .0625	20
CAP SCREW	.5 x 13 x 2	5

Fig. 21-3

11. Enter **FASTENERS** and press the tab key.

12. Enter the text shown in Fig. 21-3 into the cells. Be sure to press the tab key to advance from one cell to the next. Pick the **OK** button when you are finished and save your work.

Editing Tables

It is possible to edit the contents of a table, as well as the table itself.

Editing Text in Tables

Making changes to the text within a table is very similar to editing mtext.

1. In the table, double-click **Type** in the first cell.

This selects the cell and displays the Text Formatting toolbar.

2. Double-click **Type**, this time to highlight it.

3. Type **TYPE** in uppercase letters.

4. Repeat Steps 1 through 3 with **Size** and **Quantity**.

5. Double-click to activate (place a text cursor in) the left-most cell in the bottom (empty) row of cells.

6. Type **CAP SCREW** and press the tab key.

7. For the size, type **.5 x 13 x 1.5**; for the quantity, type **15**; pick **OK**.

8. In the docked Styles toolbar located in the upper right area of the AutoCAD window, make Fasteners the current table style.

9. From the Styles toolbar, pick the **Table Style Manager** button.

This displays the Table Style dialog box.

10. Pick the **Modify...** button.

11. In the Cell properties area, pick **Middle Left** from the dropdown located to the right of Alignment.

12. At the right of Text color, select red from the dropdown.

13. Pick the **OK** button and then the **Close** button.

The data text is now red and left-justified.

14. Undo your changes.

Changing the Table Border Lines

AutoCAD allows you to change the table itself, which is represented by border lines.

1. Select any one of the border lines.

This selects the entire table.

2. On the table's left-most vertical line, click the middle grip box and move it to the right to reduce the width of the column.

3. With the table still highlighted, right-click to display the shortcut menu and select **Size Columns Equally**.

AutoCAD adjusts the columns so they are of equal width.

4. From the docked Styles toolbar, pick the **Table Style Manager** button and pick the **Modify...** button.

5. In the Border properties area, select **0.30 mm** for Grid lineweight and **Yellow** for Grid color.

6. In the same area, pick the fourth button to make the border lines disappear.

7. Pick the first button in this area (to make the data border lines yellow), pick the **OK** button, and pick the **Close** button.

 NOTE:

If the width did not change, pick the LWT button located in the status bar.

8. Undo the changes to the border lines and save your work.

Inserting Formulas in Tables

A *formula* performs an automatic mathematical equation. You can insert simple formulas into cells in tables to calculate sums, counts, and averages. You can also define simple arithmetic expressions similar to those you can define in spreadsheet programs such as Microsoft Excel®.

1. In a blank area of the screen, create a table with two columns and five rows. Use the **Fasteners** table style. (It should be the current table style.)

2. Populate the cells with the text shown in Fig. 21-4.

3. Make the word **TOTAL** in the bottom left cell bold.

4. Select the empty cell in the lower right corner of the table.

5. Right-click and pick **Insert Formula > Sum** from the shortcut menu.

Part Count	
Item	Quantity
Bolt	8
Washer	16
Nut	8
Spacer	4
TOTAL	

Fig. 21-4

6. In response to Select first corner of table cell range, click in the first cell below Quantity.

7. In response to Select second corner of table cell range, click in the last cell in the number column next to Spacer.

The formula =Sum(B3:B6) now appears in the cell. This formula will add the contents of cells B3 through B6.

8. Pick **OK**.

The cell now contains the number 36, which is the sum of the parts listed in the table. Notice that the number in the cell is shaded, indicating that the cell contains a formula.

NOTE: ───────────────────────────

You can also type formulas directly into cells. In this example, typing =Sum(B3:B6) directly into the cell would have yielded the same result.

9. Save your work and exit AutoCAD.

Review Questions

1. What is a table in AutoCAD?

2. Describe the quickest method of editing text in a cell of a table.

3. Suppose that you created a table, but later decided to change the weight of the table's border lines. How would you go about it?

4. How many cells are in a table that contains five rows and seven columns?

5. Describe how to create a table style. Explain why this method is faster than formatting cells individually.

6. What is the purpose of a formula in a table?

7. How can you tell whether a number in a cell was entered directly or is the result of a formula?

8. Explain how you could make sure that all of the columns in a table are exactly the same width.

Challenge Your Thinking

1. Other than for parts lists and bills of materials, when might you use a table in AutoCAD?

2. AutoCAD allows you to build a table with the title and headers on the bottom and the list on top. Why do you think this option exists? When would it be useful?

Applying AutoCAD Skills

Work the following problems to practice the commands and skills you
learned in this chapter.

1. Create the door schedule shown in Fig. 21-5 using a table style that
   includes cell data that is .125 in height, column heads that are .18
   in height, and a title that is .22 in height. All text should use the
   Romans font. Name the file schedule.dwg.

DOOR SCHEDULE			
QUAN.	TYPE	SIZE	REMARKS
2	FLUSH	3–0 x 6–8	HOLLOW CORE
3	FLUSH	2–8 x 6–8	HOLLOW CORE
1	FLUSH	2–6 x 6–8	SOLID CORE, OAK
3	SLIDING	4–0 x 6–8	HOLLOW CORE

Fig. 21-5

2. Using Save As..., create a new file from schedule.dwg and name the
   file schedule2.dwg. In this new file, change the text in the door
   schedule so that it matches the schedule in Fig. 21-6. Use the tt text
   style for the column heads and title and make all of the border lines
   green.

Door Schedule			
Quantity	Type	Size	Comments
2	FLUSH	2–10 x 6–8	HOLLOW CORE
4	FLUSH	2–8 x 6–8	HOLLOW CORE
1	FLUSH	3–0 x 6–8	SOLID CORE, OAK
2	BI–FOLD	5–0 x 6–8	HOLLOW CORE

Fig. 21-6

Using Problem-Solving Skills

Complete the following activities using problem-solving skills and your knowledge of AutoCAD.

1. The frame.dwg file, which you created in "Using Problem-Solving Skills" problem 3 in Chapter 17, is shown in Fig. 21-7. (If you have not yet created the drawing, create it now.) Use the drawing and editing commands to insert windows and doors into the floor plan. Do research if necessary to find standard window and door widths so that you can create them accurately. Then create a table to show the sizes, kinds, and prices for the windows and doors. In the last column, show the number of each type of window or door. In the last row, use one or more formulas to find the total number of windows and doors.

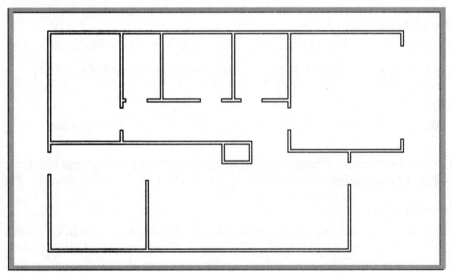

Fig. 21-7

Part 4 Project

Manufacturer's Product Table

Even in the modern electronic age, written communication remains a significant form of exchanging information. Even within a CAD technical drawing, dimensions alone often do not convey sufficient information to enable the drawing to be used as intended. A few well-chosen words can convey enough information to minimize the number of views or necessary dimensions, making the drawing cleaner in appearance.

Description

In this project, you will use written language to convey the information needed for the following scenario: A manufacturer of automotive wheels wishes to organize its product line in various ways to facilitate the task of consumers seeking to find a product that suits them. A decision has been made to organize the wheels in four different ways: the type of finish/color, the rim size (in diameter only—it is assumed each size is available in the most commonly needed widths), the material, and the price range.

Organization

Organize the following data into tables using AutoCAD's TABLE command.

- Finish/colors available include: brushed aluminum and shiny aluminum; shiny black and matte black; chromed steel and gunmetal gray steel (6 categories total)

- Rim sizes include all diameters from 16″ to 24″ in 1″ increments (9 categories total)

- Materials include aluminum, magnesium-aluminum alloy, and steel alloy (3 categories total)

Price Structure

The price structure, per wheel, is as follows: aluminum 16″ wheels start at $299, magnesium-aluminum alloy wheels at $349, and steel alloy wheels at $229. For each line of wheel, the price increases 5% (of the base price) for each incremental increase in diameter. Thus an aluminum wheel costs $299 plus 5% of $299 for a 17″ rim; an 18″ aluminum wheel would go for $299 plus 10% of $299, and so on. There is no distinction in price for color or finish.

Accompanying Drawings

Create drawings of one or more styles of wheels that might be available (again, at the direction of instructor). If you would like to see some of the many wheel designs being sold, merely do an Internet search for "automobile wheels." Figure P4-1 may also give you ideas.

Hints and Suggestions

1. Consider including instructions for consumers on how to use the information in the table to find and calculate the cost of their order.

2. AutoCAD can insert formulas directly into tables. Your instructor may therefore ask you to include, for example, the potential to calculate taxes for the purchase of one or more wheels.

3. Make sure each table and column heading is clearly labeled.

4. The newly enhanced text editing features will also make your work easier. You may, for example, wish to take advantage of the bulleting or numbering list options.

5. When illustrating the wheels, remember the convenience of using polar array.

Summary Questions

Your instructor may direct you to answer these questions orally or in writing. You may wish to compare and exchange ideas with other students to help increase efficiency and productivity.

1. Why might it be preferable to use a table instead of a list or graph to convey this information?

2. How would a company benefit by facilitating the ability of consumers to initiate and price out their own orders?

Fig. P4-1

Careers Using AutoCAD

Tim Fuller Photography

CAD Instructor

Inspiration. It's what caused Beethoven to compose amazing symphonies. It's what motivated Picasso to paint exquisite pictures. Chances are, you've been inspired at some point in your life. Inspiration can come from just about anywhere: a dream, a picture, or a person. Some of the most inspirational people you will meet in your lifetime may be your CAD instructors.

Drafting and CAD instructors work in high schools, colleges, and universities. Before becoming instructors, though, most educators gain experience in the workplace. By working in one or more areas of drafting, they acquire basic skills and obtain a thorough understanding of the profession. A CAD instructor who has progressed in this manner can draw on real-life experience in addition to material presented in a textbook to help make class practical and interesting.

Teaching vs. Industry

Although instructors' salaries typically are not quite as high as those working in industry, some additional benefits are available for educators, such as shorter workdays and summer and holiday vacations. Instructors are also rewarded by knowing that they are helping students gain skills necessary to succeed in life.

A Desire to Teach

All good instructors have one quality in common: a desire to help others succeed. Whether teaching a high school or college AutoCAD class, an instructor's positive attitude and ability to encourage students to do their best can truly help students rise to the next level. If you enjoy helping others, you may be a good candidate for a career as a CAD instructor.

However, other characteristics are also important in order to be successful in this field. Patience, social skills, and communication skills are also characteristics of good instructors in all educational fields.

Career Activities

- Research the need for CAD instructors in your area. Which high schools and community colleges offer CAD courses? How many CAD instructors currently work in the area? What are the job requirements?
- Think of one instructor you have had who made an impression on the way you think, or who encouraged you to try something new. How has this instructor helped you to grow as a person? What did you learn from him or her? Write a short essay about this inspirational person.

Chapter 22 Drawing Setup

Objectives:

- **Explain the purpose of a template file and list settings that are commonly included**
- **Choose the appropriate unit of measurement for a drawing**
- **Determine the appropriate sheet size and drawing scale for a drawing**
- **Check the current status of a drawing file**

Key Terms

drawing area
limits
scale
template file

The first 21 chapters of this book have provided a foundation for using AutoCAD. Now it's time to apply many of the pieces. As part of this effort, you will identify the scale and sheet size for a particular drawing. In doing this, you will determine the drawing units and drawing area—key elements of the drawing setup process. You will also consider settings of features such as the grid and snap that you can store in a template file and use over and over again.

Template Files

In AutoCAD, a *template file* is a file that contains drawing settings that can be imported into new drawing files. This feature is helpful to people who need to create the same types of drawings (drawings with similar settings) frequently. The purpose of a drawing template file is to minimize the need to change settings each time you begin a new drawing. Some users and companies choose to include a standard border and title block in their template files. This further shortens the time they have to spend on preparing for a new drawing.

Once you have set up a template file, subsequent drawing setups for similar drawings are fast. When you use a template, its contents are automatically loaded into the new drawing. The template's settings thus become the settings for the new drawing. Template development can include the steps shown in Table 22-1. This and the next several chapters will provide an opportunity to practice these steps in detail.

Step	Description
1.	Determine what you are going to draw (*e.g.*, mechanical detail, house elevation, etc.)
2.	Determine the drawing scale.
3.	Determine the sheet size. (Steps 2 and 3 normally are done simultaneously.)
4.	Set the drawing limits.
5.	Set the drawing area. (You will learn more about the drawing area later in this chapter.)
6.	Set the grid.
7.	ZOOM All. (This will zoom to the new drawing area.)
8.	Set the snap resolution.
9.	Enter STATUS to review the settings.
10.	Determine how many layers you will need and what information will be placed on each layer; establish the layers with appropriate colors, linetypes, lineweights, etc.
11.	Set the linetype scale (LTSCALE).
12.	Create new text styles.
13.	Set DIMSCALE, dimension text size, arrow size, etc.
14.	Store as a drawing template file.

Table 22-1

Using a Predefined Template File

AutoCAD provides many standard templates that can be used as is or modified to fit individual needs.

1. Start AutoCAD and pick the **Use a Template** button from the Startup dialog box.

AutoCAD displays a list of template files available to you.

2. Scroll down the list, and as you single-click several of them, notice those with borders and title blocks.

Notice also that the template files have a DWT file extension.

3. Find and double-click **Ansi a -named plot styles.dwt.**

The contents of the template file display on the screen with a named layout tab in the foreground. This template is set up to fit an A-size drawing sheet according to standards established by the American National Standards Institute (ANSI).

4. Pick the **Model** tab, located in the lower left area of the AutoCAD window.

As you can see, the Model tab does not display the border and title block.

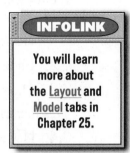

INFOLINK

You will learn more about the Layout and Model tabs in Chapter 25.

Initial Template Setup

Many AutoCAD users choose to create custom template files. We will create one also. The first step is to identify the type of object (mechanical part, building, etc.) for which the new template will be used. For this exercise, let's create a template that we could use for stair details in drawings for homes and commercial buildings.

Next, we will determine the drawing scale for the stair detail. In this sense, a *scale* is a means of reducing or enlarging a representation of an actual object or part that is too small or too large to be shown on a drawing sheet. This information will give us a basis for establishing the drawing area, linetype scale, and the scale for the dimensions. Let's use a ½″ = 1′ scale, and let's base the template on a sheet size of 8.5″ × 11″.

Standard

1. Close the current drawing (do not save changes), and pick the **QNew** button from the abbreviated Standard toolbar.

2. Pick the **Use a Wizard** button.

This begins our setup of the template file. We will create and save it as a regular drawing file. Then, when we have finished defining the settings, we will save it as a template file.

3. Read the description of the Advanced Setup wizard.

Advanced Setup uses the template file acad.dwt. So does Quick Setup.

4. Pick the **OK** button to proceed with the Advanced Setup.

Advanced Setup consists of five steps: Units, Angle, Angle Measure, Angle Direction, and Area.

Completing these steps as part of the process of creating a new file can simplify part of the template setup process.

Specifying the Unit of Measurement

Prior to Release 14, AutoCAD users were required to use either the UNITS or the DDUNITS command to set the drawing units. Both commands are still available, but now AutoCAD provides the additional option of setting the drawing units using either Quick Setup or Advanced Setup.

1. Pick the **Architectural** radio button and review the sample drawing located at the right.

2. Under **Precision**, pick the down arrow and review the list.

3. Choose the **0'0–1/16"** default setting.

This means that 1/16" is the smallest fraction that AutoCAD will display.

4. Pick the **Next** button to go to the next step.

Setting Angle Measurements

Decimal Degrees is the default setting for angle measurements. This is what we want to use for the template, so no selection is required from you. When Decimal Degrees is selected, AutoCAD displays angle measurements using decimals.

1. Under **Precision**, pick the down arrow to adjust the angle precision, and pick **0.0**.

A setting of 0.0 means that AutoCAD will carry out angle measurements to one decimal place, as shown in the sample.

2. Pick the **Next** button to proceed to the next step.

This dialog box permits you to control the direction for angle measurements. As discussed earlier in the book, AutoCAD assumes by default that 0 degrees is to the right (east). Let's not change it.

3. Pick the **Next** button.

We also established that angles increase in the counterclockwise direction by default. Do not change this, either.

4. Pick the **Next** button to proceed.

Setting the Drawing Area

The next step is to set the *drawing area*, or *limits*, of the drawing. The drawing area defines the boundaries for constructing the drawing, and it should correspond to both the drawing scale and the sheet size. The default drawing area is 12 × 9 units. Using architectural units, the drawing area is 1' × 9". (See Appendix C for a chart showing the relationships among sheet size, drawing scale, and drawing area.)

NOTE: ────────────────────────────────────

Actual scaling does not occur until you plot the drawing, but you should set the drawing area to correspond to the scale and sheet size. The drawing area and sheet size can be increased or decreased at any time using the LIMITS command. The plot scale can also be adjusted prior to plotting. For example, if a drawing will not fit on the sheet at $1/4'' = 1'$, you can enter a new drawing area to reflect a scale of $1/8'' = 1'$. Likewise, you can enter $1/8'' = 1'$ instead of $1/4'' = 1'$ when you plot.

As mentioned before, the drawing area should reflect the drawing scale and sheet size. Let's look at an example.

If the sheet size is $11'' \times 8.5''$ and the drawing scale is $1/4'' = 1'$, what is the drawing area? Since each plotted inch on the sheet would occupy 4 scaled feet, it's a simple multiplication problem: $11 \times 4 = 44$ and $8.5 \times 4 = 34$. The drawing area, therefore, would be $44' \times 34'$ because each plotted inch represents $4'$.

Since our scale is $1/2'' = 1'$, what should the drawing area be?

HINT: ────────────────────────────────────

How many $1/2''$ units would $11''$ occupy, and how many would $8.5''$ occupy?

1. Enter **22′** for the width and **17′** for the length. (See the following note.)

NOTE: ────────────────────────────────────

When entering 22′ and 17′, type the numbers exactly as you see them here; use an apostrophe for the foot mark. As discussed in Chapter 12, if you do not use a foot mark, AutoCAD assumes that the numbers are inches. You can specify inches using ″ or no mark at all. Note also that if you need to change the drawing area later, you can do so using the LIMITS command.

2. Pick the **Finish** button.

You have completed AutoCAD's Advanced Setup process.

3. Save your work in a file named **tmp1.dwg**.

Establishing Other Settings

1. Enter the **GRID** command and set it at **1'**. (Be sure to enter the apostrophe.)

The purpose of setting the grid is to give you a visual sense of the size of the objects and the drawing area. After you enter ZOOM All in the next step, the grid will fill the drawing area, with a distance of 1' between grid dots.

2. Enter **ZOOM All**.

The grid reflects the size of the sheet.

3. Enter **SNAP** and set it at **6"**.

4. Position the crosshairs in the upper right corner of the grid and review the coordinate display.

It should read exactly 22'-0", 17'-0", 0'-0". The 0'-0" is the z coordinate.

Status of the Template File

1. To review your settings up to this point, select the **Tools** pull-down menu and pick **Inquiry** and **Status**.

This enters the STATUS command. Note each of the components found in STATUS.

2. Close the window.

Tmp1 now contains several settings specific to creating architectural drawings at a scale of ½" = 1'. It will work well for the stairway detail mentioned earlier.

We have developed the basis for a template file. Technically, we have not yet created the template, because we have not stored the contents of our work in a template (DWT) file. We will do this later when we continue its development.

The next steps in creating a template file deal with establishing layers. You will create layers in the next chapter and complete the final steps for creating a template file in a subsequent chapter.

3. Save your work and exit AutoCAD.

4. Produce a backup copy of the file named tmp1.dwg.

Backup copies are important because if you accidentally lose the original (and you may, sooner or later), you will have a backup. You can produce a backup in seconds, and it can save you hours of lost work.

INFOLINK

See Chapter 8 for details on producing copies of files.

Review Questions

1. Explain the purpose and value of template files.

2. What settings are commonly included in a template file?

3. If you select architectural units, what does a precision of 0'0–¼" mean?

4. What information is used to determine the size of the drawing area?

5. If the sheet measures 22" × 16", and the scale is 1" = 10', what should you enter for the drawing area?

6. Describe the information displayed as a result of entering the STATUS command.

Challenge Your Thinking

1. Discuss unit precision in your drawings. If necessary, review the precision options listed in the Drawing Units dialog box for units and angles. (*Hint:* You can display this dialog box by entering the UNITS command.) Then describe applications for which you might need at least three of the different settings.

2. For a drawing to be scaled at ⅛" = 1' and plotted on a 17" × 11" sheet, what drawing area should you establish?

Applying AutoCAD Skills

Work the following problems to practice the commands and skills you learned in this chapter.

1. Create a new drawing using the Ansi b -named plot styles.dwt template file. Use the STATUS command to review its settings. Create another new drawing using the Iso a2 -named plot styles.dwt template file and review its settings also. What differences do you see between AutoCAD's templates for ANSI and ISO standards?

2. Establish the settings for a new drawing based on the information below, setting each of the values as indicated. Save the file as prb22-2.dwg.

   Drawing type: Mechanical drawing of a machine part
   Scale: 1″ = 2″
   Sheet size: 17″ × 11″
   Units: Decimal
   Drawing area: (You determine it.)
   Grid value: .5″
   Snap resolution: .25″

   Be sure to ZOOM All.

   Review settings with the STATUS command.

3. Establish the settings for a new drawing based on the information below, setting each of the values as indicated. Save the file as prb22-3.dwg.

   Drawing type: Architectural drawing of a house and site plan
   Scale: ⅛″ = 1″
   Sheet size: 24″ × 18″
   Units: Architectural (You choose the appropriate options.)
   Drawing area: (You determine it.)
   Grid value: 4″
   Snap resolution: 2′

   Reminder: Be sure to ZOOM All.

   Review settings with the STATUS command.

4. Consider the following drawing requirements:

   Drawing type: Architectural drawing of a detached garage
   Dimensions of garage: 32′ × 20′

   Other considerations: Space around the garage for dimensions, notes, specifications, border, and title block

   Based on this information, write the missing data for drawing setup on a separate sheet of paper. Suggest values for scale, paper size, units, drawing area, grid, and snap resolution.

5. Consider the following drawing requirements:

Drawing type: Mechanical drawing of the wheel and bearings for an in-line skate

Approximate diameter of wheel: 2.5″

Other considerations: Space for dimensions, notes, specifications, border, and title block

Based on this information, write the missing data for drawing setup on a separate sheet of paper. Suggest values for scale, paper size, units, drawing area, grid, and snap resolution.

Using Problem-Solving Skills

Complete the following activities using problem-solving skills and your knowledge of AutoCAD.

1. Your architectural office needs a drawing template setup that will accommodate the floor plans of houses measuring 16 m by 8.4 m. Create the template. In addition to the units, angles, and area, specify reasonable settings for both snap and grid. Save the template as arch1.dwt.

2. Electronic engineering requires large drawings of small parts. Set up a drawing for electronic circuits measuring $\frac{1}{4} \times \frac{3}{32}$ inches. Use decimal settings, and set the appropriate snap and grid. Save the template as elec.dwt.

Objectives:

- **Create layers with appropriate characteristics for the current drawing or template**
- **Use layers to control the visibility and appearance of objects in a drawing**
- **Change an object's properties**
- **Use object properties to filter object selections**
- **Apply a custom template file**

Key Terms

center lines	layers
filtering	palette
freeze	phantom lines
hidden lines	

AutoCAD gives you the option of separating classes of objects into layers. It may be helpful to think of AutoCAD's *layers* as transparency film that helps you organize your drawing.

One of the benefits of using layers is the ability to make them visible and invisible. For example, you can place construction lines and reference notes on a layer and then turn the layer off when you're not using them. A house floor plan could be drawn on a layer called Floor and displayed in red. The dimensions of the floor plan could be drawn on a layer called Dimension and displayed in yellow. A layer called Center could contain blue center lines.

Table 23-1 shows an example set of layers. Note the layer names, colors, linetypes, and lineweights. We will create these layers in this chapter.

NOTE:

Be sure to make a backup copy of tmp1.dwg before you begin working with it in this chapter if you have not already done so.

Layer Name	Color	Linetype	Lineweight
0	white	Continuous	Default
Border	cyan	Continuous	0.50 mm
Center	magenta	Center	0.20 mm
Dimensions	blue	Continuous	0.20 mm
Hidden	green	Hidden	0.30 mm
Notes	magenta	Continuous	0.30 mm
Objects	red	Continuous	0.40 mm
Phantom	yellow	Phantom	0.50 mm

Table 23-1

Creating New Layers

1. Start AutoCAD and open the drawing named **tmp1.dwg**.

2. Pick the **Layer Properties Manager** button from the docked Layers toolbar, or enter **LA**, the command alias for the LAYER command.

Layers

This enters the LAYER command and displays the Layer Properties Manager.

First, let's create layer Objects.

3. Pick the **New Layer** button.

This creates a new layer with a default name of Layer1. Notice that the name Layer1 is highlighted and a cursor appears in the edit box.

4. Using upper- or lowercase letters, type **Objects** and press **ENTER**.

The name Objects replaces Layer1. If you make a mistake, you can single-click the layer name and edit it.

5. Create the layers **Border, Center, Dimensions, Hidden, Notes,** and **Phantom** on your own, as listed in Table 23-1. (You will set the color, linetype, and lineweight of each layer later in this chapter.)

Changing the Current Layer

Let's change the current layer to Objects.

1. Pick **Objects** and pick the **Set Current** button. (See the note on page 308.)

Objects is now the current layer.

NOTE:

The Set Current button is the green check mark at the top of the dialog box. You can also make a layer current by double-clicking the icon in the Status column next to the layer name.

2. Pick the **OK** button.

Notice that Objects now appears in the docked Layers toolbar.

3. Save your work.

Assigning Colors

The Layer Properties Manager permits you to assign screen colors to the layers.

Layers

1. Pick the **Layer Properties Manager** button or enter **LA**.

2. Find the **Color** heading, which is located at the left of the Linetype column heading. If the Color column heading and the color names are visible, skip to Step 5. If not, proceed to Steps 3 and 4.

3. Position the pointer between the two headings until the pointer changes to a double arrow.

4. Click and drag to the right until the names of the colors appear.

5. At the right of Objects, pick the white box under the Color heading.

The Select Color dialog box appears. It contains a palette of colors available to you. *Palette* is a term used to describe a selection of colors, similar to an artist's palette of colors.

6. Select the **True Color** tab.

This offers a choice of more than 16 million colors.

7. Select the **Color Books** tab.

This allows you to specify colors using third-party color books, such as PANTONE, or user-defined color books.

8. Move the vertical slider bar up and down and select the down arrow in the Color book area to review options.

9. Select the **Index Color** tab.

10. In the bottom half of the dialog box, pick the color red (next to yellow) and pick the **OK** button.

AutoCAD assigns the color red to Objects.

11. Assign colors to the other layers as indicated in the layer listing in Table 23-1. Use the colors in the same row of colors (red, yellow, green, etc.). Cyan is light blue and magenta is purple.

12. When you have finished assigning the colors, pick the **OK** button and save your work.

Notice that the color red appears beside Objects in the Layers toolbar.

NOTE:

AutoCAD offers a command named COLOR that allows you to set the color for subsequently drawn objects, regardless of the current layer. Therefore, you can control the color of each object individually.

The ability to set the color of objects individually or by layer gives you a great deal of flexibility, but it can become confusing. It is recommended that you avoid use of the COLOR command and that its setting remain at ByLayer. The ByLayer setting means that the color is specified by the layer on which you draw the object.

The Color Control drop-down box in the docked Layers toolbar allows you to set the current color. The recommended setting is ByLayer.

13. Draw a circle of any size on the current Objects layer.

It should appear in the color red.

14. Set **Hidden** as the current layer and draw a concentric circle inside the first circle.

It should appear in the color green.

Assigning Linetypes

Various types of lines are used in drafting to show different elements of a drawing. By convention, for example, *hidden lines* (those that would not be visible if you were looking at the actual object) are shown as dashed or broken lines. *Center lines* (imaginary lines that mark the exact center of an object or feature) are shown by a series of long and short line segments. Let's look at and load some of the different linetypes AutoCAD makes available to you.

1. Display the **Layer Properties Manager**.

2. At the right of layer Hidden, and under the Linetype heading, pick **Continuous**.

This displays the Select Linetype dialog box.

3. Pick the **Load...** button.

This displays the Load or Reload Linetypes dialog box, as shown in Fig. 23-1 on page 310. AutoCAD stores the list of linetypes in a file named acad.lin, as listed in the box at the right of the File... button.

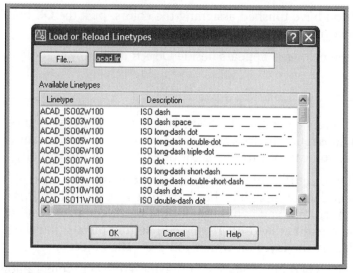

Fig. 23-1

These linetypes conform to International Standards Organization (ISO) standards. This is important because ISO is the leading organization for the establishment of international drafting standards.

At this point, you can select individual linetypes that you want to load and use in the current drawing.

4. Using the scrollbar, review the list of linetypes.

5. Find and select the **CENTER**, **HIDDEN**, and **PHANTOM** linetypes.

HINT:

Press the CTRL key when selecting them. This allows you to make multiple selections.

Hidden lines are used to show invisible edges on drawings. This is why they are called "hidden." Center lines are used to show the centers of holes, cylinders, rounded corners, and fillets. In AutoCAD, lineweights are applied to *phantom lines* based on their use. When used for cutting planes in sectional views, a thick lineweight is applied.

6. Pick the **OK** button in the Load and Reload Linetypes dialog box and notice the new list of linetypes in the Select Linetype dialog box.

7. Pick **HIDDEN** for the linetype and pick **OK**.

8. Pick **OK** in the Layer Properties Manager.

The inner circle on layer Hidden changes from a continuous line to a hidden line. However, it's unlikely that you will be able to see the hidden line until after we scale the linetypes.

9. Save your work.

NOTE: ────────────────────────────────────

You can also load linetypes using the LINETYPE command. Enter LT (for LINETYPE) and pick the Load button.

Scaling Linetypes

The LTSCALE command permits you to scale the linetypes so that they are correct for the drawing scale.

1. Enter **LTS** for the LTSCALE command.

Let's scale the linetypes to correspond to the scale of the template drawing. This is done by setting the linetype scale at ½ the reciprocal of the plot scale. When you do this, broken lines, such as hidden and center lines, are plotted to ISO standards.

NOTE: ────────────────────────────────────

As you may recall, we are creating a template file based on a scale of ½″ = 1′. Another way to express this is 1″ = 2′ or 1″ = 24″. This can be written as ¹⁄₂₄. The reciprocal of ¹⁄₂₄ is 24, and half of 24 is 12. Therefore, in this particular case, you should set LTSCALE at 12.

2. In reply to Enter new linetype scale factor, enter **12**.

The line that makes up the inner circle should now appear in the hidden linetype.

3. If you're not sure, zoom in on it.

Layers

4. Open the **Layer Properties Manager**.

5. Assign the **CENTER** linetype to layer Center and the **PHANTOM** linetype to layer Phantom.

6. Pick **OK** to close the dialog box, and save your work.

NOTE: ────────────────────────────────────

The Linetype Control dropdown box, located in the docked Properties toolbar, permits you to change the current linetype regardless of the current layer. This gives you a great deal of flexibility, but it can become confusing. It is recommended that you leave the linetype at its default setting of ByLayer.

Setting Lineweights

Lineweights are important in technical drawing. For example, object lines should be thicker so they stand out more than dimensions. Normally, object lines are thick, hidden lines and text are of medium thickness, and center lines, dimensions, and hatch lines are thin. The exact lineweights vary from industry to industry and from company to company. In Auto-CAD, the default linewidth is a thickness of .25 mm.

Layers

1. Open the **Layer Properties Manager**.

2. At the right of layer Objects, and under the Lineweight heading, pick the word **Default**.

This displays the Lineweight dialog box as shown in Fig. 23-2.

Fig. 23-2

3. Scroll down the list to review the lineweight options.

4. Pick **0.40 mm** and pick **OK**; pick **OK** to close the Layer Properties Manager.

This assigns a lineweight of .4 mm to layer Objects. This lineweight replaces the word default in the Layer Properties Manager.

5. Pick the **LWT** button in the status bar.

LWT is short for "lineweight." This turns on the display of lineweights in the drawing area.

6. Pick the **LWT** button to toggle it off; toggle it back on.

7. Assign lineweights to the other layers as shown in Table 23-1 on page 307.

8. Pick **OK** to close the dialog box, and save your work.

NOTE: ──

The Lineweight Control dropdown box in the Properties toolbar allows you to set the lineweight for objects regardless of layer. However, it is recommended that you leave the lineweights at their default setting of ByLayer to avoid confusion.

Working with Layers

Setting up the appropriate layers for a drawing gives you much more control over the drawing. It also gives you more flexibility. For example, suppose you are working on the plans for a new residence. You can set up separate layers for plumbing, electrical, and other work that is often subcontracted. By placing the plans for these elements on separate layers, you can use the same drawing to generate plans for use by each of the subcontractors. For the electrical plans, you can turn off or *freeze* the plumbing layer so that the plumbing specifications don't clutter up the electrical drawing, and so on.

Turning Layers On and Off

1. Display the **Layer Properties Manager**.

2. At the right of Objects, click the light bulb.

This toggles off layer Objects.

3. Pick the **OK** button and notice that the red circle disappears.

4. Press the spacebar to display the Layer Properties Manager.

5. Toggle on layer **Objects** by clicking the darkened light bulb.

6. Pick the **OK** button and notice that the red circle reappears.

Layers

Freezing and Thawing Layers

1. Display the **Layer Properties Manager**.

2. In the Freeze column, and at the right of Objects, pick the symbol representing the sun.

Layers

This changes the symbol to a snowflake and freezes the layer.

3. Pick the **OK** button and notice what happens to the circle.

It disappears.

4. Display the **Layer Properties Manager** again, change the snowflake into a sun by picking it, and pick the **OK** button.

The circle reappears. As you can see, freezing and thawing layers is similar to turning them off and on. The difference is that AutoCAD regenerates a drawing faster if the unneeded layers are frozen rather than turned off. Therefore, in most cases, Freeze is recommended over the Off option. Note that you cannot freeze the current layer.

Locking Layers

AutoCAD permits you to lock layers as a safety mechanism. This prevents you from editing objects accidentally in complex drawings.

1. Display the **Layer Properties Manager**.

Layers

2. At the right of Objects, click the symbol that looks like a padlock.

The lock closes, indicating that the layer is now locked.

3. Pick the **OK** button and try to edit the circle.

A locked padlock appears next to the crosshairs. AutoCAD will not permit you to edit the circle because it resides on a locked layer.

NOTE:

AutoCAD does allow you to create new objects on a locked layer. However, once you have created them, you cannot edit them.

4. Reopen the **Layer Properties Manager**, unlock layer Objects by clicking the locked padlock, and pick **OK**.

Managing Layers

There are several ways to control layers, layer properties, and the current layer in AutoCAD. The Make Object's Layer Current button allows you to change the current layer based on a selected object. The Layer Properties Manager allows you to change layer properties and delete unused layers.

Layers

1. Pick the **Make Object's Layer Current** button on the docked Layers toolbar.

Notice that the AutoCAD prompt reads Select object whose layer will become current.

2. Pick the red circle.

Objects is now the current layer.

Layers

3. Display the **Layer Properties Manager** and pick the **New Property Filter** button located in the upper left area.

This displays the Layer Filter Properties dialog box. This dialog box is available to filter layers based on name, state, color, and linetype. *Filtering* means to selectively include or exclude using specific criteria. Filtering layers is especially useful in complex drawings that contain a large number of layers.

4. Pick the **Show example** button located in the upper right area, review the layer filter examples, and then close the box.

5. Pick **Cancel** to close the Layer Filter Properties dialog box.

The New Group Filter button creates a layer filter containing layers that you select and add to the filter. The Layer States Manager button displays the Layer States Manager. It enables you to save the current property settings for layers in a named layer state that you can later restore.

6. Pick the **Objects** layer and then right-click to display the shortcut menu.

7. Select **Change Description** from the shortcut menu.

This produces a text box under the Description column.

8. Enter **Object lines** for the description.

The description can be a helpful reference when you use these layers in the future.

Focus your attention on the Plot Style and Plot headings. Colors appear under Plot Style. These are the colors you see in the drawing area. Color 1 is red, color 2 is yellow, and so on, as shown. They are grayed out and unavailable because this drawing does not use plot styles.

INFOLINK

You will learn more about plotting in Chapter 24.

Under the Plot heading, you can pick the printer icon to turn off plotting for a layer. If you turn off plotting for a layer, AutoCAD still displays the objects on that layer, but they do not plot.

9. Pick the printer icon located at the right of layer Center.

A red line appears on top of the printer, indicating that plotting has been turned off for this layer. You will learn more about plotting in the following chapter.

10. Turn on plotting for layer Center and pick **OK**.

Chapter 23: Layers and Linetypes

315

Changing and Controlling Layers

AutoCAD offers a convenient way of changing and controlling certain aspects of layers.

1. In the docked Layers toolbar, pick the down arrow to display the Layer Control dropdown box.

This displays the list of layers in the current drawing. By picking the symbols, you can quickly change the current status of the layers.

2. Pick **Center**.

Layer Center becomes the current layer.

3. Display the list again.

4. Beside Objects, click the padlock.

The padlock changes to its locked position, and layer Objects becomes locked.

5. Pick it again to unlock layer Objects.

6. Beside Hidden, pick the sun.

The sun changes to a snowflake.

7. Pick a point anywhere in the drawing area.

Layer Hidden is now frozen.

8. Display the list, pick the snowflake, and pick a point anywhere in the drawing area.

This thaws layer Hidden.

9. Make **Objects** the current layer and save your work.

Working with Objects

Now that you have created a number of layers and assigned properties to them, you have much more flexibility in working with objects in the drawing. You can control their appearance and style by placing them on appropriate layers. You can also change an object's properties and even filter objects according to their characteristics.

Changing an Object's Properties

The Properties dialog box provides a quick method of reviewing and changing the properties of an object.

1. Double-click the green circle.

This displays the Properties dialog box.

The Properties dialog box provides a lot of information on the circle, including its color, linetype, lineweight, and layer name. Suppose you created the object on the wrong layer—something you will certainly do in the future—and want to move it to a different layer.

2. At the right of Layer, pick **Hidden**.

This produces a down arrow.

3. Pick the down arrow.

This displays the list of layers you created.

4. Pick **Center**, close the dialog box, and press **ESC** to remove the selection.

The circle now resides on layer Center.

5. Using the same method, change the same circle back to layer **Hidden**.

The Properties dialog box is a handy way of quickly making changes to the objects in the drawing.

Matching Properties

The Match Properties button provides a quick way to copy the properties of one object to another.

Standard

1. From the docked Standard toolbar, pick the **Match Properties** button.

This enters the MATCHPROP command.

2. In reply to Select source object, pick the red circle.

Notice the change to the pointer.

3. In reply to Select destination object(s), pick the green circle.

This copies the properties of the first object to the second object.

4. Press **ENTER** to terminate the command.

5. Undo the last operation.

The circle reverts to its original properties.

Standard

6. Pick the **Match Properties** button again and pick the green circle.

7. Enter **S** for Settings.

This displays the Property Settings dialog box. The checked items are copied from the source object to the destination object(s), allowing you to control the properties that are copied.

8. Pick the **Cancel** button in the dialog box and press **ENTER** to terminate the command.

9. Erase both circles and save your work, but do not close the drawing.

Filtering by Object Property

The Quick Select dialog box allows you to filter a selection set based on an object's properties. You can use the Quick Select dialog box with the Properties dialog box to make changes quickly to a large number of objects.

INFOLINK

See Chapter 5 for other methods of object selection.

1. Start a new drawing using the **Quick Setup** wizard. Specify **Architectural** units and a drawing area of **17′ × 11′**.

2. Create two new layers as shown in Table 23-2 and **ZOOM All**.

3. Create the stair detail shown in Fig. 23-3. Place the risers on layer Risers and the treads on layer Treads. (The *risers* are the vertical lines, and the *treads* are the horizontal lines.)

4. Zoom in so that the stairs fill most of the screen.

5. Save your work in a file named **stairs.dwg**.

Suppose your supervisor reviews the drawing and asks you to place both the treads and the risers on the risers layer. However, you must keep the differences in lineweight and color for display purposes.

6. Select the **Tools** pull-down menu and pick the **Quick Select...** item.

This displays the Quick Select dialog box. The Properties box lists several properties by which you can select objects.

7. Pick **Layer** in the Properties selection box.

8. Pick a point in the Value box to see a list of layers, and pick **Treads**.

We will leave the rest of the values at their default values. The operator is an equal sign, and the radio button for Include in new selection set is selected. This means that all the lines in the drawing that are on layer Treads will be included in the selection set.

Layer Name	Color	Linetype	Lineweight
Risers	yellow	Continuous	0.30 mm
Treads	green	Continuous	0.60 mm

Table 23-2

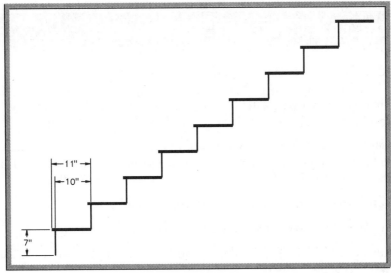

Fig. 23-3

9. Pick **OK**.

All of the stair treads are selected in a single operation.

Standard

10. Pick the **Properties** button on the Standard toolbar and pick **Layer** in the Properties dialog box.

11. Pick the down arrow to see a list of layers, and choose **Risers**.

12. Close the Properties dialog box and press **ESC** to remove the grips.

All of the stair treads are now located on layer Risers, but they are now all the same lineweight. Because they are all on the same layer, you can't select according to layer this time. However, you know that all of the treads are exactly 11 inches long. The Quick Select dialog box allows you to select them based on their length.

13. Select the **Tools** pull-down menu and pick **Quick Select...** again.

14. Scroll down the Properties selection box and pick **Length**.

15. In the Value box, enter **11** and pick **OK**.

All of the 11″ lines in the drawing—all of the treads—are selected.

Standard

16. Pick the **Properties** button on the Standard toolbar and pick **Lineweight** in the Properties dialog box.

17. Pick the down arrow and select **0.60 mm** from the dropdown list.

18. Pick **Color**, pick the down arrow, and choose green.

19. Close the dialog box, save your work, and close the file.

Applying the Custom Template

The tmp1.dwg file should still be open. This template file is nearly complete. The last few steps (12, 13, and 14 in Table 22-1 on page 298) typically involve creating new text styles and setting the dimensioning variables and DIMSCALE. All of this is covered in Chapters 26 and 27.

The template file concept may lack meaning to you until you have actually applied it. Therefore, let's convert tmp1.dwg to a template file and use it to begin a new drawing. First we will create the template file.

1. Be sure to save your work if you haven't already.

2. Select **Save As...** from the **File** pull-down menu.

3. At the right of Files of type, pick the down arrow and select **AutoCAD Drawing Template (*.dwt)**.

4. Find and select the folder with your name, enter **tmp1** for the file name, and pick the **Save** button.

The Template Description box appears.

5. Enter **Created for stair details** and pick **OK**.

6. Close **tmp1.dwt**.

7. Pick the **QNew** button in the Standard toolbar and pick the **Use a Template** button.

8. Pick the **Browse...** button.

9. Find and open the **tmp1.dwt** template file. Be sure to look in your named folder.

AutoCAD loads the contents of tmp1.dwt into the new drawing file.

10. Review the list of layers.

Look familiar? Do you see why drawing templates are of value?

11. Save the new drawing file as **staird.dwg** and exit AutoCAD.

Review Questions

1. Name at least two purposes of layers.

2. Using the Layer Properties Manager, how do you change the current layer?

3. Describe the purpose of the LTSCALE command and explain how to set it.

4. Describe a situation in which you would want to freeze a layer.

5. Name five of the linetypes AutoCAD makes available.

6. What is the purpose of locking layers?

7. If you accidentally draw on the wrong layer, how can you correct your mistake without erasing and redrawing?

8. Explain how to freeze a layer using the Layer Control dropdown box in the docked Layers toolbar.

9. Describe a way to select several similar objects using a single operation.

Challenge Your Thinking

1. When might you use the Layer Control dropdown box in the Layers toolbar instead of using the Layer Properties Manager? Explain.

2. AutoCAD allows you to use linetypes in your drawings that correspond to ISO standards. Find out more about ISO standards. When and where are they used? What is the purpose of having such standards?

Chapter 23
Review & Activities continued

Applying AutoCAD Skills

Work the following problems to practice the commands and skills you learned in this chapter.

1–2. Create two new drawings using the tmp1.dwt template file. Change the layers to match those shown in Tables 23-3 and 23-4. Name the drawings prb23-1.dwg and prb23-2.dwg. Make Objects the current layer in prb23-1.dwg, and make 0 the current layer in prb23-2.dwg.

Layer Name	State	Color	Linetype	Lineweight
0	Frozen	white	Continuous	Default
Border	On	cyan	Continuous	0.50 mm
Center	On	yellow	CENTER	0.20 mm
Dimensions	On	green	Continuous	0.20 mm
Hidden	On	yellow	HIDDEN	0.30 mm
Objects	On	red	Continuous	0.40 mm
Phantom	On	blue	PHANTOM	0.50 mm
Text	Frozen	magenta	Continuous	0.30 mm

Table 23-3

Layer Name	State	Color	Linetype	Lineweight
0	On	white	Continuous	Default
Center	Frozen	blue	CENTER	0.20 mm
Dimensions	On	yellow	Continuous	0.20 mm
Electrical	On	cyan	Continuous	0.30 mm
Found	On	magenta	DASHED	0.40 mm
Hidden	On	blue	HIDDEN	0.30 mm
Notes	On	yellow	Continuous	0.30 mm
Plumbing	Frozen	white	Continuous	0.30 mm
Title	Frozen	magenta	Continuous	0.30 mm
Walls	On	red	Continuous	0.40 mm

Table 23-4

Chapter 23: Layers and Linetypes

3. Create a new drawing from scratch. Set up the layers shown in Table 23-5. Then create the slide shown in Fig. 23-4 on layer Object. Place the center lines on layer Center, positioning them as shown in the illustration. Change back to the Object layer and create the hole (circle) for the slide so that its center point is at the intersection of the two center lines. To finish the drawing, offset the hole by .2 unit to the outside and trim the center lines to the outside circle. Erase the temporary trim circle, and save the drawing as slide.dwg.

Layer Name	Color	Linetype	Lineweight
Object	white	Continuous	0.40 mm
Center	blue	CENTER2	0.20 mm
Dims	red	Continuous	Default

Table 23-5

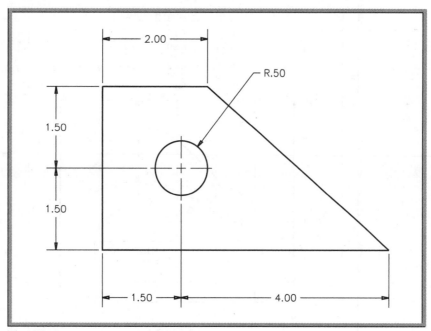

Fig. 23-4

4. Create a drawing template for use with an A-size drawing sheet. Use a scale of $1/4'' = 1'$. Use architectural units with a precision of $1/16''$. Set up the following layers: Floor, Dimensions, Electrical, Plumbing, and Furniture. Name the template ch23tmp.dwt.

5. Using the ch23tmp.dwt template file you created in the previous problem, create a simple floor plan for a house. Make the floor plan as creative and as detailed as you wish. Place all the objects on the correct layers.

6. The graph in Fig. 23-5 shows the indicated and brake efficiencies as functions of horsepower for a small engine. Reproduce the graph as follows: Using a suitable scale, draw the grid and plot the given points as shown; then draw a spline through each set of points. Place the border, title block, and curves on a layer named Visible; the grid and point symbols on layer Grid; and the text on layer Text. Trim the grid around the text and arrows, and trim the curves and grid out of the symbols. Use the appropriate text justification to align the axis numbers and titles properly.

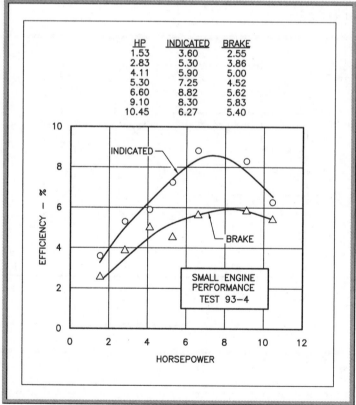

Fig. 23-5

Problem 6 courtesy of Gary J. Hordemann, Gonzaga University

Chapter 23
Review & Activities

Using Problem-Solving Skills

Complete the following activities using problem-solving skills and your knowledge of AutoCAD.

1. Create the drawing of the washing machine spacer from the graph paper sketch sent to the drafting department by engineering (Fig. 23-6). Each square represents .25 inch. Add a title to the drawing. Include a note saying that the material is SAE 1010 carbon steel. Make the appropriate drawing setup and create the necessary layers. Add center lines for all holes. Save the drawing as ch23spacer.dwg.

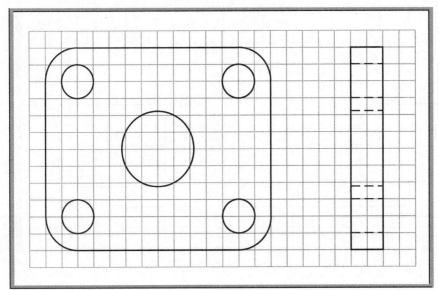

Fig. 23-6

2. The chief engineer created a sketch of the step block for a bowling
   alley pin-setter (Fig. 23-7). Create the drawing. Each square represents 5 cm. Add a title to the drawing and these notes: 1) break
   all sharp edges, 2) fillets and rounds 5 cm, and 3) material is UNS
   S30451 stainless steel. Make the appropriate drawing setup, and
   create the necessary layers. Save the drawing as ch23pinsetter.dwg.

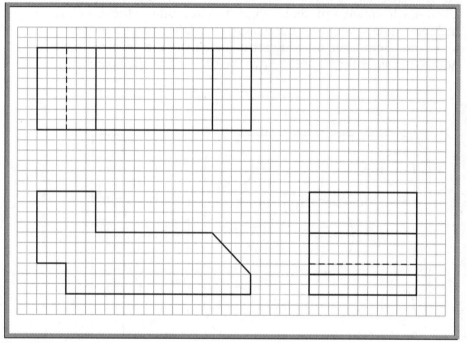

Fig. 23-7

Chapter 23: Layers and Linetypes

Chapter 24 Plotting and Printing

Objectives:

- **Preview a plot**
- **Adjust plotter settings**
- **Plot an AutoCAD drawing to scale**

Key Terms

hidden lines   portrait
landscape   printer
plot scale   rendered
plot style   wireframe
plotter

CAD users plot and print drawings using plotters and printers. Many years ago, there was a distinct difference between a plotter and a printer, but today, the distinction has blurred to the extent that there's now little difference between the two. Both use the same basic technology—usually inkjet or laser—and both print black, grayscale, or color onto paper and other sheet materials. Some people still refer to devices that handle large sheets, such as 36" × 48" and even larger, as *plotters*. They think of *printers* as smaller, tabletop devices handling sheets up to 11" × 17". Note that this is not a universally accepted difference, however. The terms *plotter* and *printer* and *plotting* and *printing* are used interchangeably in this book.

Let's create and plot a set of stairs.

1. Start AutoCAD and open the file named **staird.dwg**.

This file, which is based on the tmp1.dwt template file, was created in the previous chapter.

2. On layer **Objects**, draw the stair step shown in Fig. 24-1. Use polar tracking and enter the lengths at the keyboard. (The LWT button should *not* be depressed.)

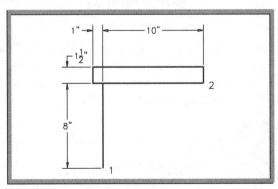

Fig. 24-1

327

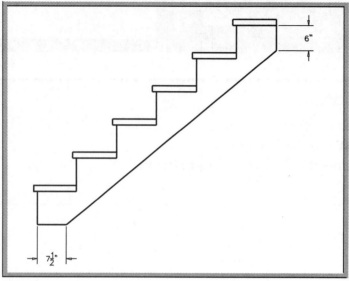

Fig. 24-2

3. Copy the objects five times to produce the stairs as shown in Fig. 24-2. Be sure to add the lines that make it complete.

4. **ZOOM All** and save your work.

Previewing a Plot

AutoCAD allows you to preview a plot before you send it to the plotter. This feature allows you to catch mistakes before you spend the time and supplies to create the actual plot.

Standard

1. From the docked Standard toolbar, pick the **Plot** button or enter the **PLOT** or **PRINT** command at the keyboard.

The Plot dialog box appears, as shown in Fig. 24-3.

In the illustration, notice that Dell Photo AIO Printer 922 is the current printer/plotter. The device you see in this area will probably be different. AutoCAD stores plotter settings for each configured plotter, so other settings may also differ.

NOTE:

If no plotter names appear in the upper left area named Printer/plotter, you must configure one before you proceed with the following steps. Page 334 provides information on how to configure plotter devices.

Note the preview button located in the lower left area of the dialog box.

2. Pick the **Preview...** button.

This feature gives you a preview of how the drawing will appear on the sheet, as shown in Fig. 24-4.

3. Use the pick button to zoom in on the drawing.

4. Right-click to display a pop-up menu, and pick **Pan** to move around in the drawing.

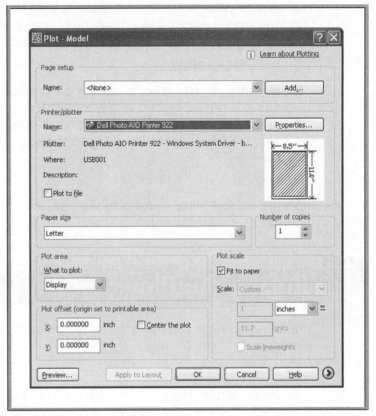

Fig. 24-3

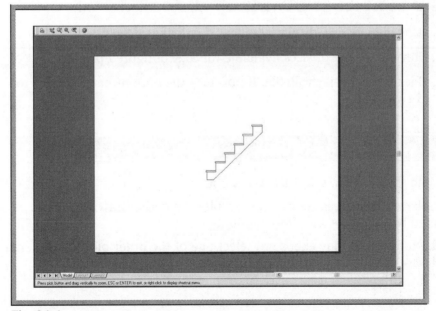

Fig. 24-4

This enables you to examine the drawing closely before you plot it. Plotting large drawings can take considerable time and supplies, so spotting errors before you plot can save time and supplies.

5. Press **ESC** or **ENTER**.

6. If your computer is connected to the configured plotter or printer and it is turned on and ready, continue with Step 7. If it is not, prepare the device. If you do not have access to an output device, pick the **Cancel** button and skip to the next section, titled "Tailoring the Plot."

7. Pick **OK**.

Since we did not make changes to the plot settings before plotting, the plot may be incorrect.

Tailoring the Plot

The following steps consider sheet size, drawing scale, and other plotter settings and parameters.

Standard

1. Pick the **Plot** button.

2. In the bottom right corner, pick the arrow (**>**) to expand the dialog box to include other options.

Specifying Sheet Size

1. Under Paper size, display the list of options.

2. Review the list of sizes available for the configured printer.

3. Pick **Letter** if it is available. If not, pick the option that is closest to 11″ × 8.5″.

Defining the Plot Area

1. In the Plot area, select the **Limits** option.

We selected Limits because we want to plot the entire drawing area as defined by the 22′ × 17′ drawing limits.

In the future, you may choose to select one of the other plotting options. Table 24-1 explains what each option means.

Option	Function
Display	Plots the current view
Extents	Similar to ZOOM Extents: plots the portion of the drawing that contains objects
Limits	Plots the entire drawing area
Window	Plots a window whose corners you specify; use the Window... button to specify the window to plot

Table 24-1

Setting the Drawing Scale

1. Focus your attention on the Plot scale area of the dialog box.

The *plot scale* is the scale at which the drawing is plotted to fit on a drawing sheet.

2. Uncheck the Fit to paper check box.

The scale options become available.

3. Pick the down arrow next to Scale: and review the list of options.

Focus on these values. As you may recall, the scale for the stairs drawing is ½″ = 1′.

4. Select **1/2″ = 1′-0″**.

The numbers 1 and 24 (or 0.5 and 12) are entered automatically in the inch and units boxes because 1 inch = 24 units is another way to express our plot scale.

When Plot object lineweights is available and checked, AutoCAD scales the lineweights in proportion to the plot scale. This setting controls the LWSCALE system variable.

 NOTE:

You can manage the list of scales you use for viewports, page layouts, and plotting in the Edit Scale List dialog box. To open the list, pick Scale List... from the Format pull-down menu or type the command SCALELISTEDIT. Scales can be added, deleted, and modified, and you can rearrange the scale list to display the most commonly used scales at the top.

Setting the Drawing Orientation

Notice the illustration to the right of the Portrait and Landscape radio buttons in the Drawing orientation area. In *portrait* orientation, the drawing is positioned so that "north" or "up" falls on the narrower edge of the paper. In *landscape*, the wide edge of the paper is north.

1. Pick each radio button while watching the illustration to see the difference between portrait and landscape.

2. In the Drawing orientation area, pick **Landscape**.

Reviewing Other Settings

Let's look briefly at the other settings available from the Plot dialog box.

1. Focus on the Page setup area and pick the **Add** button.

This button enables you to save the current settings in the Plot dialog box to a named page setup. When plotting in the future, you can select the named page setup to retrieve all of the settings associated with it. You can modify the page setup using the Page Setup Manager.

2. Focus your attention on the Printer/plotter area and notice the Plot to file check box.

When the Plot to file check box is checked, AutoCAD sends plot output to a file rather than to a device and creates a PLT file type.

3. Pick the **Properties...** button located in the Printer/plotter area.

This displays the Plotter Configuration Editor, which presents information relevant to the device.

4. Pick the **Cancel** button to close the Plotter Configuration Editor.

5. Now focus your attention on the Plot offset area of the dialog box.

It specifies an offset of the plotting area from the left corner of the sheet.

6. Focus your attention on the Plot style table (pen assignments) area.

The Plot style table (pen assignments) area allows you to create and edit plot style tables. A *plot style* in AutoCAD is a collection of property settings saved in a plot style table.

Focus your attention on the Shaded viewport options area. This area specifies how shaded and rendered viewports are plotted and determines their resolution. A viewport is a bounded area that displays some portion of a drawing. You will learn more about viewports in the following chapter.

7. Select the down arrow located to the right of Shade plot.

- As displayed plots objects the way you see them on the screen.

- Wireframe plots objects in wireframe regardless of the way you see them on the screen. *Wireframe* is the representation of a 3D object using lines and curves to show its boundaries.

- Hidden plots objects with hidden lines removed regardless of the way you see them on the screen. In this context, *hidden lines* are those lines and curves of a 3D model that are hidden from view.

- Rendered plots objects as rendered regardless of the way you see them on the screen. A *rendered* 3D object is one that contains shading to make it look more realistic.

8. Select the down arrow located to the right of Quality.

- Draft sets rendered and shaded objects to plot as wireframe.
- Preview sets rendered and shaded objects to plot at a maximum of 150 dots per inch (dpi).
- Normal sets rendered and shaded objects to plot at a maximum of 300 dpi.
- Presentation sets rendered and shaded objects to plot at the current device resolution (maximum of 600 dpi).
- Maximum sets rendered and shaded objects to plot at the current device resolution with no maximum resolution.
- Custom sets rendered and shaded objects to plot at the resolution setting specified in the DPI box, but not greater than the current device's maximum resolution.

9. Focus on the DPI area, which may be grayed out.

DPI specifies the dpi for shaded and rendered views, up to the maximum resolution of the plotting device. This option is available if you select Custom in the Quality box.

10. Focus on the Plot options area.

This area of the dialog box specifies options for lineweights, plot styles, shaded plots, and the order in which objects are plotted.

- The Plot in background option designates that the plot is processed in the background. You can also turn on this option in the Options dialog box.
- Plot object lineweights indicates whether lineweights assigned to objects and layers are plotted.
- Plot with plot styles specifies whether plot styles that have been applied to objects and layers are plotted.
- Plot paperspace last plots model space first. Normally, paper space is plotted first. You will learn about model space and paper space in the following chapter.
- Hide paperspace objects designates whether the HIDE operation applies to objects in the paper-space viewport.

- Plot stamp on places a plot stamp on a specified corner of each drawing. Plot stamp settings are specified in the Plot Stamp dialog box.
- Save changes to layout saves changes made in the Plot dialog box to the layout.

11. Pick the **Cancel** button.

Plotting the Drawing

Let's preview and plot the drawing.

1. Pick the **Preview...** button.

2. Review the preview and then press **ESC**.

3. Prepare the device for plotting.

4. Pick the **OK** button to initiate plotting.

After plotting is complete, examine the output carefully. The dimensions on the drawing should measure correctly using a $\frac{1}{2}'' = 1'$ scale.

Adding and Configuring a Plotter

Adding a new plotter for AutoCAD is an elaborate process that should involve an instructor or systems administrator.

1. Pick **Options...** from the **Tools** pull-down menu to display the Options dialog box.

2. Select the **Plot and Publish** tab.

3. Pick the down arrow located in the upper left area to display a list of currently available output devices, but do not select a new one.

4. Pick the **Add or Configure Plotters...** button.

5. Double-click **Add-A-Plotter Wizard** and read the information that displays in the Add Plotter - Introduction Page.

6. Pick the **Next** button.

7. Read the information in the box and then pick the **Cancel** button.

8. Close the Plotters window.

9. To learn more about adding and configuring plotters, pick the **Help** button and read the information presented in the Help window.

10. Close any windows that are still open and exit AutoCAD without saving.

Review Questions

1. Explain why a plot preview is useful.

2. In the Plot dialog box, what is the purpose of the Plot to file check box?

3. Briefly describe each of the following plot options.
   Limits
   Extents
   Display
   View

4. The drawing plot scale for a particular drawing is 1 = 4″. What does the 1 represent and what does the 4″ represent?

5. Describe how you would add a new printer or plotter device.

Challenge Your Thinking

1. Explore ways of sending PLT files to a printer or plotter.

2. Suppose you wanted to plot all of your drawings to PLT files and store them automatically in the folder of your choice. How can you set the default folder for PLT files?

Applying AutoCAD Skills

Work the following problems to practice the commands and skills you learned in this chapter. In problems 1 and 2, prepare to plot a drawing using the information provided. Choose any drawing to plot.

1. Paper size: Letter
   Units: Inches
   Drawing orientation: Landscape
   Plot scale: 1 inch = 2 units
   Plot with object lineweights
   Do not plot with plot styles
   Shade Plot: As Displayed
   Quality: Normal
   Number of copies: 1
   Do not plot to file
   Perform a preview

2. When starting a drawing from scratch, select Metric.
   Paper size: A4
   Units: Millimeters
   Drawing orientation: Landscape
   Plot area: Limits
   Plot scale: 1 mm = 10 units
   Plot with lineweights
   Number of copies: 1
   Perform a preview
   Plot to file

3. Open slide.dwg (problem 3 in Chapter 23). Zoom in so that the slide fills most of the screen. Enter the PLOT command and choose the Limits option in the Plot area portion of the dialog box. Plot the drawing. Then create two more plots, one with the Extents radio button selected, and one with the Display radio button selected. Compare the drawings.

4. Choose and plot a drawing that you created in an earlier chapter. Consider the scale and drawing area so that dimensions measure correctly on the plotted sheet. Text and linetypes should also measure correctly on the sheet. For example, ⅛″ text should measure ⅛″ in height.

Using Problem-Solving Skills

Complete the following activities using problem-solving skills and your knowledge of AutoCAD.

1. The R&D department has requested a full-size plot of the step block for the bowling alley pin-setter you created in Chapter 23. Adjust the plot area and scale accordingly, and plot the drawing.

2. The engineering department has requested a half-size plot of the washing machine spacer that you created in Chapter 23. Make the necessary adjustments and plot the drawing.

Chapter 25 Multiple Viewports

Objectives:

- **Create and use multiple viewports in model space**
- **Create and use viewports in paper space**
- **Position objects in paper space viewports**
- **Edit, position, and plot paper space layouts**

Key Terms

layout   paper space
model space   viewports

Viewports in AutoCAD are portions of the drawing area that you define to show a specific view of a drawing. By default, AutoCAD begins a new drawing using a single viewport. By defining and using multiple viewports, you can see more than one view of a drawing at once.

AutoCAD includes two different environments—*model space* and *paper space*—for working with objects. When you create a new drawing without using a predefined layout template, AutoCAD presents a single viewport in model space, because this is where most drafting and design work is done.

Each new drawing also has two default layout tabs. A *layout* is an arrangement of one or more views of an object on a single sheet. When you pick a layout tab, AutoCAD automatically switches to paper space. In general, paper space is used to lay out, annotate, and plot one or more views of an object or 3D model. After you have set up the paper space parameters, AutoCAD positions model space objects against a white "paper" background to show you how they will appear on the printed sheet.

Viewports in Model Space

You can apply viewports to both model space and paper space. In model space, you can use viewports to draw and edit in more than one view at a time. The magnification of each viewport can be set individually, so viewports provide capabilities that would otherwise be impossible.

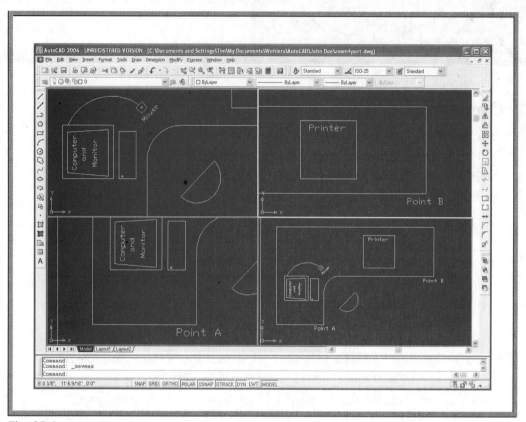

Fig. 25-1

After creating a drawing in model space, you can create floating viewports in paper space to display different views of the drawing. Because they float, you can easily position them for plotting. Depending on your needs, you can set options that determine what is plotted and how the viewports fit on the sheet.

Figure 25-1 gives an example of applying multiple viewports in model space to zoom.dwg. Notice that each viewport is different both in content and in magnification.

Creating Viewports in Model Space

Viewports are controlled with the VPORTS (or VIEWPORTS) command.

1. Open the drawing file named **zoom.dwg**.

2. Enter **ZOOM All** to make the drawing fill the screen.

3. Enter the **VPORTS** command or select **Viewports** and **New Viewports...** from the **View** pull-down menu.

4. Under **Standard viewports**, click on each of the options.

Each viewport option appears in the Preview area.

5. Pick **Four: Equal** and pick the **OK** button.

AutoCAD produces four viewports of equal size, each with an identical view of the drawing.

6. Move the pointer to each of the four viewports.

The crosshairs appears only in the viewport with the bold border. This is the current (active) viewport.

7. Move to one of the three nonactive viewports.

An arrow appears in place of the crosshairs.

8. Press the pick button on the pointing device.

This viewport becomes the current one.

Using Viewports in Model Space

Let's modify zoom.dwg using the viewports.

1. Refer to Fig. 25-1 and create four similar viewports using AutoCAD's zoom and pan features.

 HINT: _____

> Make one of the viewports current, and then zoom and pan. Repeat this process for the other three viewports.

2. Make the lower left viewport current.

3. Enter the **LINE** command and pick point **A**. (Refer to Fig. 25-1 for point A.)

4. Move to the upper right viewport and make it current.

Notice that the LINE command is now active in this viewport.

5. Pick point **B** and press **ENTER**.

View all four viewports. The line represents the edge of a hard surface for the chair.

So you see, you can easily begin an operation in one viewport and continue it in another. Any change you make is reflected in all viewports. This is especially useful when you are working on large drawings with lots of detail.

Let's move the printer from one viewport to another.

6. With the upper right viewport current, select the printer, and pick a grip box.

7. Right-click and pick **Move** from the shortcut menu.

8. Pick a corner of the printer as the base point.

9. Move to the lower left viewport and make it current. (Make sure ortho is off.)

10. Place the printer in the open area on the table by picking a point at the appropriate location.

Notice that the printer location changed in the other viewports.

Viewport Options

Let's combine two viewports into one.

1. From the **View** pull-down menu, pick **Viewports** and **Join**.

2. Choose the upper left viewport in reply to Select dominant viewport.

3. Now choose the upper right viewport in reply to Select viewport to join.

As you can see, the Join option enables you to expand—in this case, double—the size of a viewport.

4. Enter **VPORTS**, pick **Single**, and press **ENTER**.

The screen changes to single viewport viewing. This single viewport is inherited from the current viewport at the time you selected Single.

AutoCAD also allows you to subdivide current viewports into two or more additional viewports.

5. Enter the **VPORTS** command, pick **Three: Right**, and pick **OK**.

6. Make the upper left viewport the current one.

7. Redisplay the Viewports dialog box and pick **Two: Vertical**.

8. Under Apply to, located in the lower left corner, pick the down arrow and select **Current Viewport**, and pick **OK**.

As you can see, AutoCAD applies Two: Vertical to the current viewport.

9. Try the remaining viewport options on your own. Practice drawing and editing using the different viewport configurations.

 ## NOTE:

The REGEN command affects only the current viewport. If you are using multiple viewports and you want to regenerate all of them, you can use the REGENALL command.

10. Save your work and close the drawing file, but do not exit AutoCAD.

Viewports in Paper Space

When you are creating a layout in AutoCAD, you can consider viewports as objects with a view into model space that you can move and resize. By default, AutoCAD presents a single viewport in the paper space layout.

1. Open the file named **staird.dwg**.

2. Right-click the **Layout1** tab.

This produces a menu that allows you to name and rename the current layout.

3. Select **Rename** from the menu, enter **My Layout**, and pick **OK**.

AutoCAD renames the layout tab.

4. With the pick button, pick the **My Layout** tab.

When you pick a layout tab, a sheet with margins displays, reflecting the paper size of the currently configured plotter and printable area of the sheet.

In paper space, you can view the exact size of the sheet and see how the drawing will appear on the sheet when you plot. The dashed line represents the plotting boundary, and the solid line represents the single viewport. (AutoCAD's paper space defaults to a single viewport if you do not specify more than one.)

The paper space icon, located in the lower left corner of the drawing area, replaces the standard coordinate system icon. The paper space icon is present whenever paper space is the current space. The coordinate system icon is present whenever model space is the current space.

In the status bar, notice that PAPER replaced MODEL, indicating that you are now in paper space. MODEL appears when you are in model space.

5. Pick the viewport (solid red border).

As you can see, a viewport in paper space is an object.

6. Move the viewport a short distance.

The stair detail moves too, because it belongs to the viewport.

Standard

7. Undo the last step.

Plotting a Single Viewport in Paper Space

Plotting a viewport containing an object in paper space is different than plotting the same object in model space. Before plotting to scale, you must fit the objects in the viewport using the ZOOM command.

1. In the status bar, click **PAPER** so that it now reads **MODEL**.

The outline of the viewport becomes bold and the coordinate system icon returns, indicating that you are now in model space within the layout.

2. Enter the **ZOOM** command and enter the **Scale** option.

3. For the scale factor, enter **1/24xp**.

The "1 to 24" reflects the scale of the drawing. As you may recall, we established a scale of ½" = 1' for tmp1.dwt, which we used for the stair detail. You can also express the scale as 1" = 24". When you enter a value such as 1/24 followed by xp, AutoCAD specifies the scale relative to the paper space scale. (The term xp means "times paper space scale.") If you enter .5xp, AutoCAD displays model space at half the scale of the paper space scale.

4. In the status bar, change **MODEL** to **PAPER**.

5. Pick the **Plot** button from the docked Standard toolbar.

Standard

Notice under Plot area that Limits is not available; Layout appears in its place.

6. Produce the following settings:

    Paper size: Letter or 11" × 8.5"
    Drawing orientation: Landscape
    Plot area: Layout
    Scale: 1:1
    1 inch = 1 unit

The plot scale is 1=1 because we do not want to scale the layout up or down. We did that in Step 3. This is one distinct difference between plotting from model space and plotting from paper space.

7. Perform a full preview and then press **ESC**.

8. Make sure that your plotter is ready, and pick the **OK** button to initiate plotting.

AutoCAD plots the stairs to scale.

9. Save your work and close the drawing file.

Adding Viewports in Paper Space

You can add viewports in paper space using a method similar to the one you used in model space, but you must be in paper space. Let's set up a drawing with multiple paper space viewports.

1. Create a new drawing and pick the **Quick Setup** wizard.

2. Pick **Next** to accept Decimal units.

3. Enter **11"** for the width and **8.5"** for the length, pick the **Finish** button, and **ZOOM All**.

Layer Name	Color	Linetype	Lineweight
Objects	red	Continuous	0.40 mm
Border	blue	Continuous	0.60 mm
Vports	magenta	Continuous	0.20 mm

Table 25-1

4. Create the layers shown in Table 25-1, making **Vports** the current layer.

5. Set snap at **.5"** and create a new text style named **romans** using the **romans.shx** font. Use the default settings.

6. Save your work in a file named **pspace.dwg**.

7. Pick the **Layout1** tab to enter paper space.

8. Erase the viewport that appears on layer Vports.

9. From the **View** pull-down menu, pick **Viewports** and **4 Viewports**, and press **ENTER** to accept the **Fit** default value.

AutoCAD inserts four equally sized viewports on layer Vports.

NOTE:

You can insert viewports of any polygonal shape. From the View pull-down menu, select Viewports and Polygonal Viewport.

10. On the status bar, change **PAPER** to **MODEL**.

Coordinate system icons appear in each of the four viewports, showing that you are now in model space.

Objects in Viewports

You will better understand the benefits of using viewports in paper space when you create an object.

1. Make **Objects** the current layer.

2. Enter the **THICKNESS** system variable and set it to **1**.

The THICKNESS system variable enables you to specify the thickness of an object in the z direction, resulting in a three-dimensional (3D) object.

INFOLINK

Chapters 37-46 contain more information about <u>3D objects</u>.

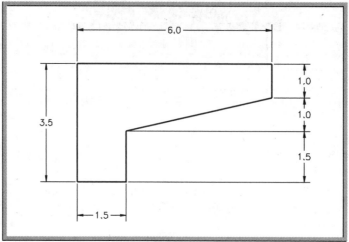

Fig. 25-2

3. In the upper left viewport, draw the object shown in Fig. 25-2.

An identical view of the object appears in all four viewports.

Creating Four Individual Views

Taking advantage of AutoCAD's viewports, let's create four different views of the solid object.

1. Make the lower left viewport the current one by picking a point inside the viewport.

2. Display the **View** toolbar.

3. View the object from the front by picking the **Front View** button from the toolbar.

4. Enter **ZOOM** and **1**.

View

This scales the viewport to an apparent size of 1. Entering 2 would make it twice the size.

5. Make the lower right viewport the current one.

6. View the object from the right side by picking the **Right View** button.

7. Enter **ZOOM** and **1**.

8. Make current the upper right viewport and view the object from above, in front, and to the right by picking **SE Isometric View.**

9. Enter **ZOOM** and **1**, and save your work.

View

View

The screen should look similar to the one in Fig. 25-3, with the exception of the border and title block.

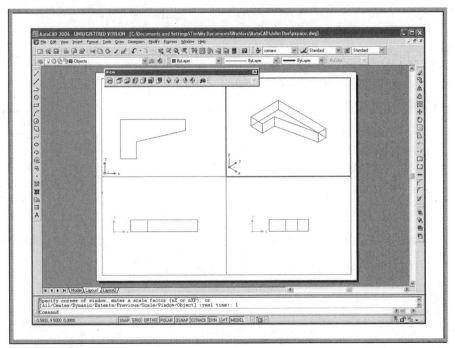

Fig. 25-3

Editing Objects

When you are working in paper space, you cannot edit objects created in model space. Likewise, when you are in model space, you cannot edit objects created in paper space.

1. In the status bar, change **MODEL** to **PAPER**.

Little appears to change except for the coordinate system icons.

2. Attempt to select the object in any of the four viewports.

As you can see, you cannot select it because it was created in model space.

Working with Paper Space Viewports

One of the big advantages of working with paper space viewports is that they allow you to print more than one view of an object on a single sheet of paper. Paper space is used to arrange views and embellish them for plotting, while model space is used to construct and modify the objects that make up the model or drawing.

Editing Viewports

Viewports in paper space are treated much like other AutoCAD objects. They can be moved and even erased, but not in model space. Only the views (objects) themselves can be edited in model space.

1. Enter the **UNDO** command and the **Mark** option.

2. Select one of the lines that make up one of the four viewports, and move the viewport a short distance toward the center of the screen.

3. Erase one of the viewports.

4. Scale one of the viewports by **.75**.

5. Switch to model space within the layout.

6. Attempt to move, erase, or scale a viewport.

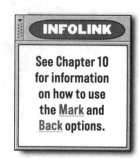

INFOLINK

See Chapter 10 for information on how to use the Mark and Back options.

NOTE: ─────────────

Do not double-click a viewport because this will cause you to enter paper space.

You can't do it, because you are no longer in paper space.

7. Enter **UNDO** and the **Back** option.

Plotting Multiple Viewports in Paper Space

One of the benefits of paper space is multiple viewport plotting. (You cannot plot more than one model space viewport at a time.)

1. Freeze layer **Vports** and make **Border** the current layer.

The lines that make up the viewports should now be invisible.

2. Switch to model space within the layout.

The viewport lines are invisible in model space also.

3. Switch to paper space and draw a border and basic title block similar to those in Fig. 25-4.

4. Plot the layout at a scale of **1 = 1**.

INFOLINK

Refer to Chapter 24 for more information about plotting a drawing.

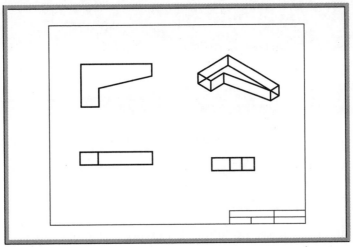

Fig. 25-4

Positioning Viewports

It is possible that the current position of the border and views did not plot perfectly. Even if they did, make adjustments to the location of the border, title block, and views by following these steps.

1. Thaw layer **Vports**.

2. Move the individual viewports to position them better in the drawing. It is normal for them to overlap.

3. Edit the size and location of the border and title block if necessary.

4. Freeze layer **Vports**, save your work, and replot the drawing.

5. Close the View toolbar and exit AutoCAD.

Review Questions

1. Explain the difference between model space and paper space.

2. How can using multiple viewports in model space help you construct drawings?

3. How do you make a viewport in model space the current viewport?

4. Explain how you would join two viewports in model space.

5. What option should you enter to obtain two viewports in the top half of the screen and one viewport in the bottom half of the screen?

6. If you are working in paper space and you discover that you need to change an object that was created in model space, what must you do before you can make the change? Why?

7. Explain why you may want to edit viewports in paper space.

8. What is the main benefit of plotting in paper space?

Challenge Your Thinking

1. Find out how many viewports you can have at one time in AutoCAD. Would you want to use that many? Why? Explain the advantages and disadvantages of using multiple model space viewports in your drawings.

2. Experiment with viewports created in model space and paper space. Is it possible to create more than one viewport in model space, then import the model space viewports into a paper space viewport? Explain.

Chapter 25
Review & Activities

Applying AutoCAD Skills

Work the following problems to practice the commands and skills you learned in this chapter.

1. Create each of the viewport configurations shown in Fig. 25-5.

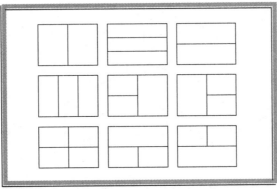

Fig. 25-5

2. Open the db_samp.dwg file in AutoCAD's Sample folder and save it in your named folder as Fremont.dwg. The current tenant of this building wants to reconfigure the office by moving a printer island to an area of unused offices.
   a. Create two viewports side by side (use the Vertical option).
   b. Make the right viewport active, and zoom in so that the building almost fills the viewport. Then zoom in further on the printer island on the right side of the building.
   c. Make the left viewport active, and again zoom in so that the building almost fills the viewport. Then zoom in on the empty offices in a horizontal line at the bottom of the viewport (office numbers 6156, 6158, 6162, and 6164). This will be the new location of the printer island.
   d. Erase the five offices, including the wall that lies against the outside wall.
   e. Using the two viewports as necessary, move the entire printer island to the space formerly occupied by the offices. (Note: When you select the printer island, do not include the gray H or the lines above and below it. It is a structural I-beam that helps support the building.)
   f. Return to a single viewport and save the drawing.

3. Using viewports in paper space, create the top, front, right side, and SE isometric view of the security clip shown in Fig. 25-6. Create a border and title block, and plot the multiple views on a single sheet.

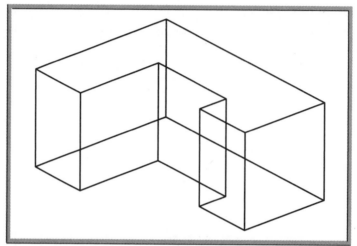

Fig. 25-6

 Using Problem-Solving Skills

Complete the following activities using problem-solving skills and your knowledge of AutoCAD.

1. Draw the locking receptacle (Fig. 25-7) for the alarm system to be installed in the administration building. Create Three: Right viewports in paper space, and show the top and front views. In the large right viewport, show the SW isometric view. Plot the views.

2. Draw the hexagonal locking pin that fits into the alarm system receptacle in the administration building from problem 1. The pin, shown in Fig. 25-8, is 2″ long and measures 2.5″ across the flats. Use three viewports in paper space, and show the same views that you used for the locking receptacle in problem 1. Plot the views.

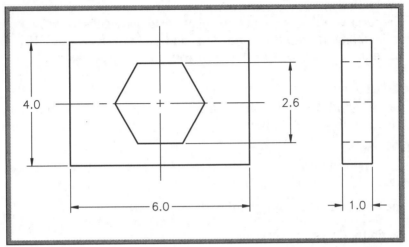

Fig. 25-7

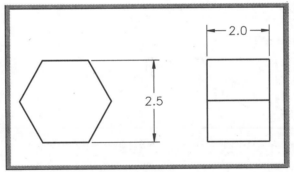

Fig. 25-8

Part 5 Project

Designing a Bookend

Creating and working with multiple viewports allows the AutoCAD user greater flexibility in preparing, viewing, and plotting drawings. This exercise requires the use of both model space and paper space.

Description

Design a bookend based primarily on one or more initials of your name. Figure P5-1 shows an example of a bookend created using the initials L.B.

Create or modify the design to have a thickness (Z axis) of 2 or more inches. Actual dimensions are not critical, but they should approximate the height and width of a small book.

Keep the following suggestions in mind as you work.

1. Begin by drawing everything in model space.

2. Switch to paper space and set up four viewports.

Fig. P5-1

3. Show the plan view in the lower left viewport, the top view in the upper left viewport, the right view in the lower right viewport, and an isometric view in the upper right viewport.

4. Place each view on its own layer. Assign each layer a different color.

5. Zoom each viewport appropriately.

Hints and Suggestions

1. Thickness gives the piece a three-dimensional look.

2. Remember as you work that lines and objects created in model space cannot be altered in paper space, and vice versa.

3. Only paper space allows adjustment and manipulation of multiple viewport plotting. Some trial and error may be required to achieve the desired look.

Summary Questions

Your instructor may direct you to answer these questions orally or in writing. You may wish to compare and exchange ideas with other students to help increase efficiency and productivity.

1. What factors determine the drawing setup? (How do you determine the "right" choices?) Why can't all drawings successfully use the same setup?

2. What variables can be modified on each layer?

3. Discuss some advantages of being able to draw on and individually control separate layers.

4. Compare and contrast model space with paper space.

5. What is useful about working in multiple viewports?

Careers Using AutoCAD

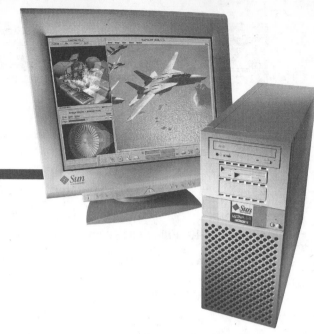

Courtesy Sun Microsystems

Aeronautical Drafter

When planning a trip of substantial distance, travelers sometimes choose to use a form of air-based transportation. Airplanes and helicopters are able to avoid traffic jams and road construction delays, and they can travel at much faster speeds than cars. When considering that aircraft actually allow people to fly through the sky instead of using the paved roads we have become accustomed to, it is easy to feel amazed at the thought of traveling by air.

Making Flight Possible

Aeronautical drafters play an important role in the process of manufacturing passenger planes, helicopters, military jets, and even missiles. They work with aeronautical engineers to create plans and specifications for aircraft and for the parts needed to build them. Opportunities for employment are available with aircraft manufacturing companies or in government positions, such as the U.S. Department of Defense or NASA.

Continuous Learning

The desire to continue learning is useful for employees who work in aeronautics. An ambitious drafter in this field may eventually become an aeronautical engineer, although additional schooling is necessary to advance to a higher position. The ability to learn on the job and to work well with others is important. Aeronautical engineers may choose to specialize in a specific area of design, such as heating and cooling systems or air and gas pressure systems.

Education and Training

Regardless of the subspecialty, aeronautical drafters must have a strong background in science and mathematics. They must also have solid experience in drafting and in using a CAD program such as AutoCAD, CATIA, or Pro/Engineer. In addition, many companies and particularly governmental jobs require some kind of drafting or CAD certification.

 Career Activities

- Write a report about the various occupations involved in the aeronautical industry, from aeronautical drafters to engineers and astronauts. What other drafting specialties are involved? What kinds of training does each specialty require?

- On the Internet or in your local or school library, research common job requirements for aeronautical drafters. How might you begin preparing for a career in aeronautical drafting now?

Objectives:

- Set up a text style for dimensions
- Produce linear dimensions using dimensioning commands and shortcuts
- Dimension round shapes, curves, and holes
- Dimension angles
- Determine the need for and use baseline and ordinate dimensioning when appropriate

Key Terms

baseline dimensions   dimension line
datum   extension lines
datum dimensions   linear dimensions
dimensioning   ordinate dimensions

Technical drawings lack meaning without information that communicates size. Drafters and designers use a method known as *dimensioning* to describe the size of features on a drawing. In this and the two chapters that follow, you will discover the wide range of dimensioning options, settings, and styles that AutoCAD makes available to you for mechanical, architectural, and other types of drafting and design work.

Figure 26-1 on page 356 shows a drawing of a part that fits into an injection mold (mold insert). We will dimension the drawing, but first we need to prepare the drawing file.

1. Start AutoCAD and use the **Quick Setup** wizard to establish decimal units and a drawing area of 11″ × 8.5″.

2. Create two new layers using the information in Table 26-1.

3. Set layer **Objects** as the current layer.

Layer Name	Color	Linetype	Lineweight
Objects	red	Continuous	0.50 mm
Dimensions	green	Continuous	0.20 mm

Table 26-1

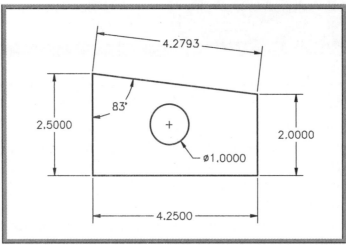

Fig. 26-1

4. Set snap at **.25** and grid at **.5**; **ZOOM All**.

5. Produce the drawing shown in Fig. 26-1 on layer **Objects**, but omit the dimensions. From the lower left corner of the drawing, position the center of the hole **2″** in the positive *x* direction and **1.25″** in the positive *y* direction.

6. Save your work in a file named **dimen.dwg**.

Setting the Dimension Text Style

AutoCAD uses text styles for its dimension text. AutoCAD's romans font is suitable for dimension text and is popular among companies that use AutoCAD. This font resembles the Gothic lettering used extensively in hand-produced drawings. For most drafting applications, a text height of .125″ (¹⁄₈″) is standard practice in industry. Some companies use taller text, such as .150″. Whichever height you use, it's important that you use a consistent text height throughout the drawing or set of drawings.

1. Create a new text style named **rom** using the **romans.shx** font. Accept the default settings for the new style.

 HINT: ─────────────────────────────────────

Enter the STYLE command by entering ST.

Dimension

2. Display the **Dimension** toolbar and pick the **Dimension Style** button.

This displays the Dimension Style Manager dialog box. We will use it here, but you will learn much more about it in the following two chapters.

3. Pick the **Modify...** button.

4. Pick the **Text** tab.

5. At the right of Text style, pick the down arrow and select **rom**.

6. Pick **OK** and then pick **Close**.

Creating Linear Dimensions

Linear dimensions are those with horizontal, vertical, or aligned dimension lines. A *dimension line* is the part of the dimension that typically contains arrowheads at each of its ends, as shown in Fig. 26-2.

1. Set layer **Dimensions** as the current layer.

2. Pick the **Linear** button from the Dimension toolbar.

Dimension

This enters the DIMLINEAR command.

3. In reply to Specify first extension line origin, pick one of the endpoints of the mold insert's horizontal edge.

4. In reply to Specify second extension line origin, pick the other end of the horizontal line.

5. Move the crosshairs downward to locate the dimension line **1″** away from the object and press the pick button.

The dimension's *extension lines* are the lines that extend from the object to the dimension line.

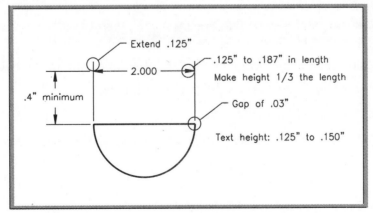

Fig. 26-2

Dimensioning Overall Size

Let's dimension the vertical edges of the object.

1. Press the spacebar to reenter the DIMLINEAR command.

2. This time, when AutoCAD asks for the first extension line origin, press **ENTER** or the spacebar.

The crosshairs changes to a pick box.

3. Pick any point on the left edge of the object.

4. Move the crosshairs to the left and pick a point **1″** from the edge to place the dimension.

Let's dimension the other vertical edge, but this time we will use AutoCAD's Quick Dimension feature.

Dimension

5. Pick the **Quick Dimension** button from the Dimension toolbar or enter the **QDIM** command.

As you can see, a pick box replaces the crosshairs. AutoCAD asks that you select geometry to dimension.

6. Pick the right edge of the object and right-click for **ENTER**.

7. Move the crosshairs to the right and pick a point for the dimension.

Notice that adding the last two dimensions required only three clicks for each dimension.

NOTE:

You can create more than one dimension at a time using the Quick Dimension method. However, the results are often unpredictable.

Dimensioning Inclined Edges

Let's dimension the inclined edge by aligning the dimension to the edge.

Dimension

1. Pick the **Aligned** button from the Dimension toolbar.

This enters the DIMALIGNED command.

2. Press **ENTER**, pick the inclined edge, and place the dimension.

Dimensioning Round Features

Now let's dimension the hole. For mechanical drafting, you should use diameter dimensions for features such as holes and cylinders. Use radius dimensions for features such as fillets and rounds (rounded outside corners).

Dimension

1. Pick the **Diameter** button from the Dimension toolbar.

This enters the DIMDIAMETER command.

2. Pick a point anywhere on the hole. (You may need to turn off snap.)

The dimension appears.

3. Move the crosshairs around the circle and watch what happens.

Notice that you have dynamic control over the dimension.

4. Pick a point down and to the right as shown in Fig. 26-1 (page 356).

Dimensioning Angles

Let's dimension the angle as shown in Fig. 26-1.

Dimension

1. Pick the **Angular** button from the Dimension toolbar.

This enters the DIMANGULAR command.

2. Pick both edges that make up the angle.

3. Move the crosshairs outside the mold insert and watch the different possibilities that AutoCAD presents.

4. Pick a location for the dimension arc inside the drawing as shown in the illustration.

5. Save your work. Also, save as a template file in your named folder. Name the file **dimen.dwt** and enter **Dimensioning Practice** for the template description.

6. Close the file.

Dimensioning Arcs

Now we will modify our injection mold insert by adding an arc. Then we will dimension the arc in two ways.

1. Select **dimen.dwt** as the template file to begin a new drawing.

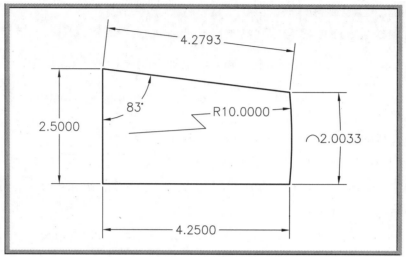

Fig. 26-3

2. Save the drawing and name it **dimenarc.dwg**.

3. Erase the right edge of the part, the center circle, and the two associated dimensions.

4. Set layer **Objects** as the current layer.

5. Pick the **Arc** button from the docked Draw toolbar, or enter **A** at the keyboard.

6. Create the arc as shown in Fig. 26-3 by following this procedure:
   - Pick the lower right corner of the part as the start point of the arc.
   - Type **E** for End and pick the upper right corner of the part as the second endpoint.
   - Type **R** for Radius, move the crosshairs to the far left side of the screen, and type **10**.

7. Set layer **Dimensions** as the current layer.

Dimension

8. Pick the **Arc Length** button from the Dimension toolbar.

This enters the DIMARC command.

9. Pick the arc you just created.

The dimension appears. Notice the arc symbol at the beginning of the dimension. This indicates that the dimension is the length of an arc and not a linear dimension.

10. Move the crosshairs to the right and pick a location for the dimension.

The radius of this arc is much larger than our part. To dimension large-radius arcs and circles, we use the DIMJOGGED command.

Chapter 26: Basic Dimensioning

11. Pick the **Jogged** button from the Dimension toolbar.

12. Pick the arc.

13. In reply to Specify center location override, pick a point inside the part near its left edge.

14. Specify a location for the dimension line; then specify a location for the jog. Undo and recreate the jogged dimension until yours looks like the dimension in Fig. 26-3.

The jog in the dimension line indicates that the center point of the dimension is not the true center point of the arc.

15. Save your work and close the file.

Using Other Types of Dimensioning

The dimensioning process you have followed so far in this chapter works for most objects. However, other types of dimensioning are better suited for some objects and engineering drawing applications. In many cases, they produce a cleaner, less confusing drawing.

Baseline Dimensioning

Baseline dimensions are progressive, each starting at the same place, as shown in Fig. 26-4. Baseline dimensioning is most useful on machine drawings where precision is critical. Measuring from a single reference dimension reduces the chance of errors that can accumulate from a "stack" of dimensions.

1. Select **dimen.dwt** as the template file to begin a new drawing.

2. Save the drawing and name it **base.dwg**.

3. Erase the drawing and dimensions and create the sheet-metal drawing shown in Fig. 26-4. Place the sheet-metal part on layer **Objects**. Fill the top half of the screen, and estimate the dimensions.

4. Make **Dimensions** the current layer.

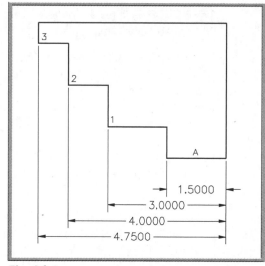

Fig. 26-4

5. Pick the **Linear** button and dimension line A. It is important that you pick line A's right endpoint first.

6. Pick the **Baseline** button.

This enters the DIMBASELINE command.

7. Pick points 1, 2, and 3 in order.

8. Press **ENTER** twice and save your work.

The Quick Dimension method also works with baseline dimensioning.

9. Erase the dimensions on the drawing.

10. Pick the **Quick Dimension** button or enter **QDIM**.

11. Select the entire drawing, press **ENTER**, and read the options AutoCAD presents.

12. Move the crosshairs downward, but do not pick a point.

13. Enter **B** for Baseline and pick a point about 1″ below the right corner of the sheet-metal part.

This establishes the location of the first dimension.

HINT:

For the procedure to work correctly, you must enter the All option to select the entire drawing. Do not use a regular or crossing window.

14. Place the dimensions.

15. Save your work.

Ordinate Dimensioning

Ordinate dimensions, also known as *datum dimensions*, are similar to baseline dimensions in that both use a datum, or reference, dimension. A *datum* is a surface, edge, or point that is assumed to be exact. A basic difference between baseline and ordinate dimensions is their appearance. Baseline dimensioning uses conventional dimension lines, whereas ordinate dimensions show absolute coordinates. Both types of dimensioning are especially useful when producing machine parts. Using a datum reduces the chance of error buildup caused by successive dimensions.

1. From the **File** pull-down menu, pick **Save As...** to create a new file named **ordinate.dwg**.

2. Create a new layer named **Orddim**, assign the color magenta to it, make **Objects** the current layer, and freeze layer **Dimensions**.

Chapter 26: Basic Dimensioning

3. Mirror the sheet-metal part to the right (see Fig. 26-5) and delete old objects. Center the part on the screen.

4. On the Objects layer, create a hole with a diameter of **.5**. Place the hole **.5** unit down and **.5** unit to the right of the top left corner.

5. Use **COPY** to create four additional holes. Place the holes as shown in Fig. 26-5.

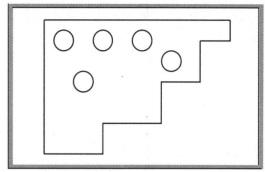

Fig. 26-5

6. Create a layer named **Center**, and on this layer add center marks to each of the holes. To do this, pick the **Center Mark** button on the Dimension toolbar. Now we're ready to define the datum, or reference dimension.

Dimension

7. Make **Orddim** the current layer.

8. Pick the **Ordinate** button.

Dimension

This enters the DIMORDINATE command.

9. In reply to the Specify feature location prompt, pick the upper left corner of the sheet-metal part.

The Xdatum option measures the absolute X ordinate along the X axis of the drawing. Also, it determines the orientation of the leader line and prompts you for its endpoint. This defines the location of the datum.

10. Enter **X** for Xdatum.

11. Pick a point about **1″** above the object to make the ordinate dimension appear.

12. Reenter the command and pick the upper left corner of the sheet-metal part again.

13. Enter **Y** for Ydatum and pick a point about **1″** to the left of the object to make the ordinate dimension appear.

The Ydatum option measures the absolute Y ordinate along the Y axis of the drawing. This defines the location of the datum.

Your drawing should now look similar to the one in Fig. 26-6 on page 364, although it will not include the 3.0000 ordinate dimension.

NOTE:

The ordinate values may differ in your drawing, depending on where in the drawing area you placed the drawing.

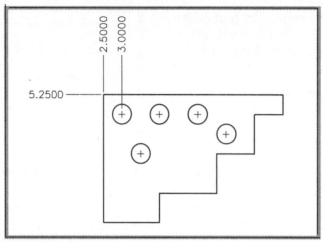

Fig. 26-6

Dimension

14. Pick the **Ordinate** button.

15. Snap to the center of the hole nearest to the upper left corner, as shown in Fig. 26-6.

16. Pick a point about **1″** from the top of the sheet-metal part. (Snap should be on.)

The X-datum ordinate dimension appears.

17. Repeat Steps 14 through 16 to create X-datum ordinate dimensions for the remaining holes.

18. Create the Y-datum ordinate dimensions on your own using Fig. 26-7 as a guide.

When you finish, your drawing should look similar to the one in Fig. 26-7.

19. Save your work and exit AutoCAD.

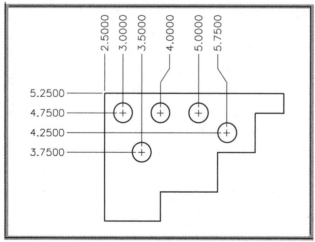

Fig. 26-7

Chapter 26: Basic Dimensioning

Review Questions

1. How do you specify a text style for dimension text?

2. Describe the alternative to specifying both endpoints of a line when dimensioning the entire length of an edge.

3. Which dimension button do you use to dimension inclined lines? angles?

4. Which dimension buttons do you use to dimension fillets, rounds, and holes? Explain when you would use each button.

5. What does a jog in a radius dimension represent?

6. Describe baseline dimensioning. On what types of drawings is it useful?

7. Describe ordinate dimensioning. On what types of drawings is it useful?

Challenge Your Thinking

1. Experiment with altering a linear dimension using its grips. Try to change the location of the dimension line by selecting and moving a grip located at the end of an extension line that is closest to the object. Is it possible? Explain.

2. Drafters generally classify dimensions in two broad categories: size dimensions and location dimensions. What is the difference between them? Are both necessary on every drawing you create? Write a paragraph explaining your findings.

3. What is the relationship between the height of text saved in a text style and the height saved in a dimension style?

Applying AutoCAD Skills

Work the following problems to practice the commands and skills you learned in this chapter. Begin a new drawing for each problem. Save your work.

1–2. Create the blocks shown in Figs. 26-8 and 26-9. Place object lines on a layer named Objects and dimensions on a layer named Dimensions. Create a new text style using the romans.shx font. Approximate the location of the holes. If you have access to the drawing files on the Instructor Resource CD, refer to block1.dwg and block2.dwg.

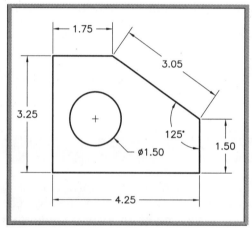

Fig. 26-8

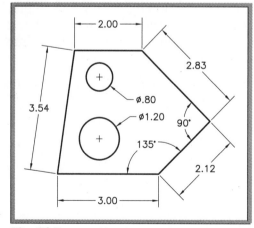

Fig. 26-9

3. Open the slide.dwg file that you created at the end of Chapter 23 and dimension it. When you finish, your drawing should look like the one in Fig. 26-10. Save the drawing as slide2.dwg.

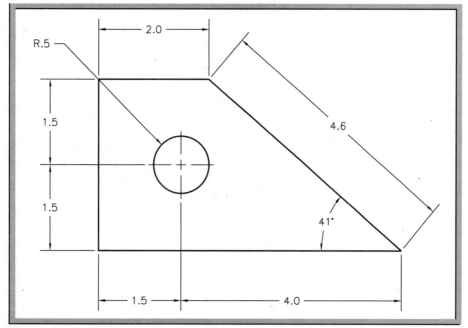

Fig. 26-10

4. Create the mounting bracket drawing shown in Fig. 26-11. Place the bracket on layer Objects. Then dimension the bracket on a layer named Dimensions. If you have access to the files on the Instructor Resource CD, refer to the file bracket1.dwg.

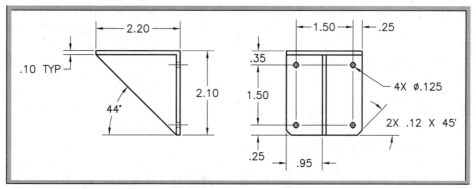

Fig. 26-11

Problem 4 courtesy of Joseph K. Yabu, Ph.D., San Jose State University

Chapter 26: Basic Dimensioning

5. Open the rockerarm.dwg file that you created at the end of Part 1 on page 98. Place the rocker arm on a layer named Objects. Then dimension the part on a layer named Dimensions. See Fig. 26-12. Save the file as rockerarmdim.dwg.

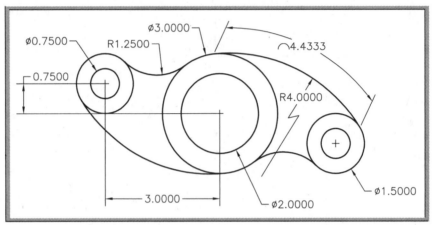

Fig. 26-12

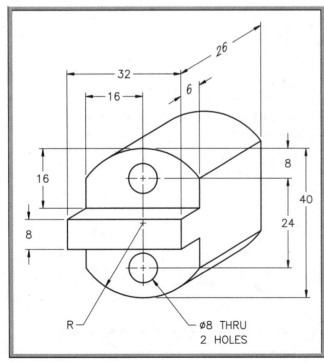

Fig. 26-13

6. Draw the front view of the shaft lock shown in Fig. 26-13. Create appropriate layers for the dimensions and linetypes. Use the romans.shx font to create a new text style named roms for the dimension text. Dimension the front view, placing the dimensions on layer Dimensions. Show all pertinent dimensions. Also, draw and dimension the right-side view. Resize the drawing area as necessary to accommodate the dimensions of the part. If you have access to the files on the Instructor Resource CD, refer to the file shaftloc.dwg.

Using Problem-Solving Skills

Complete the following activities using problem-solving skills and your knowledge of AutoCAD.

1. The architect for your firm has given you the preliminary elevation drawing shown in Fig. 26-14. Select an appropriate template file, complete the title block, and draw the front elevation; include dimensions. Since the drawing was made by an architect and not a drafter, make any changes necessary to comply with correct dimensioning practices. Save the drawing as elev1.dwg.

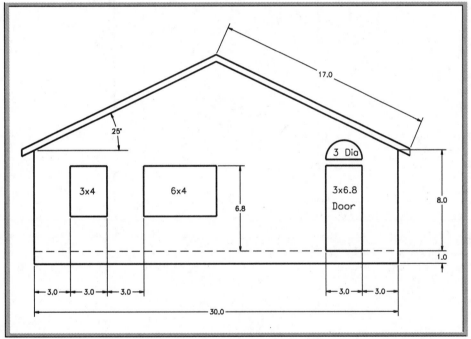

Fig. 26-14

2. Your manager has given you the sketch of a shaft support shown in Fig. 26-15 and told you that it is not to scale; however the following description applies. The base is 2″ square and ¼″ thick. The boss (lipped area) is in the center of the base, has an outer diameter of 0.5″, a thickness of ¼″, and a through hole with a .35″ radius. The four corner holes are each .15″ in diameter and their center lines are offset from the horizontal and vertical sides by .35″. Draw two views of the shaft support from this description and dimension the drawing correctly.

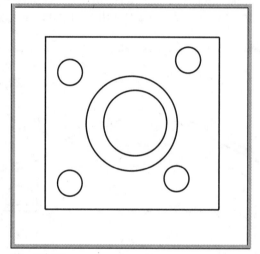

Fig. 26-15

Chapter 26: Basic Dimensioning

Chapter 27 Advanced Dimensioning

Objectives:

- Create a dimension style and apply it to a drawing and template file
- Adjust the dimension text, arrowheads, center marks, and scale in a dimension style
- Edit individual dimensions using a shortcut menu
- Use AutoCAD's associative dimensioning feature to update dimensions on a rescaled drawing
- Apply dimension styles to a drawing template file

Key Terms

associative dimensions   leaders
dimension styles   spline leader

AutoCAD permits you to create *dimension styles*, which consist of a set of dimension settings. Dimension styles control the format and appearance of dimensions and help you apply drafting standards. They define the format of dimension lines, extension lines, arrowheads, and center marks. They also define the appearance and position of dimension lines and text. Using a dimension style, you can format and define the precision of the linear, radial, and angular units.

Before we can apply a dimension style to something meaningful, we need to produce a drawing.

1. Start AutoCAD and select the **tmp1.dwt** file.

2. On layer **Objects**, create the top view drawing of the redwood deck shown in Fig. 27-1 on page 372. Do not add dimensions at this time.

The round object represents an outdoor table. It was included in the drawing to ensure that the deck is of sufficient size and shape to accommodate a table of this size.

3. Save your work in a file named **deck.dwg**.

HINT:

For best results, begin in the upper right corner and work counterclockwise to draw the edges of the deck. Add the inside rounded corner using the FILLET command.

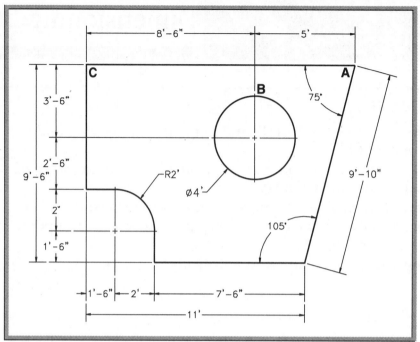

Fig. 27-1

Creating a Dimension Style

Dimension styles provide an easy method of managing the appearance of dimensions. This saves time and effort and helps you produce professional drawings. Note that drafting standards vary from industry to industry and even among companies in the same industry.

1. Display the **Dimension** toolbar.

2. On layer Dimensions, place a horizontal dimension along the bottom edge of the deck.

As you can see, the dimension text, arrowheads, etc., are much too small. We need to make some adjustments.

3. Erase the dimension.

4. From the Dimension toolbar, pick the **Dimension Style** button.

Dimension

 NOTE: ————————————————————

This button is also available in the docked Styles toolbar.

This displays the Dimension Style Manager dialog box. As you can see, Standard is the default dimension style. The sample drawing shows how the dimensions will appear using this style.

5. Pick the **New...** button.

6. For New Style Name, enter **My Style**.

7. Pick the down arrow next to Start Width and pick **Standard**.

This selection means that we will start with all of the settings saved in the Standard style.

8. After Use for, pick the down arrow.

You can apply a dimension style to the types of dimensions included in this list.

9. Select **All dimensions** and pick the **Continue** button.

The New Dimension Style dialog box appears, showing several tabs across the top. This is where you change the settings to store in the dimension style.

10. Pick the **OK** button.

Notice that My Style now appears in the list of styles, but Standard is still the current style.

11. Pick the **Close** button.

12. In the Dimension or Styles toolbar, pick the down arrow, and pick **My Style**.

My Style is now the current dimension style.

13. Save your work.

Changing a Dimension Style

Currently, My Style is identical to the Standard style. We will need to customize the new My Style dimension style for our particular drawing. This means changing the size of the arrowheads and center marks, the style and size of the text, and the overall scale of the dimensions.

Arrowhead Size

Dimension

1. Pick the **Dimension Style** button and pick the **Modify...** button.

2. Pick the **Symbols and Arrows** tab unless it is already in front.

You can make adjustments to the arrowheads in the left side of the dialog box.

3. Using the up and down arrows at the right of Arrow size, change the value to ⅛″ and watch closely at how the arrowheads in the sample drawing change to reflect the new value. They may be difficult to see.

Sizing Center Marks or Center Lines

Notice that center marks currently appear at the center of the circle in the dialog box's sample drawing. Below Arrow size, in the area titled Center marks, you can make changes to the center marks. Use center marks when space is not available for complete center lines. You may also need to make changes in this area when a company style requires one or the other.

1. At the right of Size, click the down arrow until it drops to a value of **0″** and then click the up arrow and stop at ¹⁄₁₆″.

2. Select the **Line** radio button in the Center marks area and watch to see how the sample drawing changes to reflect the new center lines.

This causes complete center lines to appear at the centers of circles and arcs, as shown in the sample drawing.

Tailoring Dimension Text

Let's create a new text style and height for our dimension style.

1. Pick the **Text** tab.

2. In the upper left area titled Text appearance, pick the button containing the three dots.

This causes the familiar Text Style dialog box to appear.

3. Create a new style named **roms** using the **romans.shx** font and the default settings.

4. After closing the Text Style dialog box, pick the down arrow at the right of Text style and select **roms**.

Roms is now the text style for My Style, as shown in the sample drawing.

5. In the Text height edit box, change the value to ⅛″.

As a reminder, the dimension text should be .125″ to .150″ in height.

Setting the Units

1. Pick the **Primary Units** tab.
2. At the right of Unit format, change the setting to **Architectural**.
3. At the right of Precision, change the value to **0'-0"**.

The sample drawing changes to reflect the unit format and precision.

Scaling the Dimensions

1. Pick the **Fit** tab and focus your attention on the area titled Scale for dimension features.

This is where you set the scale for dimensions. The value should be the reciprocal of the plot scale. When creating tmp1.dwt, we determined that the scale would be ½" = 1', which is the same as ¹⁄₂₄. The reciprocal of ¹⁄₂₄ is 24.

2. Double-click the number in the edit box and change it to **24**.

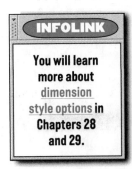

INFOLINK

You will learn more about dimension style options in Chapters 28 and 29.

NOTE:

You could also enter DIMSCALE at the AutoCAD prompt and enter 24.

As you have discovered by now, there are many other options. We have changed only those needed for the dimension style we will use to dimension the redwood deck.

3. Pick **OK**; pick **Close**.
4. Save your work.

Applying a Dimension Style

Now that you have made several adjustments to the dimension style, you are ready to dimension the drawing. Layer Dimensions should be the current layer.

Dimension

1. Begin by dimensioning the table and rounded corner. Use the Diameter button for the table and the Radius button for the filleted corner. Place the dimensions as shown in Fig. 27-1 (page 372).

Dimension

Now that you have center lines, you can use them to locate the centers of the table and rounded corner using dimensions.

Dimension

2. Pick the **Linear** button from the Dimension toolbar.

3. For the first and second extension line origins, snap to points A and B, and then place the dimension **1.5′** from the edge of the deck.

HINT: _____

Use grid or snap to determine the 1.5′ distance.

Dimension

4. Pick the **Continue Dimension** button, snap to point C, and press **ENTER** twice to terminate the command.

5. Using a similar method, add the remaining horizontal and vertical dimensions.

HINT: _____

When locating the center of the rounded corner with dimensions, use the end of the center lines for the extension line origins, similar to the way you picked point B. Separately, using the grips method of moving dimension text, move the 9′-6″ text of the vertical dimension so that it is lower than the 2′-6″ dimension beside it.

6. Add the aligned and angular dimensions.

Your dimensioned drawing of the redwood deck should now look similar to the one in Fig. 27-1.

7. Save your work.

8. If you have access to a plotter or printer, plot the drawing.

Editing a Dimension

Using AutoCAD's grips, you can change the location of dimension lines and dimension text.

1. Pick one of the linear dimensions.

Five grip boxes appear.

2. Pick the grip box located at the center of the dimension text.

3. Move the text up and down and back and forth, and then pick a new point.

4. Press **ESC**.

By selecting a dimension and right-clicking it, you can make other useful changes.

5. Select one of the linear dimensions and right-click.

A shortcut menu appears.

6. Rest the pointer on **Dim Text position**.

These options permit you to adjust the location of the dimension text.

7. Try each of them, and pick **Home text** last.

As you can see, it's easy to tailor individual dimensions, regardless of the current dimension style.

8. Select one of the linear dimensions and right-click.

9. Rest the pointer on **Dim Style**.

This allows you to save a new style or apply an existing one.

10. Select **Standard**.

The dimension uses settings stored in the Standard dimension style. The arrowheads and dimension text are there, but you can't see them because they are so small. The value of the dimension scale in Standard is 1, and we set it to 24 in My Style. This means that the text, arrowheads, and other dimensioning features are 24 times larger in My Style.

11. Change the dimension's style back to **My Style**.

12. Save your work.

Leaders

Leaders are lines with arrowheads at one end that are used to point out a particular feature in a drawing. Radial and diameter dimensions usually use leaders. The plastic retainer in Fig. 27-2 provides an example of a leader.

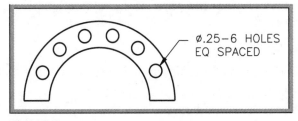

Fig. 27-2

1. Pick the **Quick Leader** button on the Dimension toolbar.

This enters the QLEADER command.

2. Pick point A (shown in Fig. 27-1, page 372).

3. Move the crosshairs up and to the right about **1.5'** and pick a point.

4. Press **ENTER**.

AutoCAD asks for the text width.

5. Enter **1'**.

6. In reply to Enter first line of annotation text, enter **USE REDWOOD**.

7. Press **ENTER** twice to display the text and terminate the command.

A leader appears with the note USE REDWOOD. Notice that the width of the note does not exceed 1'.

Using the Text Editor with Leader Text

You can also use AutoCAD's multiline text editor to produce notes for leaders.

1. Undo the addition of the leader.

2. Repeat Steps 1 through 5 from the previous section.

3. When AutoCAD asks you to enter the first line of annotation text, press **ENTER**.

4. Enter **USE REDWOOD** and pick **OK**.

The leader and note appear.

Spline Leaders

A *spline leader* is similar to a regular leader, except the leader line can consist of a spline curve, giving you flexibility with its shape. Spline leaders permit you to produce leaders where space is tight on a drawing.

1. Pick the **Quick Leader** button and enter **S** for Settings.

The Leader Settings dialog box appears.

2. On your own, review all of the parts of this dialog box, including each of the three tabs.

3. Pick the **Leader Line & Arrow** tab.

4. In the Arrowhead area, pick the down arrow to review the list of arrowhead options.

5. In the Leader Line area, pick the **Spline** radio button.

6. In the Number of Points area, pick the **No Limit** check box.

Now AutoCAD will not restrict the number of points you can use to create a spline leader.

7. Pick **OK**.

8. Pick three points anywhere on the screen to form a complex leader.

9. Press **ENTER** and **3'** for the text width.

10. Enter **THIS IS A SPLINE LEADER**.

11. Press **ENTER** twice to display the text and terminate the command.

12. Undo the addition of the spline leader.

13. Save your work.

Working with Associative Dimensions

By default, AutoCAD creates associative dimensions. *Associative dimensions* are dimensions that update automatically when you change the drawing by stretching, scaling, and so on.

1. Using the **SCALE** command, scale the drawing by **1.1**.

 HINT:

In reply to Select objects, enter all. Use 0,0 for the base point and 1.1 for the scale factor.

Notice that the dimensions changed automatically. This becomes useful when you need to change a drawing after you've dimensioned it.

2. Using the **STRETCH** command, stretch the rightmost part of the drawing a short distance to the right. (When selecting, use a crossing window.)

The dimensions change accordingly.

3. Undo Steps 1 and 2.

4. Save your work.

Adding Dimension Styles to Templates

Now you know how to perform the remaining steps (12 through 14) in creating a template file (see Table 22-1 on page 298). Let's apply these steps to the tmp1.dwt template.

1. Open **tmp1.dwt**.

HINT:

When the Select File dialog box appears, you will need to select Drawing Template File (*.dwt) at the right of Files of type.

2. Create a new dimension style using the following information.
   - Name the dimension style **Preferred**. In the Create New Dimension Style dialog box, start with **Standard** and apply the new style to all dimensions.
   - In the **Primary Units** tab, set Unit format to **Architectural** and Precision to **0'-0"**.
   - In the **Symbols and Arrows** tab, set the dimension arrow size to 1/8". Pick the **Line** radio button in the Center marks area and set the size of the center marks to 1/16".
   - In the **Text** tab, create a new text style named **roms** using the **romans.shx** font and use the default values.
   - After you have closed the Text Style dialog box, make **roms** the current style for Preferred. Then set the text height at 1/8".
   - In the **Fit** tab, set Scale for dimension features to **24**.

You could make other changes to the dimension style, but let's stop here.

3. Pick **OK** and close the dialog boxes.

4. In the Dimension or Styles toolbar, pick the down arrow and select **Preferred**.

5. Save your work and close the file.

The tmp1.dwt file is now ready for use with other new drawing files. As you use template files, feel free to tailor them to your specific needs.

6. Close the Dimension toolbar and exit AutoCAD.

Review Questions

1. Describe the purpose of dimension styles. Why might you need to create one?

2. How do you determine the dimensioning scale?

3. Explain the difference between center marks and center lines. When should you use each?

4. Explain the use of the Continue Dimension button in the Dimension toolbar.

5. When you select a dimension and right-click, AutoCAD displays a menu. Name and briefly describe the three dimension-related options that are available to you.

6. When might you use a spline leader?

7. What is the advantage of using associative dimensions?

8. What is the fastest way to change the position of a dimension line?

Challenge Your Thinking

1. In this chapter, you placed object lines on one layer and dimensions on another. Discuss the advantages of placing dimensions on a separate layer. Are there any disadvantages to placing dimensions on a separate layer? Explain.

2. Describe the changes you would need to make to tmp1.dwt to create a template for a B-size sheet at the same drawing scale.

Chapter 27
Review & Activities continued

Applying AutoCAD Skills

Work the following problems to practice the commands and skills you learned in this chapter.

1. Begin a new drawing and establish the following drawing settings. Store as a template file. (Name it tmp2.dwt.)

   Units: Engineering
   Scale: 1″ = 10′ (or 1″ = 120″)
   Sheet size: 17″ × 11″
   Drawing Area: You determine the drawing area based on the scale and sheet size.
   Grid: 10′
   (Reminder: Be sure to enter ZOOM All.)
   Snap: 2′
   Layers: Set up as shown in Table 27-1.

Layer Name	Color	Linetype	Lineweight
Thick	red	Continuous	0.50 mm
Thin	green	Continuous	0.25 mm

Table 27-1

Create a new dimension style for the template using the following information.

- Name the dimension style D001. In the Create New Dimension Style dialog box, start with Standard and apply the new style to all dimensions.
- Set the unit format to Engineering and the precision to 0′-0.00″.
- Set the dimension arrow size to .125″.
- Use center marks and set their size to .1″.
- Create a new text style named arial using the Arial font. Use the default settings and make arial the current style for NewStyle. Set the text height at .125″.
- Set the dimension scale to the proper value.

Chapter 27: Advanced Dimensioning

2–3. Use the template file you created in problem 1 to create the hotel recreation areas shown in Figs. 27-3 and 27-4.

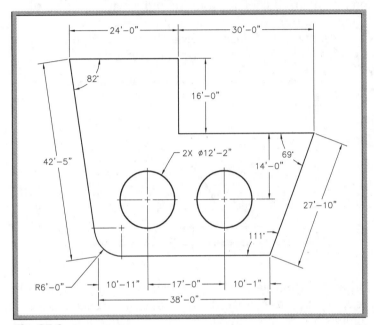

Fig. 27-3

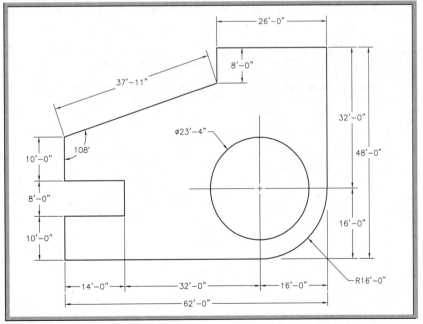

Fig. 27-4

4. Create and identify the radii of the sheet metal elbow shown in Fig. 27-5. First draw the perpendicular lines, and then create two arcs: one with a radius of 1.5 and the other with a radius of 3. Then dimension the 90° angle and add the notes. Before you create the leaders, open the Dimension Style Manager and change the arrowheads to dots. Add the text as shown, and save the drawing as elbow1.dwg.

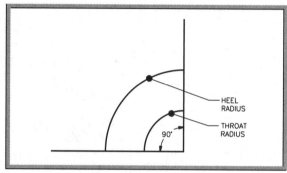

Fig. 27-5

5. Draw the front view of the shaft shown in Fig. 27-6. Use the Quick Setup method to change the drawing area to accommodate the dimensions of the drawing. Place the visible lines on a layer named Visible using the color green. Create a layer named Forces using the color red. Draw and label the four load and reaction forces, using leaders and a text style based on the romans.shx font. Create another layer named Dimensions using the color cyan. Dimension the front view, placing the dimensions on layer Dimensions.

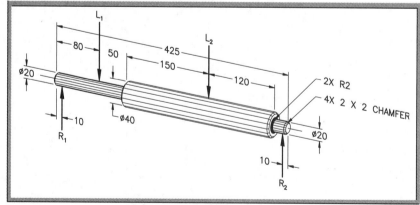

Fig. 27-6

Problem 5 courtesy of Gary J. Hordemann, Gonzaga University

6. Draw and dimension the front view of the metal casting shown in Fig. 27-7 on page 386. Use the following drawing and dimension settings, and place everything on the proper layer.

Units: Decimal with two digits to the right of the decimal point
Drawing Area: 11 × 8.5
Scale: 1:1
Grid: .1
Snap: .05
LTSCALE: Start with .5
Layers: Set up as shown in Table 27-2.

Layer Name	Color	Linetype	Lineweight
Visible	red	Continuous	0.50 mm
Dimensions	blue	Continuous	0.25 mm
Text	magenta	Continuous	0.30 mm
Center	green	CENTER	0.20 mm

Table 27-2

Create a new dimension style using the following information:
- Name the dimension style using a name of your choice. Choose a name that will help you remember the purpose of this dimension style. In the Create New Dimension Style dialog box, start with Standard and apply the new style to all dimensions.
- Set the unit format to Decimal and the precision to 0.00.
- Set the dimension arrow size to .125.
- Set the size of the center marks to .1 and use center lines.
- Create a new text style using a name of your choice. Use the romans.shx font, apply default settings, and make it the current style. Set the text height at .125.
- Set the dimension scale to the proper value.

Problem 6 courtesy of Gary J. Hordemann, Gonzaga University

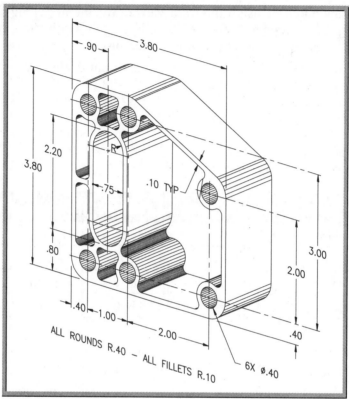

Fig. 27-7

7. Draw and dimension the front view of the steel plate shown in Fig. 27-8. Use the following drawing and dimension settings to set up the metric drawing, and place everything on the proper layers. Also draw and dimension the top and right-side views, assuming the plate and boss thicknesses to be 10 and 5 respectively.

Units: Decimal with no digits to the right of the decimal point
Drawing Area: 280 × 216
Scale: 1:1
Grid: 10
Snap: 5
LTSCALE: Start with 10
Layers: Set up as shown in Table 27-3.

Create a new dimension style using the following information:

• Name the dimension style using a name of your choice. In the Create New Dimension Style dialog box, start with Standard and apply the new style to all dimensions.

Layer Name	Color	Linetype	Lineweight
Visible	red	Continuous	0.60 mm
Dimensions	blue	Continuous	0.20 mm
Text	magenta	Continuous	0.20 mm
Center	green	CENTER	0.20 mm
Hidden	magenta	HIDDEN	0.30 mm

Table 27-3

- Set the unit format to Decimal and the precision to 0.
- Set the dimension arrow size to 3.
- Set the size of the center marks to 1.5 and use center lines.
- Create a new text style using a name of your choice. Use the romans.shx font, apply default settings, and make it the current style. Set the text height at 3.
- Set the dimension scale to the proper value.

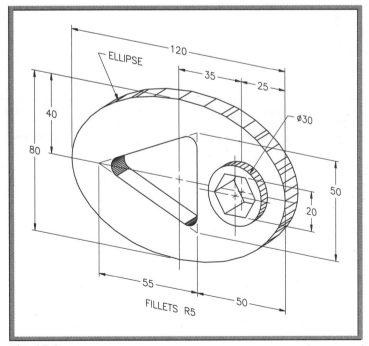

Fig. 27-8

Problem 7 courtesy of Gary J. Hordemann, Gonzaga University

Using Problem-Solving Skills

Work the following problems to practice the commands and skills you learned in this chapter.

1. Your architectural drafting trainee showed you the wall corner shown in Fig. 27-9. Notice that the trainee dimensioned the drawing in a mechanical drafting style. Redraw the corner and set the correct dimension size and units, change the arrowheads to architectural ticks (short diagonal lines), and move the dimensions above the dimension line to conform to your company's style. Save the drawing as framing1.dwg.

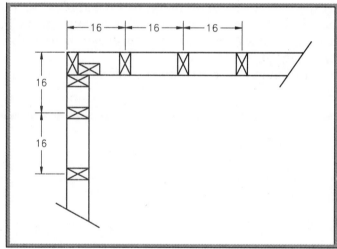

Fig. 27-9

2. As the set designer for a theatrical company, you are creating a
template for a cityscape. The director has given you a sketch of
the general building landscape she would like (Fig. 27-10). Draw
the template and dimension it from the datum plane. Each square
equals 9 inches. Save and plot the drawing using the Extents plotting
option.

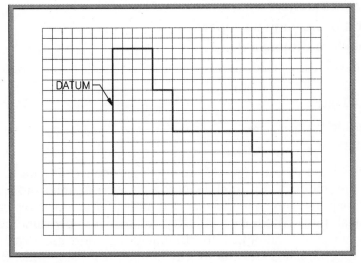

Fig. 27-10

Chapter 28 Fine-Tuning Dimensions

Objectives:

- Adjust the appearance of dimension lines, arrowheads, and text
- Control the placement of dimension text
- Adjust the format and precision of primary and alternate units
- Edit a dimension's properties
- Explode a dimension

Key Terms

alternate units
primary units

Most industries that use AutoCAD apply dimensioning differently. The architectural industry, for example, uses architectural units and fractions, whereas manufacturers of consumer and industrial products typically use decimal units. Within each industry, companies apply standards, some more rigidly than others. In manufacturing, most use a variation of ANSI or ISO standards and enforce the standard company-wide. Because thousands of companies use AutoCAD worldwide, the software must accommodate a wide range of industries and company standards. This chapter focuses on how you can fine-tune dimensions to meet a particular company's style or standard.

Dimension Style Options

AutoCAD offers a wealth of options for tailoring a dimension style to a specific need.

- Start AutoCAD and open **deck.dwg**.

- From the **File** pull-down menu, select **Save As...** and enter **deck2.dwg** for the file name.

Dimension

- Open the **Dimension** toolbar and pick the **Dimension Style** button.

The current dimension style is My Style, as shown in the upper left corner of the dialog box. In the last chapter, we covered the purpose of the New... and Modify... buttons. The Set Current button permits you to make another style current. The Override... button enables you to override one or more settings in a dimension style temporarily. The Compare... button allows you to compare the properties of two dimension styles or view all the properties of one style.

Lines

1. Pick the **Modify...** button and pick the **Lines** tab unless it is already in front.

Focus your attention on the Dimension lines area. Color, Linetype, and Lineweight permit you to change the color, linetype, and lineweight of dimension lines. In most cases, it is best not to change them. Extend beyond ticks specifies a distance to extend the dimension line past the extension line when you use oblique, architectural tick, integral, and no marks for arrowheads. Baseline spacing sets the spacing between the dimension lines when they're outside the extension lines. Dim line 1 suppresses the first dimension line, and Dim line 2 suppresses the second dimension line, as shown in Fig. 28-1.

2. Make changes to the settings in the Dimension lines area and notice how the sample drawing changes.

Focus on the area labeled Extension lines. Color and Lineweight allow you to change the color and lineweight of the extension lines. Extend beyond dim lines specifies a distance to extend the extension lines beyond the dimension line. Offset from origin specifies the distance to offset the extension lines from the origin points that define the dimension. Suppress prevents the display of extension lines. Fixed length extension lines allows you to specify a fixed length for all extension lines. Ext line 1 suppresses the first extension line, and Ext line 2 suppresses the second extension line. Suppressing an extension line is useful when you want to dimension to an object line. In this case, the object line would serve the same purpose as the extension line.

3. Make changes to the settings in the Extension lines area and notice how the sample drawing changes.

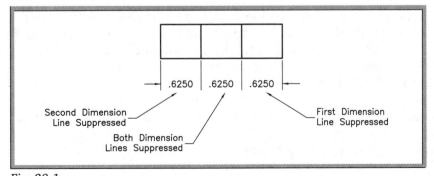

Fig. 28-1

Symbols and Arrows

1. Pick the **Symbols and Arrows** tab.

Concentrate on the area titled Arrowheads. First sets the arrowhead for the first end of the dimension line. Second sets the arrowhead for the second end of the dimension line. In most cases, you would use the same arrowhead style for both ends of a dimension, but AutoCAD gives you flexibility in case you need to make them different. Leader sets the arrowhead for leaders.

2. To the right of First, pick the down arrow and review the types of arrowheads available.

Some companies and individual architects use ticks instead of arrows. Oblique is a less bold variation of a tick. Integral is the name of a curved tick. It would be rare to use None, but AutoCAD gives you that option.

As you learned in the previous chapter, Arrow size displays and sets the size of the arrow. You also learned how to change the appearance of center marks in the Center marks area.

3. Make changes to the settings in the Arrowheads area and notice how the sample drawing changes.

Text

1. Pick the **Text** tab.

In the Text appearance area, you are familiar with Text style and Text height. Text color permits you to change the color of the dimension text. In most circumstances, it is best to leave it alone. Fraction height scale sets the scale of fractions relative to dimension text by multiplying the value entered here by the text height. Draw frame around text draws a frame around the dimension text. This option would be useful to companies that frame certain notes in a drawing.

2. Make changes to the settings in the Text appearance area and notice the changes in the sample drawing.

The Text placement area allows you to control the placement of text. Vertical controls the vertical justification of dimension text along the dimension line. Horizontal controls the horizontal justification of dimension text along the dimension line and the extension line. Offset from dim line sets the text gap (distance) around the dimension text when the dimension line is broken to accommodate the dimension text. If the space between the dimension text and the end of the dimension lines becomes too large or small, use this option to adjust the space. It is rare for a company or even a government agency to require a specific distance.

3. Make changes to the settings in the Text placement area and notice the changes in the sample drawing.

The Text alignment area permits you to change the orientation (horizontal or aligned) of dimension text, whether it is inside or outside the extension lines. Horizontal places text in a horizontal position. Aligned with dimension line aligns the text with the dimension line. ISO Standard aligns text with the dimension line when text is inside the extension lines. If the text is outside the extension lines, it aligns it horizontally.

4. Experiment with the text alignment options.

Fit

1. Pick the **Fit** tab.

Focus on the Fit options area. The fit options permit you to control the placement of dimension text and arrows when there is not enough space to place them inside the extension lines.

2. Select each of the fit options and notice how they affect the placement of dimension text and arrows.

The Text placement area allows you to control the placement of text when it's not in the default position.

3. Try each of the placement options.

As you discovered in Chapter 27, the Scale for dimension features area controls the scaling of dimensions. When you pick Scale dimension to layout, AutoCAD determines a scale factor based on the scaling between the current model space viewport and paper space.

In the Fine tuning area, if you check Place text manually, AutoCAD allows you to place the text at the position you specify. If you check Draw dim line between ext lines, AutoCAD draws dimension lines between the measured points even when the arrowheads are outside the extension lines.

4. Experiment with these two options.

Primary Units

The *primary units* are those that appear by default when you add dimensions to a drawing. The term *primary* is used by AutoCAD to distinguish them from alternate dimensions, which are discussed in the next section.

1. Pick the **Primary Units** tab.

This tab establishes the format and precision of the primary dimension units, as well as prefixes and suffixes for dimension text.

You were introduced to the Linear dimensions area in Chapter 27, so you know the purpose of Unit format and Precision. Fraction format controls the format for fractions. Decimal separator sets the separator for decimal formats. Options include a period (.), comma (,), or space ( ). Round off sets rounding rules for dimension measurements for all dimension types except angular. Prefix and Suffix permit you to enter text or use control codes to display special symbols such as the diameter symbol.

2. Experiment with the settings in the Linear dimensions area.

Under Measurement scale, Scale factor establishes a scale factor for all dimension types except angular.

Zero suppression controls the suppression of zeros. Some companies do not use a leading zero, for example. One widely accepted rule of thumb is to include the leading zero in drawings dimensioned using the metric system, but to exclude it when dimensioning using the English system (feet and inches). Space (or lack of it) for the dimension may also be a consideration.

A distance of 0.2000 becomes .2000 when you check Leading. A distance of 2.0000 becomes 2 when you check Trailing. A distance of 0'-2¼" becomes 2¼" when you check 0 feet, and a distance of 2'-0" becomes 2' when you check 0 inches.

Concentrate on the area labeled Angular dimensions. Units format sets the angular units format, and Precision sets the precision for the units. The Zero suppression area controls the suppression of zeros in angular dimensions.

3. Make changes to the settings in the Angular dimensions area.

Alternate Units

Alternate units are a second set of distances inside brackets in dimensions. The most common reason for using alternate units is to show both inches and millimeters, so millimeters are shown by default when you activate alternate units. However, the conversion factor is customizable, as you will see, so that you can change the alternate units to meet needs for specific drawings.

1. Pick the **Alternate Units** tab.

2. Pick the **Display alternate units** check box in the upper left corner.

This displays alternate units in the sample drawing. The alternate units, in millimeters, are in brackets.

Unit format and Precision in the Alternate units area are self-explanatory. Multiplier for alt units sets the multiplier to use as the conversion factor between primary and alternate units. Since an inch is equal to 25.4 millimeters, the default value is 25.4. Round distances to establishes the rounding rules for alternate units for all dimension types except angular. Prefix and Suffix permit you to include a prefix and suffix in the alternate dimension text.

3. Experiment with the settings in the Alternate units area.

The Zero suppression area is the same as the one for Primary units.

In the Placement area, After primary value places the alternate units after the primary units. Below primary value places alternate units below the primary units.

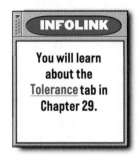

INFOLINK

You will learn about the Tolerance tab in Chapter 29.

4. Pick **Below primary value** to compare the difference between the two options.

5. Pick the **Cancel** button; pick the **Close** button to close the Dimension Style Manager dialog box.

Editing a Dimension

AutoCAD offers additional options for editing a dimension.

1. Pick the **Dimension Edit** button from the Dimension toolbar.

AutoCAD enters the DIMEDIT command and displays several options.

Dimension

Home moves dimension text to its default position. New allows you to change dimension text using the Multiline Text Editor. Rotate rotates dimension text. Oblique makes the extension lines of linear dimensions oblique instead of perpendicular to the dimension line.

2. Press **ESC** to end the command.

Changing a Dimension's Properties

1. Select any linear dimension.

2. Right-click and select **Properties** at the bottom of the shortcut menu.

AutoCAD displays the Properties dialog box.

3. If information is not displayed under the General tab, click the down arrow located to the right of General.

AutoCAD provides a list of the dimension's properties.

4. Change the color and lineweight to new values of your choice.

5. Open the Text tab by clicking the down arrows located to the right of Text.

This displays the settings associated with the dimension text.

6. Change the height to ¾₆″.

7. Open each of the other tabs, review their contents, and then close them.

You can also flip individual arrows on dimensions.

8. Select the 9′-10″ aligned dimension on the right side of the deck.

9. Rest the crosshairs on one of the arrows and right-click.

10. Pick **Flip Arrow** from the shortcut menu.

11. Repeat for the other arrow.

As you can see, you can easily custom-tailor the properties of an individual dimension. In most circumstances, all of the dimensions on a drawing should use a consistent style, but AutoCAD gives you the option of changing the individual elements of a particular dimension if necessary.

12. Close the dialog box.

Exploding a Dimension

1. Attempt to erase a single element of any dimension on the drawing, such as the dimension text or an extension line.

AutoCAD selects the entire dimension because it treats an associative dimension as a single object.

Modify

2. Press the **ESC** key and pick the **Explode** button from the docked Modify toolbar.

This enters the EXPLODE command.

3. Select any dimension and press **ENTER**.

AutoCAD explodes the dimension, enabling you to edit (move, erase, trim, etc.) the individual parts of the dimension.

4. Select the dimension.

As you can see, it is now possible to select individual parts of the dimension. Sometimes, it is necessary to explode a dimension to edit it because it is impossible to change it any other way. Use this method, however, as the last alternative because exploded dimensions lose their associativity and you cannot control their appearance using a dimension style.

5. Close the Dimension toolbar.

6. Exit AutoCAD without saving your changes.

Review Questions

1. Why is it important for AutoCAD to offer so many settings that control the appearance of dimensions?

2. What are alternate units?

3. Can you rotate dimension text? Describe a drawing situation in which you might need to do this.

4. What is the purpose of the New option of the DIMEDIT command?

5. How do you adjust the angle of a dimension's extension lines? Why might you choose to make this adjustment?

6. Will the Home option of the DIMEDIT command work with a dimension that is not an associative dimension? Explain.

7. What is the fastest way of reviewing and changing the properties of an individual dimension?

8. Under what circumstances would you want to suppress one or both dimension lines? Explain.

Challenge Your Thinking

1. A drawing of a complex automotive part has been completed with AutoCAD using the English inch system. Without redimensioning the drawing, how could you change it to show metric units only?

2. Contact two different companies in your area that use CAD and find out what their company dimensioning standards are. How do the two companies compare? How might any differences be explained?

Applying AutoCAD Skills

Work the following problems to practice the commands and skills you learned in this chapter.

1. Create a new drawing named prb28-1.dwg. Plan for a drawing scale of 1″ = 1″ and sheet size of 11″ × 17″. After you apply the following settings, create and dimension the drawing of an aircraft door hinge component shown in Fig. 28-2. In addition to saving prb28-1.dwg as a drawing file, save it as a template file.

Snap:	.25
Grid:	1
Layers:	Create layers to accommodate multiple colors, linetypes, and lineweights.
Linetype scale:	.5
Font:	romans.shx
Dimension style name:	Heather
Dimension scale:	1
Dimension text height:	.16
Arrowhead size:	.16
Center mark size:	.08
Mark with center lines?	Yes

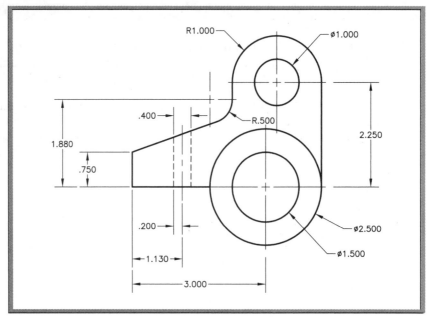

Fig. 28-2

2. Draw and dimension the front view of the oscillating follower shown in Fig. 28-3 on page 400. Use the drawing settings shown below. Draw the top and right-side views, assuming the follower to have a thickness of 8. This is a metric drawing.

Units: Decimal with no digits to the right of the decimal
Drawing area: 280 × 216
Grid: 10
Snap: 5
Linetype scale: Start with 10
Text style: Use the romans.shx font to create a new text style
                  and set the height at 3.
Layers: Set up layers as shown in Table 28-1.

Layer Name	Color	Linetype	Lineweight
Visible	red	Continuous	0.60 mm
Dimensions	blue	Continuous	0.20 mm
Text	magenta	Continuous	0.35 mm
Center	green	CENTER	0.20 mm
Hidden	magenta	HIDDEN	0.35 mm

Table 28-1

Create two dimension styles named Textin and Textout. Use Textin for those dimensions in which the text is inside the dimension lines, and Textout for the few radial dimensions in which the text should be outside the extension lines. Use the Standard settings except for the following:

Arrowhead size: 3
Text size: 3
Distance to extend extension line beyond dimension line: 1.5
Distance the extension lines are offset from origin points: 1.5
Distance around the dimension text: 1.5
Color of dimension text: magenta

Set the dimension style to Textin and Textout as appropriate for each of the two styles.

Problem 2 courtesy of Gary J. Hordemann, Gonzaga University

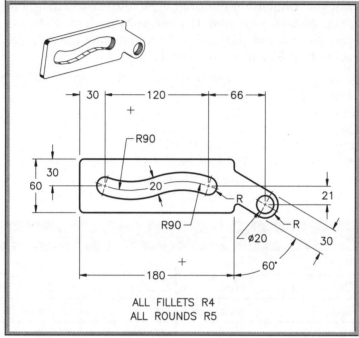

Fig. 28-3

3. Dimension the gasket as shown in Fig. 28-4. Set the arrowheads to Integral and the text font to Times New Roman. Display alternate units below the primary values and in the same unit format. Save the drawing as ch28gasket.dwg.

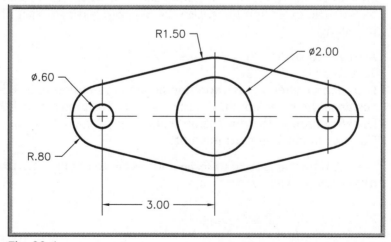

Fig. 28-4

4. Create a new drawing using the prb28-1.dwt template file and make the changes shown in Fig. 28-5. Note that the 2.250 diameter is now 2.500. Stretch the top part of the hinge component upward .250 unit and let the associative dimension text change on its own.

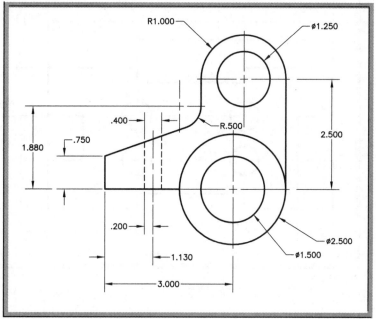

Fig. 28-5

 ## Using Problem-Solving Skills

Work the following problems to practice the commands and skills you learned in this chapter.

1. Architects are noted for their individuality. Open the elevation drawing (elev1.dwg) from Chapter 26. Redimension the elevation, including text, with the following style modifications. Save the drawing as ch28elev-rev1.dwg.

   Text font name: Trekker
   Arrowheads: Architectural tick
   Dimension Lines: Extend beyond ticks $^1/_8$
   Vertical Text Placement: Outside

2. Dimension the shim shown in Fig. 28-6. Set the font for Times New Roman. Apply Override as necessary. Save the drawing as ch28shim.dwg.

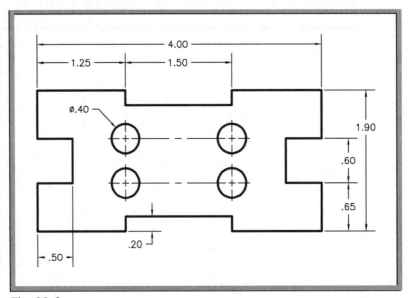

Fig. 28-6

Careers Using AutoCAD

Landscape Designer

Parks and outdoor recreation centers are perfect locations for picnics, Frisbee games, and other social activities. The next time you visit your local park, try to take notice of the surroundings: healthy green grass, immaculate flower beds, a ball diamond or two, a playground, and perhaps even a water fountain for kids to splash in. The design and placement of such items, all of which make up a visually appealing outdoor atmosphere, require strategic planning.

Working Together

Landscape designers typically work closely with a landscape architect. While architects supply practical and artistic views of how they want to arrange outdoor items, landscape designers possess the knowledge to create the actual plans for the design scheme. Before construction begins, the landscape designer must prepare detailed drawings for landscape plans that result in landscapes that are both attractive and economical.

Employment Opportunities

Employment opportunities for landscape designers are plentiful. Large development corporations, private sectors, and government organizations such as cities, counties, and even government agencies need people with specialized landscape design skills. Landscape designers create plans for golf courses, sports facilities, and college campuses. They also design drainage systems, outdoor lighting systems, and irrigation systems for farmlands, estates, and airports.

Tim Fuller Photography

All landscape designers must be able to produce accurate drawings based on conceptualizations and rough sketches provided by architects. They should also be prepared to assist the landscape architect with the actual design work. It is important for landscape designers to understand mathematical and measurement concepts, since the plans must be drawn to scale.

Career Activities

- Interview someone who can tell you about employment of landscape designers in your area. Are apprenticeships available?
- Has employment in this field been steady over the past few years? What are the forecasts for future employment?
- Find out more about the educational requirements for landscape designers. Do you think they should also have some knowledge of ecology and science? Why or why not?

Chapter 29 Tolerancing

Objectives:

- Apply the symmetrical, deviation, limits, and basic methods of tolerancing
- Insert surface controls on a drawing
- Create geometric characteristic symbols and feature control frames that follow industry standard practices for geometric dimensioning and tolerancing
- Add material condition symbols, datum references, data identifiers, and projected tolerance zones to drawings in accordance with industry standards

Key Terms

basic dimension
deviation tolerancing
feature control frames
geometric characteristic
   symbols
geometric dimensioning
   and tolerancing (GD&T)
limits tolerancing

material condition symbols
Maximum Material
   Condition (MMC)
projected tolerance zone
surface controls
symmetrical deviation
symmetrical tolerancing
tolerance

AutoCAD permits you to add tolerances to dimensions. A *tolerance* specifies the largest variation allowable for a given dimension. Tolerances are necessary on drawings that will be used to manufacture parts because some variation is normal in the manufacturing process.

This chapter applies AutoCAD's tolerancing capabilities and provides a basic introduction to *geometric dimensioning and tolerancing (GD&T)*. This is a system by which drafters and engineers define tolerances for the location of features (such as holes) of a part. The guidelines in this chapter reflect the standards that have been adopted by the American National Standards Institute (ANSI).

Methods of Tolerancing

Tolerances can be shown in more than one way, so AutoCAD provides many options for different kinds and styles of tolerances. The method used depends on the individual drawing. In some cases, you can place the tolerance information in a note or title block. However, when many different tolerances apply to different parts of the drawing, you may need to specify them individually with the dimension information.

1. Start AutoCAD and use the **dimen.dwt** template file to create a new drawing.

2. Display the **Dimension** toolbar.

Dimension

3. Pick the **Dimension Style** button from the toolbar and pick the **New...** button.

4. Create a new dimension style named **toler**. Start with **Standard** and use it for all dimensions.

5. Pick the **Text** tab and change the text height to **0.125**.

6. Pick the **Primary Units** tab and set Precision to **0.000**.

7. Pick the **OK** button; pick the **Close** button.

8. Make **toler** the current dimension style.

9. Save your work in a file named **toler.dwg**.

10. Pick the **Dimension Style** button from the toolbar and pick the **Modify...** button.

11. Pick the **Tolerances** tab to display the tolerance options.

Focus your attention on the area titled Tolerance format. Notice that many of the settings are grayed out and not available. This is because Method is set to None. Method refers to the approach used to calculate the tolerance. The four methods that AutoCAD makes available, along with an example, are shown in Fig. 29-1.

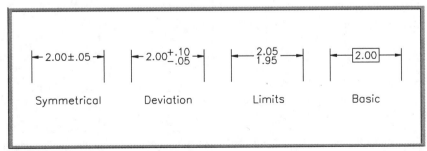

Fig. 29-1

The Symmetrical option permits you to create a plus/minus expression tolerance. AutoCAD applies a single value of variation to the dimension measurement. In other words, the upper and lower tolerances are equal. Deviation allows you to add different values for the upper and lower tolerance for cases in which the two are not equal. Limits adds a limit dimension with a maximum and a minimum value—one on top of the other. Basic creates a plain dimension inside a box. A *basic dimension* is one to which allowances and tolerances are added to obtain limits.

12. After Method, pick each of the options and notice how the sample drawing changes.

Precision sets the number of decimal places for tolerances. Precision varies with the tolerance precision and the numbering system you use. If you are using the Imperial system (inches, feet, etc.), the basic dimension should have the same number of decimal places as the tolerance. For example, a 1.5-inch hole with a tolerance of ±.005 should be written as 1.500 ±.005. Note that leading zeros (zeros to the left of the decimal point) are often not used with Imperial units.

Upper value sets the maximum or upper tolerance value, and Lower value sets the minimum or lower value. Scaling for height sets the current height for tolerance text. Vertical position controls text justification for symmetrical and deviation tolerances. Zero suppression works the same way as it does for primary and alternate units.

The settings in the Alternate unit tolerance area sets the precision and zero suppression rules for alternate tolerance units. Zero suppression works the same way for tolerancing as it does for primary and alternate units.

13. Experiment with each of the options in the Tolerance tab and notice how their settings affect the sample drawing.

14. Pick the **Cancel** button to clear all changes that you made.

Symmetrical Tolerancing

In *symmetrical tolerancing* (also called *symmetrical deviation*), the dimension specifies a plus/minus (±) tolerance in which the upper and lower limits are equal.

1. Pick the **Modify...** button and pick the **Tolerances** tab.

2. Set Method to **Symmetrical** and Precision to **0.000**.

3. Set Upper value to **0.005**.

Lower value is not available because it is the same as the upper value by definition in symmetrical tolerancing.

4. Under Zero suppression, check both **Leading** and **Trailing**, and do not change anything else.

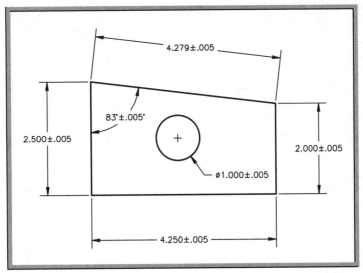

Fig. 29-2

5. Pick **OK**; pick **Close**.

Tolerances do not appear because the toler dimension style was created after the drawing was completed, and toler has not yet been applied to the drawing.

6. Select the drawing and dimensions.

7. Make **toler** the current dimension style and press **ESC** to remove the grips.

Your drawing should look similar to the one in Fig. 29-2.

Deviation Tolerancing

Sometimes the upper and lower tolerances for an object are unequal. AutoCAD's *deviation tolerancing* method allows you to set the upper and lower values separately.

1. Open the **Dimension Style Manager** and display the **Tolerances** tab of the Modify Dimension Style dialog box.

2. Change Method to **Deviation**, set Lower value to **0.007**, and leave everything else the same. Pick **OK**.

3. Close the Dimension Style Manager and notice how the drawing changes.

Your drawing should now look similar to the one in Fig. 29-3 on page 408.

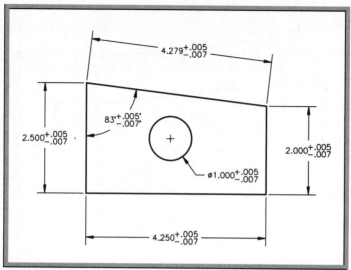

Fig. 29-3

Limits Tolerancing

In *limits tolerancing*, only the upper and lower limits of variation for a dimension are shown. The basic dimension is not shown.

1. Open the **Dimension Style Manager** and display the **Tolerances** tab in the Modify Dimension Style dialog box.

2. Change Method to **Limits** and pick **OK**.

3. Close the Dimension Style Manager and notice how the drawing changes.

Your drawing should now look similar to the one in Fig. 29-4.

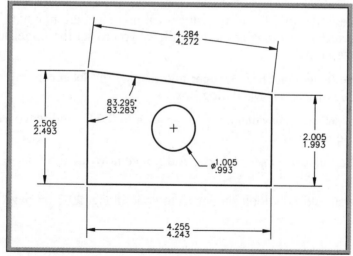

Fig. 29-4

When you select the Basic method, AutoCAD draws a box around the dimension. As you may recall, a basic dimension is one to which allowances and tolerances are added to obtain limits. Therefore, you should use the Basic method to present theoretically exact dimensions.

1. In the Dimension Style Manager, display the **Tolerances** tab of the Modify Dimension Style dialog box.

2. Change Method to **Basic** and pick **OK**.

3. Close the Dimension Style Manager and notice how the drawing changes.

GD&T Practices

With AutoCAD, you can create geometric characteristic symbols and feature control frames that follow industry standard practices for geometric dimensioning and tolerancing (GD&T). *Geometric characteristic symbols* are used to specify form and position tolerances on drawings. *Feature control frames* are the frames used to hold geometric characteristic symbols and their corresponding tolerances.

Surface Controls

When a feature control frame is not associated with a specific dimension, it usually refers to a surface specification regardless of feature size. For this reason, such frames are often known as *surface controls*.

In Chapter 27, you created simple and moderately complex leaders using the QLEADER command. You can also use QLEADER to create surface controls.

1. Create a new layer named **GDT** and assign the color magenta to it. Make it the current layer and freeze layer Dimensions.

2. Pick the **Quick Leader** button from the Dimension toolbar and press **ENTER** to accept the Settings default.

Dimension

This displays the Leader Settings dialog box.

3. Pick the **Annotation** tab, **Tolerance** radio button, and **OK**.

4. In reply to Specify first leader point, pick the midpoint of the inclined line.

HINT: ───────────────────────────────

Use the Midpoint object snap.

5. Pick a second point about **1** unit up and to the right of the first point.

6. Press **ENTER**.

The Geometric Tolerance dialog box appears, as shown in Fig. 29-5.

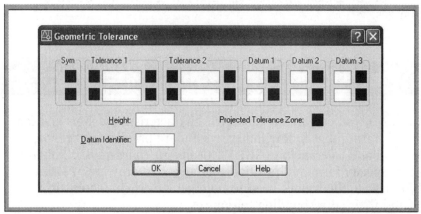

Fig. 29-5

This dialog box permits you to create complete feature control frames by adding tolerance values and their modifying symbols. Focus your attention on the area labeled Sym.

7. Pick the top black box under Sym.

This displays the Symbol dialog box, as shown in Fig. 29-6.

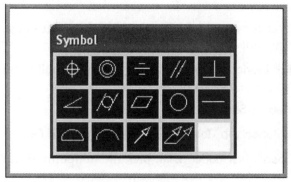

Fig. 29-6

The dialog box contains 14 geometric symbols that are commonly used in GD&T.

INFOLINK

See Appendix D for more information about geometric symbols.

8. Pick the flatness symbol (the third symbol on the second row).

AutoCAD adds the symbol to the black box you picked under Sym.

9. Under Tolerance 1, in the top white box, type **.020** for the tolerance value.

10. Pick **OK**.

The feature control frame appears as a part of the leader, as shown in Fig. 29-7. The feature control frame specifies that every point on the surface must lie between two parallel planes .020 apart.

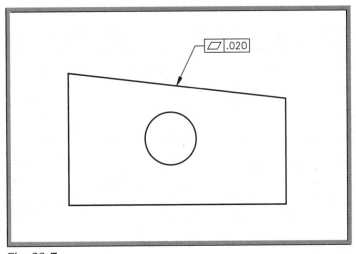

Fig. 29-7

Geometric Characteristic Symbols

To associate geometric symbols with specific dimensions, you must first create the dimension using the appropriate dimensioning command. Then you use the TOLERANCE command to add the tolerancing information.

Dimension

1. Pick the **Diameter Dimension** button from the toolbar and pick a point anywhere on the hole.

2. Pick a second point to position the dimension as shown in Fig. 29-8 on page 412.

Let's change the method of tolerance from Basic to Limits.

3. Select the new dimension and pick the **Properties** button from the Standard toolbar.

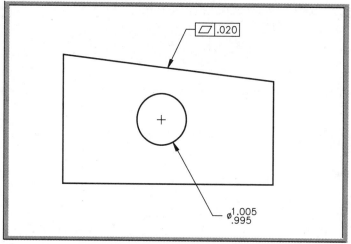

Fig. 29-8

4. In the Properties dialog box, click the down arrows at the right of Tolerances, change Tolerance display from **Basic** to **Limits**, and close the Properties dialog box.

5. Press **ESC** to remove the selection.

Your drawing should now look similar to the one in Fig. 29-8.

6. Pick the **Tolerance** button from the Dimension toolbar.

Dimension

This enters the TOLERANCE command and displays the Geometric Tolerance dialog box.

7. Under Sym, pick the top black box and pick the true position symbol located in the upper left corner of the Symbol box.

8. Under Tolerance 1, pick the black box located in the upper left corner.

The diameter symbol appears.

9. Pick it again, and again, causing it to disappear and reappear.

As you can see, this serves as a switch, toggling the diameter symbol on and off.

10. In the top white box under Tolerance 1, type **.010** for the tolerance value.

Material Condition Symbols

The black box located at the right of the white text box specifies material condition. *Material condition symbols* are used in GD&T to modify the geometric tolerance in relation to the produced size or location of the feature. In this case, we will use the symbol for *Maximum Material Condition (MMC)*. MMC specifies that a feature, such as a hole or shaft, is at its maximum size or contains its maximum amount of material.

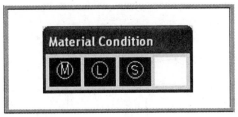

Fig. 29-9

11. At the right of the same white box, pick the black box.

This displays the Material Condition dialog box, as shown in Fig. 29-9.

12. Pick the first symbol, which is the symbol for Maximum Material Condition.

The symbol appears in the Geometric Tolerance dialog box.

Datum References and Datum Identifiers

AutoCAD allows you to specify datum reference information on three different levels (primary, secondary, and tertiary).

13. In the top white boxes under Datum 1, Datum 2, and Datum 3, enter **A**, **B**, and **C**, respectively.

You can also associate a material condition with any or all of the datum references. To do this, you would pick the black box to the right of the datum boxes and select the appropriate material condition symbol. For the current drawing, do not use material condition symbols.

Datum Identifier near the bottom left of the dialog box allows you to enter a feature symbol to identify the datum.

14. In the box located to the right of Datum Identifier, enter **-D-**.

The Geometric Tolerance dialog box on your screen should now look like the one in Fig. 29-10.

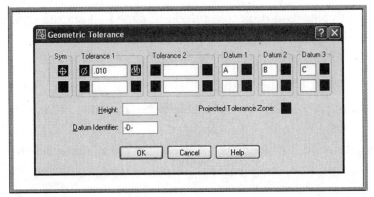

Fig. 29-10

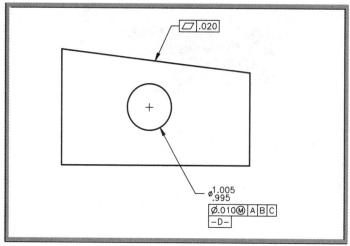

Fig. 29-11

The black and white boxes in the second row permit you to create a second feature control frame. You would need to use a second feature control frame when the feature has two geometric characteristics that are of dimensional importance.

15. Pick the **OK** button and position the feature control frame as shown in Fig. 29-11.

16. Save your work.

Projected Tolerance Zone

You can specify projected tolerances in addition to positional tolerances to make a tolerance more specific. The *projected tolerance zone* controls the height of the extended portion of a perpendicular part. To display a projected tolerance, you would pick the black box next to Projected Tolerance Zone in the Geometric Tolerance dialog box. Then, under the Tolerance 1 area of the dialog box, enter a value for the height. Doing so creates a projected tolerance zone in the feature control frame.

Experiment with placing a projected tolerance zone in a feature control frame by following the steps below.

1. Enter the **UNDO** command and **Mark** option.

2. Create another feature control frame.

3. Pick the black box next to Projected Tolerance Zone and enter a height in the Height box.

4. Place a dimension to see how the projected tolerance appears.

5. Enter the **UNDO** command and specify the **Back** option.

6. Close the Dimension toolbar and exit AutoCAD.

Review Questions

1. What is the purpose of adding tolerances to a drawing?

2. When is it necessary to include tolerances on a drawing?

3. Give an example for each of the following tolerancing methods.
    a. Symmetrical
    b. Deviation
    c. Limits

4. Which tolerancing method should you choose in AutoCAD if the upper tolerance value of a dimension is different from the lower value?

5. What is a surface control? How can you create a surface control in AutoCAD?

6. How is the TOLERANCE command similar to the QLEADER command?

7. What is a geometric characteristic symbol?

8. Explain how you would include material condition symbols in feature control frames.

9. How would you include a second feature control frame below the first one?

Challenge Your Thinking

1. With AutoCAD, you can quickly produce feature control frames, complete with geometric characteristic symbols. Investigate ways of editing them.

2. When a second feature control frame is added to a dimension, how does this affect the meaning of the first feature control frame?

3. Investigate material condition symbols and their meanings. In what way do material conditions affect dimensions?

Applying AutoCAD Skills

Work the following problems to practice the commands and skills you learned in this chapter.

1. Create the two-view drawing of the plug according to the dimensions shown in Fig. 29-12A. Use a drawing area of 100,80, and name the drawing plug.dwg. Dimension it as shown in Fig. 29-12B to show the runout of the plug. (*Runout* is the form and location of a feature relative to a datum.) Notice that the limits are different on the two holes.

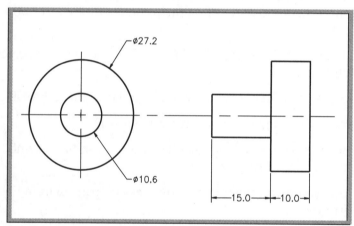

Fig. 29-12A

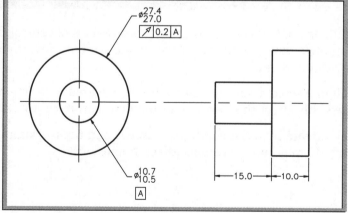

Fig. 29-12B

2. Create a new drawing named ch29gasket.dwg and draw the gasket shown in Fig. 29-13. Dimension the gasket using limits tolerancing, and specify an upper value of .02. Create a feature control frame to show a true position diameter tolerance of .025 at maximum material condition.

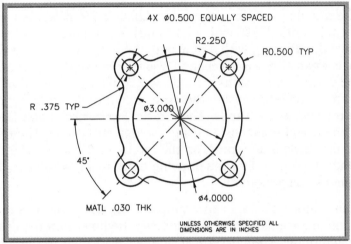

Fig. 29-13

3. Create a new drawing named hinge.dwg and draw the aircraft door hinge component shown in Fig. 29-14. Dimension the hinge using symmetrical tolerancing and specify tolerances as shown.

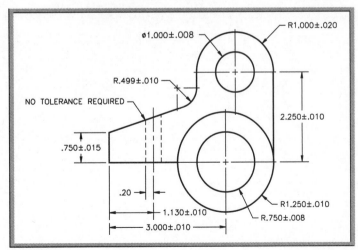

Fig. 29-14

Using Problem-Solving Skills

Complete the following activities using problem-solving skills and your knowledge of AutoCAD.

1. The drawing you dimensioned in Chapter 26, slide2.dwg, will be used in a new CAD textbook to illustrate dimensioning styles. Save the drawing with a new name of slide3.dwg. Create two new layers named Symmetrical and Deviation. Freeze the layer on which the current dimensions were drawn. Make the Symmetrical layer current and redimension the drawing with tolerances expressed in the symmetrical style. Then freeze Symmetrical, make Deviation current, and redimension using the deviation style. The tolerance on the angle is ±15 units, on the radius +0.02° and −0.00°, and all others ±.05 unit. Plot the drawing twice to show the two different dimensioning methods.

2. Your editor also wants an example of GD&T with limits dimensioning. Save slide3.dwg as slide4.dwg. Redimension the drawing to maintain the 3.0-unit vertical side perpendicular to the 5.5-unit base within 0.02 unit. Save the drawing.

Chapter 30 Calculations

Objectives:

- Find the exact coordinates of points
- Calculate the distance between specific points in a drawing
- Calculate area and circumference of objects in a drawing
- Use AutoCAD's on-line geometry calculator to place objects at precise points in a drawing
- Use **QuickCalc** to perform arithmetic, algebraic, and trigonometric calculations
- Display information from AutoCAD's drawing database on objects and entire drawings
- Divide an object into equal parts
- Place markers or points at specified intervals on an object

Key Terms

area	integer
circumference	perimeter
delta	vector
drawing database	

With the power of the computer, AutoCAD can perform measurement and calculation tasks that would take significant time to do manually. An example is calculating miles of a chain-link fence on a drawing.

The drawing in Fig. 30-1 on page 420 shows an apartment complex with parking lots, streets, and trees. With AutoCAD, you can calculate the square footage of the parking lot and the distance between parking stalls on such a drawing.

Performing Calculations

AutoCAD includes commands that help you find the coordinates of specific points on a drawing, calculate the distance between two points, calculate the area of an object, and perform other geometric calculations.

1. Start AutoCAD and start a new drawing using the **dimen.dwt** template file.

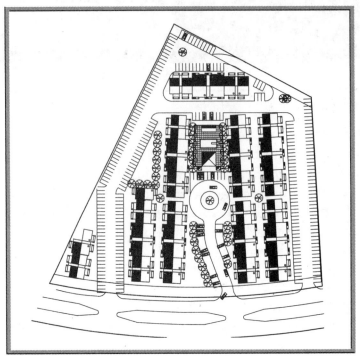

Fig. 30-1

2. Erase the drawing and dimensions.

3. Create the drawing of the end view of a shaft with a square pocket machined into it (Fig. 30-2). Use the following guidelines:
   - Create the end view on the layer named Objects.
   - Use the sizes shown, but omit dimensions.
   - The shaft and pocket share the same center point.
   - Use **RECTANG** to create the square pocket.

4. Save your work in a file named **calc.dwg**.

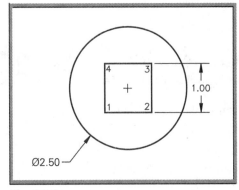

Fig. 30-2

Locating Points

Let's find the coordinates of point 1.

Inquiry

1. Open the **Inquiry** toolbar and pick the **Locate Point** button.

AutoCAD enters the ID command.

2. In reply to Specify point, pick point 1.

AutoCAD displays the coordinates of point 1.

3. Determine the coordinates of point 2.

Calculating Distances

The DIST command calculates the distance between two points.

Inquiry

1. From the same toolbar, select the **Distance** button.

AutoCAD enters the DIST command.

2. Pick points 1 and 3.

In addition to the distance, AutoCAD calculates angle and delta information. If you have studied geometry, you may be familiar with the use of the term *delta* to mean "change." Delta X is the distance, or change, in the X direction, and Delta Y is the distance in the Y direction from one point to the other.

Calculating Area

The AREA command determines the area and perimeter of several different object types. *Area* is the number of square units (inches, acres, miles, etc.) needed to cover an enclosed two-dimensional shape or surface. *Perimeter* is the distance around a two-dimensional shape or surface.

Inquiry

1. Select the **Area** button from the Inquiry toolbar.

AutoCAD enters the AREA command.

2. Enter **O** for Object and pick the square pocket.

AutoCAD displays the area and perimeter of the pocket. Suppose we want to know the area of the end of the shaft minus the pocket.

3. Reenter **AREA** and enter **A** for Add.

This puts the command in Add mode.

4. Enter **O** for Object and pick the shaft.

AutoCAD displays 4.9087 for the area and 7.8540 for the circumference. (*Circumference* is the distance around a circle.) Notice that AutoCAD is still in Add mode and is asking you to select objects.

5. Press **ENTER**.

6. Enter **S** for Subtract.

Now the command is in Subtract mode.

7. Enter **O** for Object and select the pocket.

AutoCAD subtracts the area of the pocket from the area of the end of the shaft and displays the result (3.9087).

8. Press **ENTER** twice to terminate the command.

Using the On-Line Geometry Calculator

With the CAL command, AutoCAD offers an on-line geometry calculator that evaluates vector, real, and integer expressions. A *vector* is a line segment defined by its endpoints or by a starting point and a direction in 3D space. An *integer* is any positive or negative whole number or 0. You can use the expressions in any AutoCAD command that requests points, vectors, or numbers.

1. Enter the **CAL** command.

2. Type **(3*2)+(10/5)** at the keyboard and press **ENTER**.

As in basic algebra, the parts in parentheses are calculated first. The asterisk (*) means to multiply and the forward slash means to divide the numbers.

AutoCAD calculates the answer as 8.0.

You can also use CAL with object snaps to calculate the exact placement of a point. For example, suppose you wanted to add a feature to the shaft that required starting a polyline halfway between the center of the shaft and the lower left corner of the machined pocket.

3. Enter the **PLINE** command.

4. Enter **'CAL**. (Note the leading apostrophe.)

By preceding the command with an apostrophe, you have entered the command transparently. In other words, when you complete the CAL command, the PLINE command will resume.

5. Type **(cen+end)/2** and press **ENTER**.

6. Pick any point on the shaft.

7. Pick either line near point 1.

AutoCAD calculates the midpoint between the shaft's center and point 1 and places the first point of the polyline at that location.

8. Press **ENTER** to terminate the PLINE command.

9. Undo the polyline and save your work.

Using QuickCalc

The QUICKCALC command allows you to perform a full range of mathematical and trigonometric calculations, as well as determine graphic information such as the location of a point or the length of a line.

Fig. 30-3

1. Open the **deck.dwg** file.

2. From the **Tools** pull-down menu, pick **QuickCalc**, or type **QUICKCALC** at the keyboard.

This displays the QuickCalc window, as shown in Fig. 30-3. The buttons across the top of the window allow you to find the coordinates of a point, the distance between two points, the angle of a line, and the intersection point of two lines.

3. Pick the **Get Coordinates** button and select the upper right corner of the deck.

The x,y,z coordinates of the point are displayed.

4. Pick the **Clear** button.

This clears the display of the calculator.

5. Pick the **Distance Between Two Points** button.

6. Pick the upper left and lower right corners of the deck.

The distance between the two corners of the deck is displayed (14'-6-3/8").

7. Pick the **Clear** button.

Now focus your attention on the QuickCalc Number Pad.

NOTE:

You may have to drag the bottom line of the calculator down to enlarge the window so that you can see the entire calculator.

8. Experiment with the calculator by entering mathematical equations, just as you would with an ordinary calculator.

Notice the equations and solutions are stored in a history list at the top of the QuickCalc window. More functions are located below the Number Pad in the Scientific area.

9. Pick the arrow to the right of Scientific and review the trigonometric and algebraic functions available there.

Now focus your attention on the Units Conversion area. QuickCalc allows you to obtain values for different units of measurement.

10. Pick the arrow to the right of Units Conversion to display the Units Conversion area.

11. Convert an area of 15 square feet to square meters.

HINT:

To access the Square feet and Square meters conversion options, you must first select Area in the Units type dropdown box.

In the Variables section, you can use existing values and functions or create your own for use in calculations.

12. Close the QuickCalc window and close the deck.dwg file without saving.

Displaying Database Information

AutoCAD maintains an internal database on every drawing. The *drawing database* contains information about the types of objects in the drawing, defined layers, and the current space (model or paper). It also contains specific numerical information about individual objects and their placement in the drawing. This information can be useful to drafters and programmers using AutoCAD for a variety of tasks.

Listing Information About Objects

1. Pick the **List** button from the Inquiry toolbar.

2. Pick any point on the shaft and press **ENTER**.

The AutoCAD Text window displays the object type, layer, space, center point, radius, circumference, and area.

3. Close the window.

4. Select the shaft.

5. Right-click and pick **Properties** from the shortcut menu.

AutoCAD displays most of the same information, but does not give the space. The benefit of this list is that you can make changes to the object, such as changing its circumference and area.

6. Pick the text field to the right of Area.

Notice the calculator button that appears on the right side of the Area line. This provides you with another way to open the QuickCalc window.

7. Pick the calculator button to open the QuickCalc window.

8. In the Properties window, change the area of the shaft to **3**.

AutoCAD redraws the shaft with an area of 3 units.

9. Close the Properties and QuickCalc windows and undo Step 8.

Editing Objects Mathematically

You can apply the same power that AutoCAD uses to calculate coordinates, areas, and circumferences to edit objects with mathematical precision.

Dividing an Object into Equal Parts

The DIVIDE command divides an object, such as the end of the shaft, into a specified number of equal parts.

1. From the **Draw** pull-down menu, pick **Point** and **Divide**.

This enters the DIVIDE command.

2. Select the shaft.

3. Enter **20** for the number of segments.

It appears as though nothing happened. Something did happen: the DIVIDE command divided the end of the shaft into 20 equal parts using 20 points; you just can't see them. Here's how to use them.

4. Turn off snap and object snap and enter the **LINE** command.

5. Enter the **Node** object snap. (Node is used to snap to the nearest point.)

6. Move the crosshairs along the shaft and snap to one of the nodes.

7. Snap to the center of the shaft.

8. Enter the **Node** object snap again and snap to another point on the shaft.

9. Terminate the **LINE** command.

10. Save your work.

Displaying the Points

AutoCAD's DDPTYPE command is used to control the appearance of points. Let's use it to make the points on the shaft visible.

1. Enter **DDPTYPE**.

AutoCAD displays the dialog box shown in Fig. 30-4.

2. Pick the first point style in the second row—the one with a circle and a point at its center—and pick **OK**.

AutoCAD places a small circle at each of the 20 equally spaced points as shown in Fig. 30-5.

3. Display the same dialog box (press the spacebar).

4. Experiment with other point styles and adjust the point size.

5. Display the **Point Style** dialog box again and set the style to the last one in the top row.

6. Pick **OK** to close the dialog box.

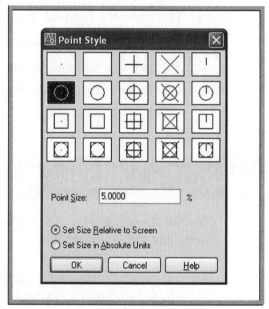

Fig. 30-4

Placing Markers at Specified Intervals

The MEASURE command is similar to DIVIDE except that MEASURE does not divide the object into a given number of equal parts. Instead, the MEASURE command allows you to place markers along the object at specified intervals.

1. From the **Draw** pull-down menu, pick **Point** and **Measure**.

This enters the MEASURE command.

2. Select one of the two lines.

3. Enter **.3** unit.

AutoCAD adds points spaced .3 unit apart.

4. Further experiment with MEASURE.

5. Set the point style to a single dot.

6. Close the Inquiry toolbar, save your work, and exit AutoCAD.

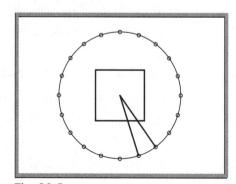

Fig. 30-5

Review Questions

1. Which AutoCAD commands are used to find coordinate points?

2. What information is produced with the AREA command?

3. What information is produced with the LIST command?

4. How do you calculate the perimeter of a polygon?

5. How do you find the circumference of a circle?

6. Explain how you control the appearance of points.

7. Explain the difference between the DIVIDE and MEASURE commands. Under what conditions would you use each of these commands?

Challenge Your Thinking

1. Describe one drafting situation in which you might need to use the DIVIDE command and one in which you would prefer to use the MEASURE command.

2. Investigate the difference between AutoCAD's points created when you use the POINT command and those created when you use the DIVIDE and MEASURE commands. Write a paragraph comparing and contrasting them.

Applying AutoCAD Skills

Work the following problems to practice the commands and skills you learned in this chapter.

Draw the views of the fasteners in Figs. 30-6 through 30-8 at the sizes indicated. Omit text and dimensions. Use the appropriate commands to find the information requested. Write their values on a separate sheet of paper. If you have access to the Instructor Resource CD, refer to the files calcprb1.dwg, calcprb2.dwg, and calcprb3.dwg. These files correspond to Figs. 30-6, 30-7, and 30-8, respectively.

1. Draw the top view of the Phillips-head screw shown in Fig. 30-6 according to the dimensions shown. Then find the following information:

   - location of point A
   - distance between points A and B
   - area of the polygon
   - perimeter of the polygon

2. Draw the top view of the nut shown in Fig. 30-7 according to the dimensions shown. Then find the following information:

   - distance between A and B
   - distance between B and C
   - area of hole
   - circumference of hole
   - area of nut
   - perimeter of nut

3. What information does AutoCAD list for arc A on the screw shown in Fig. 30-8?

4. List the information AutoCAD provides for line B on the screw.

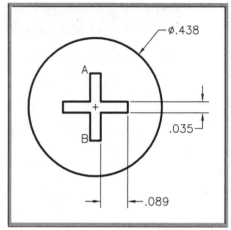

Fig. 30-6

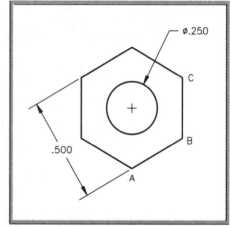

Fig. 30-7

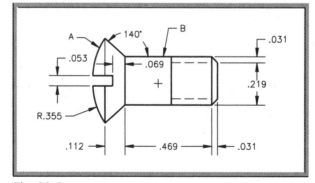

Fig. 30-8

5. On the screw shown in Fig. 30-8, divide line B into five equal parts. Make the points visible.

6. On arc A in Fig. 30-8, place markers along the arc at intervals of .02 unit. If the markers are invisible, make them visible.

7. Create the floor plan shown in Fig. 30-9 according to the dimensions given. Then find the information requested below the floor plan. This floor plan is also available on the Instructor Resource CD as flplan.dwg.

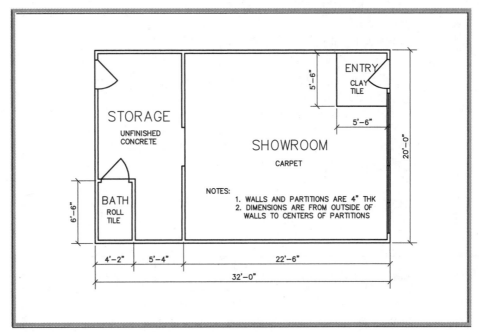

Fig. 30-9

Find the following information in square feet:

• area of showroom carpet
• area of entry clay tile
• area of bathroom roll tile

(continued on page 430)

Problem 7 courtesy of Mark Schwendau, Kishwaukee College

As you may know from your math courses, 1 square yard contains 9 square feet. Calculate the area of the carpet, entry clay tile, and bathroom roll tile as square yards using the Units Conversion area of the QuickCalc window.

- square yards of carpet
- square yards of clay tile
- square yards of roll tile

Calculate the distance between opposite corners of the following areas.

- showroom
- entry area
- bathroom

8. Create the top view of the nut shown in Fig. 30-10 according to the dimensions given. Then find the information requested below the illustration. This drawing is also available on the Instructor Resource CD as 1-8UNC2B.dwg.

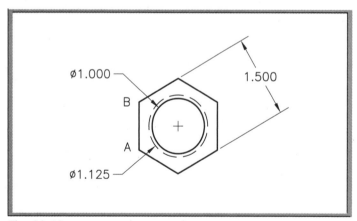

Fig. 30-10

Find the following information:

- distance between A and B
- area of the minor diameter of the thread
- area of the major diameter of the thread
- circumference of the minor diameter
- circumference of the major diameter
- area of top surface of the nut

Chapter 30
Review & Activities

- area of top surface of the nut minus the minor diameter
- area of top surface of the nut minus the major diameter
- Calculate the average of the last two answers using the QUICKCALC command.

Using Problem-Solving Skills

Complete the following activities using problem-solving skills and your knowledge of AutoCAD.

1. Create a new drawing using the Quick Setup wizard. Specify a drawing area of 300′ × 200′. Draw two horizontal lines of different lengths. Place one near the top of the drawing area and the other near the bottom. Then connect their ends to form a rectangular area.

2. The area you created in problem 1 represents a real estate plot. As a paralegal for an attorney, you have been asked to divide this plot into five equal lots. Determine how to do this and carry out your plan. (*Hint:* consider using the DIVIDE command and the Node object snap.) To avoid future conflict, the attorney also wants you to verify the size of each lot. Determine the area and perimeter of each lot, and prepare a statement for the attorney explaining why the areas are the same, but the perimeters are not. Save the drawing as realestate.dwg.

Part 6 Project

Designer Eyewear

Part 6 reviewed the extensive dimensioning and calculating abilities of AutoCAD. In this exercise, you will be asked to design and draw an object and then to dimension it as specified. To review, traditional dimensions generally inform of size and location. Tolerance dimensions also yield information about characteristics or features of the item (*e.g.,* How round is the hole? How flat is the surface?).

Description

Your task is to design a pair of eyeglasses. The actual style (appearance) and size are up to you. Whether latest fashion fad or very old-fashioned, the eyeglasses should be designed to be appropriate for the human face. (Feel free to inquire whether your instructor will allow a design for some other creature, real or imagined.) In any event, the chosen design must be revealed in two or three views and fully dimensioned. Of particular interest will be dimensioning information concerning the minimum and maximum tolerance permitted for the frame and lenses (such as ±.01). The nose pad(s) or nosepiece may be separate or part of the frame itself. You may wish to measure your own eyeglasses or those of family or friends, or you may have someone actually measure your face to gain some insight with respect to realistic sizes. Figures P6-1 and P6-2 may also give you some design ideas.

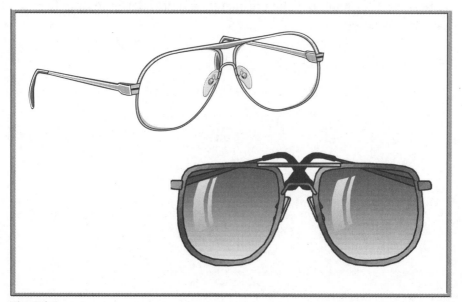

Fig. P6-1

Fig. P6-2

Hints and Suggestions

1. Consider showing a sketch of your idea to the instructor for feed-back before investing significant time and effort into the three-view CAD version. Guidance gained early is likely to save more time than constructive criticism given upon completion.

2. Don't spend too much time on exact sizes for the eyeglasses, but once drawn, dimension details thoroughly and accurately.

3. Include dimensions for the angles of the earpieces (end piece) as well as radii for the curvature of the lenses.

4. Consider drawing one half of the eyeglasses; then mirror the drawing to complete the pair.

Summary Questions

Your instructor may direct you to answer these questions orally or in writing. You may wish to compare and exchange ideas with other students to help increase efficiency and productivity.

1. Why might accurate dimensions be needed for one drawing and not for another?

2. Name some purposes for which tolerance dimensioning (min/max) might be considered of vital importance.

3. Once created, a dimension style can be a great timesaver. In what ways might it save time?

4. How do you determine which dimensions are needed on a drawing?

Careers
Using AutoCAD

Stone/Donovan Reese

Pipeline Drafter

Although we don't always realize it, we sometimes take modern-day conveniences for granted. If we want a drink of water, all we have to do is go to the kitchen sink and help ourselves to a glass. If we feel like taking a bath, we can just walk into the bathroom and fill the tub. The water we use for these purposes doesn't just magically appear, of course. Someone has taken the time to strategically plan a way to direct water from a main source to our homes. Thanks to people like pipeline drafters, we are able to enjoy one of our most important natural resources by merely turning a nozzle.

A Knack for Design

In addition to finding ways to channel water into homes, pipeline drafters are responsible for designing piping systems to move substances in refineries, oil and gas fields, sanitation systems, and chemical plants. The drawings they create are used for construction and layout of parts. The dimensions of pipes and the connecting units must adhere to certain specifications. An accurately designed system is efficient in its duty to transport substances from place to place. It also resists leakages or damage that may be potentially harmful to the environment.

Analyzing the Situation

Pipeline drafting requires some knowledge of the materials used to build pipeline systems and the advantages and disadvantages of each. Metals and plastics must be adequate in terms of quality and safety standards. Analytical skills are necessary to determine which materials are best suited for a particular system in a particular environment. Drafters frequently consult with surveyors, scientists, and engineers to ensure proper design, so communication skills are also an important part of the job, as well.

Career Activities

- Write a report about the work of a pipeline drafter. What are the advantages and challenges of this type of drafting?
- Does this kind of work fit with your goals and hopes for the future? Explain.
- Why is it important for pipeline drafters to consult with other individuals when designing piping transportation systems? With what types of people might a pipe drafter expect to work?

Objectives:

- Create a group
- Add and delete objects from a group
- Edit the group name and description
- Make a group selectable or unselectable
- Reorder the objects in a group

Key Terms

group
selectable

A *group* in AutoCAD is a named set of objects. A circle with a line through it, for instance, could become a group. Groups save time by allowing you to select several objects with just one pick. This speeds up editing operations such as moving, scaling, and erasing these objects.

Creating a Group

The GROUP command permits you to create a named selection set of objects.

1. Start AutoCAD and open the drawing named **calc.dwg**.

2. Using **Save As...**, create a new drawing named **groups.dwg** and erase all objects.

3. Create the drawing of the pulley wheel shown in Fig. 31-1 using the following guidelines:

   - Create a layer named **Hidden** and assign the HIDDEN linetype and a light gray color to it. Place the hidden line on this layer.
   - Place the remaining objects on layer Objects.
   - Omit the dimensions, including the center line.
   - Set LTSCALE to **.5**.

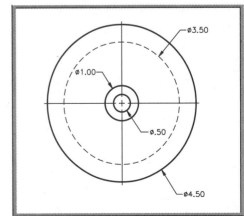

Fig. 31-1

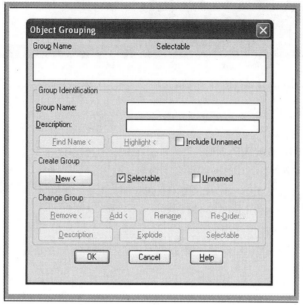

Fig. 31-2

4. Enter the **GROUP** command.

This displays the Object Grouping dialog box, as shown in Fig. 31-2.

5. For Group Name, enter **wheel** using upper- or lowercase letters.

6. Enter **Cast aluminum** for Description.

Focus your attention on the Create Group area of the dialog box. Notice that the Selectable check box is checked.

7. Pick the **New** button, select all four objects, and press **ENTER**.

When the dialog box returns, notice that WHEEL is listed under Group Name. Yes appears under Selectable because the Selectable box was checked when you picked the New button. *Selectable* means that when you select one of the group's members (*i.e.*, one of the circles), AutoCAD selects all members in the group.

8. Pick the **OK** button.

9. Pick any one of the objects.

Notice that AutoCAD selected all four circles.

10. Move the group a short distance and then press **ESC**.

11. Lock layer Hidden.

12. Pick one of the objects again and then move the group a short distance.

The gray hidden-line object did not move because it is on a locked layer.

13. Undo the move and unlock layer Hidden.

Changing Group Properties

The Object Grouping dialog box offers several features for changing a group's definition and behavior. For example, you can add and delete objects from the group, change the group's name or description, and re-order the contents (objects) in the group. The dialog box also allows you to explode the group, which deletes the group definition from the file.

Adding and Deleting Objects

1. Enter the **GROUP** command.
2. Select **WHEEL** in the Group Name list box.
3. Pick the **Highlight** button.

This highlights all members of the group to show you at a glance which objects are members of the group. This could be particularly useful when you're editing a complex drawing containing many objects and groups.

4. Pick the **Continue** button.

Focus your attention on the Change Group area of the dialog box.

5. Pick the **Remove** button and pick the gray hidden line.

AutoCAD removes this object from the group.

6. Press **ENTER**.
7. Select the **Highlight** button again.

Notice that the hidden-line circle is not highlighted; it is no longer part of the group.

8. Pick **Continue**.
9. Pick the **Add** button, pick the gray hidden-line circle, and press **ENTER**.

This adds the circle back to the group.

Changing the Group Name and Description

1. In the Group Name edit box, change the group name to **pulley** and pick the **Rename** button.

PULLEY appears in the Group Name list box near the top of the dialog box.

2. Rename the group back to **wheel**.
3. In the Description edit box, change the group description from Cast aluminum to **Cast magnesium**.

4. Pick the **Description** button (just above the OK button).

This causes the group description to update, as indicated at the bottom of the dialog box.

Making a Group Selectable

1. In the Change Group area, pick the **Selectable** button.

In the upper right area under Selectable, notice that Yes changed to No.

2. Pick the **OK** button and select one of the objects in the group.

As you can see, the group is no longer selectable. Grips appear only for the specific object you select.

3. Reenter the **GROUP** command, pick **WHEEL** in the Group Name list box, pick the **Selectable** button again, and pick **OK**.

The group is selectable once again.

Reordering the Objects

AutoCAD numbers objects in the order in which you select them when you create the group. In the Order Group dialog box, AutoCAD allows you to change the numerical sequence of objects within a group. Reordering can be useful when you are creating tool paths for computer numerical control (CNC) machining.

1. Enter the **GROUP** command.

2. Pick the **Re-Order...** button to make the Order Group dialog box appear, as shown in Fig. 31-3.

3. Select **WHEEL** under Group Name.

This makes the grayed-out items become available.

4. Pick **OK**; pick **OK** again.

5. Save your work and exit AutoCAD.

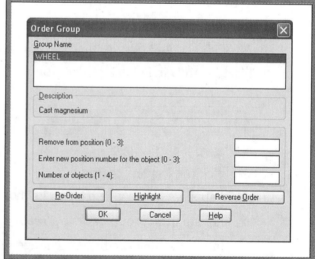

Fig. 31-3

Review Questions

1. What is a group?

2. How can groups help AutoCAD users save time?

3. When creating a new group, what is the purpose of entering the group description?

4. Explain how you would add and delete objects from a group.

5. Is it possible to remove the selectable property from a group but still retain the group definition in the drawing? Explain.

6. Under what circumstances might you need to reorder the objects in a group?

Challenge Your Thinking

1. Experiment with object selection using a group name. Note that you can enter group, followed by the group name, at any Select objects prompt for moving, copying, scaling, etc. When might this feature be useful? Explain.

2. Under what circumstances might you want to make a group unselectable? Explain.

Applying AutoCAD Skills

Work the following problem to practice the commands and skills you learned in this chapter.

1. Perform the following tasks.
   a. Load the calc.dwg file from the previous chapter and use Save As... to create a file named stop.dwg. If the points are visible, make them invisible.
   b. Create a group named WIDGET. Enter New product design for the description. The group should consist of all four objects.
   c. Change the name of the group to WATCH and the group description to Stop watch design. Remove the two lines from the group.

d. Move the group to a new location, away from the two lines. Set the text height to .2 and add 12:30 inside the square. Add this text to the group. Your drawing should look similar to the one in Fig. 31-4.

Fig. 31-4

Using Problem-Solving Skills

Complete the following activities using problem-solving skills and your knowledge of AutoCAD.

1. The exterior trim design of your company's administration building has not yet been approved. Open db_samp.dwg in AutoCAD's Sample folder and save it as AdmBldg.dwg in the folder with your name. Make a group of each of the three exterior entrances (two sides in turquoise and the front in yellow). Move these groups to various locations to find a better arrangement. You may also remove the five-sided structure at the rear of the building to be replaced by a possible entry. Save the design you think is the best, and be prepared to explain your choice.

2. Your company has been hired to create a sales brochure for a new line of bicycles. The bicycle will be available in several colors, and your customer wants the brochure to show the bicycle in each color. Research typical bicycle shapes and designs. Then create a bicycle design of your own. Use commands such as PLINE, SOLID, and FILL to make the body (frame) of the bicycle solid green. Save the drawing as bikesales.dwg. Then, using the GROUP and COPY commands, create additional copies of the bicycle and change the body colors to show a version in six colors that you think would look good on a bicycle. Use the entire Color palette to make your color selections.

Objectives:

- Create and insert blocks
- Rename, explode, and purge blocks
- Organize and insert blocks using the **DesignCenter** and **Tool Palettes** windows
- Add commands to a tool palette
- Insert a drawing file into the current drawing
- Create a drawing file from a block
- Copy and paste objects from drawing to drawing

Key Terms

block
copy and paste
purge

With AutoCAD, you seldom need to draw the same object twice. Using blocks, you can define, store, retrieve, and insert symbols, components, and standard parts without the need to recreate them. A *block* is a collection of objects that you can associate together to form a single object. Blocks are especially useful when you are creating libraries of symbols and components.

Blocks are similar to groups, but the two are different. A block definition is stored as a single, selectable object in a drawing file that you can insert, scale, and rotate. A group is a named selection set of objects.

Working with Blocks

The BLOCK command allows you to combine several objects into one and then store and retrieve it for later use.

1. Start AutoCAD and start a new drawing from scratch using **Imperial** units.

2. Draw the top view of a flathead screw as shown in Fig. 32-1 on page 442. Omit the dimensions.

3. Save your work in a file named **blks.dwg**.

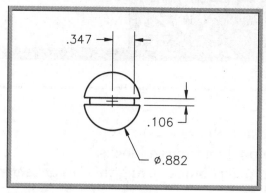

Fig. 32-1

Creating a Block

1. Pick the **Make Block** button from the docked Draw toolbar, or enter the **BLOCK** command.

Draw

This displays the Block Definition dialog box, as shown in Fig. 32-2.

2. In the Name text box, type **flat**.

3. In the Base Point area, pick the **Pick point** button.

4. Snap to the center point of the head of the screw.

This defines the base point for subsequent insertions of the screw head.

5. In the Objects area, pick the **Convert to block** radio button and then pick the **Select objects** button.

6. Select the entire screw head and press **ENTER**.

In the Objects area, AutoCAD displays 6 objects selected. The Retain option retains the selected objects as distinct objects in the drawing after you create the block. Convert to block converts the selected objects to a block instance in the drawing after you create the block. Delete deletes the selected objects from the drawing after you create the block. (The block definition remains in the drawing database.)

Fig. 32-2

7. In the Settings area, review the options available for the Block unit.

8. Set Block unit at **Unitless**, do not add a description, and pick the **OK** button.

You have just created a new block definition.

Inserting a Block

Draw

1. Pick the **Insert Block** button from the docked Draw toolbar or enter the **INSERT** command.

This displays the Insert dialog box, as shown in Fig. 32-3.

Flat is listed after Name because it is the only block definition in the drawing. Picking the down arrow would list additional blocks, if any existed, in alphabetical order.

The Insertion point area permits you to enter specific coordinates for the block's insertion. Specify On-screen, which should be checked, allows you to specify the insertion point on screen. The Scale area allows you to scale the block now or on the screen when you insert it. The Rotation area enables you to specify a rotation angle in degrees now or when you insert the block. The Block Unit area is not selectable; it displays the unit and scale factor used when the block was created. The Explode check box explodes the block and inserts the individual parts of the block.

2. Under Scale, pick the **Specify On-screen** check box.

3. Under Rotation, pick the **Specify On-screen** check box.

4. Pick the **OK** button.

AutoCAD locks the insertion point of the block onto the crosshairs.

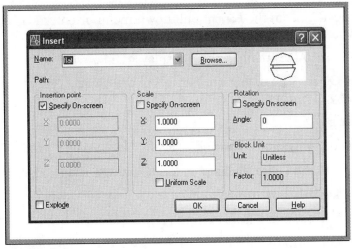

Fig. 32-3

5. Pick an insertion point anywhere on the screen.

6. Move the crosshairs up and down, back and forth.

As you can see, you can drag the X and Y scale factors.

7. Enter **1.5** for the X scale factor and **1.5** for the Y scale factor.

You can also drag the rotation into place, or you can enter a specific degree of rotation.

8. Enter **45** for the rotation angle.

9. Insert additional instances of the block using different settings in the Insert dialog box.

Editing an Inserted Block

1. Try to erase one of the lines from the first inserted block and then press **ESC**.

You cannot erase part of a block because a block is a single object.

Modify

2. Pick the **Explode** button from the docked Modify toolbar, pick the inserted block, and press **ENTER**.

3. Now try to erase a line from the inserted block.

As you can see, exploding a block returns it to its component parts. Exploding a block with the EXPLODE command is the same as checking the Explode check box in the Insert dialog box when inserting the block.

Renaming a Block

The RENAME command lets you rename previously created blocks.

1. Pick **Rename...** from the **Format** pull-down menu or enter **RENAME** at the keyboard.

This enters the RENAME command and displays the Rename dialog box.

Under Named Objects, the dialog box lists all of the object types in AutoCAD that you can rename.

2. Pick **Blocks**.

AutoCAD lists the only block, flat, under Items.

3. Pick **flat** and then enter **flathead** in the box at the right of Rename To.

4. Pick the **Rename To** button.

Flathead is now listed under Items.

5. Pick the **OK** button.

6. Save your work.

In the future, if you need to rename blocks, dimension styles, layers, linetypes, text styles, UCSs, viewports, or views, use the RENAME command.

Purging Blocks

The PURGE command enables you to selectively delete, or *purge*, any unused named objects, such as blocks. Purging named objects reduces drawing file size.

1. Enter the **PURGE** command.

AutoCAD displays the Purge dialog box and shows the different object types that you can purge. If the drawing database contained unused blocks, the Purge button would not be grayed out. In the future, you may choose to purge unused blocks, dimension styles, layers, and other object types.

2. Pick the **Close** button.

DesignCenter and Tool Palettes

The DesignCenter is a window that allows you to browse, find, and insert content, such as blocks and hatches. The Tool Palettes window offers tabbed areas that provide an easy method of organizing, sharing, and placing blocks and hatches.

Inserting Blocks from DesignCenter

Standard

1. Pick the **DesignCenter** button from the Standard toolbar.

This displays the DesignCenter window.

2. Click each of the tabs to familiarize yourself with DesignCenter.

3. Pick the **Open Drawings** tab and double-click **Blocks**.

Notice that the flathead block was added automatically to DesignCenter.

4. Click and drag the block named flathead from DesignCenter to the drawing area.

As you can see, DesignCenter offers an alternative for inserting blocks.

5. Right-click and drag the flathead block into the drawing area and pick **Insert Block...** from the pop-up menu.

This displays the Insert dialog box.

6. Make sure that **Specify On-screen** is checked under Insertion point, Scale, and Rotation and pick **OK**.

7. Pick an insertion point.

8. Enter **.75** for the X and Y scale factors and **45** for the rotation angle.

9. After inserting the block, resize DesignCenter to make it smaller, but leave it open.

Inserting Blocks from a Tool Palette

Standard

1. Pick the **Tool Palettes Window** button from the Standard toolbar.

2. Right-click the title bar of the Tool Palettes window and select **New Palette** from the menu.

3. In the text box, type **Fasteners** and press **ENTER**.

This creates a new (but empty) tool palette named Fasteners.

4. From DesignCenter, click and drag the block named flathead to the new palette.

The flathead block appears in the tool palette.

5. Close DesignCenter.

6. From the new tool palette, click and drag the flathead block into the drawing area and release the pick button.

As you can see, this is a fast and easy way of inserting blocks.

7. Insert another flathead block into the drawing area from the tool palette; and then another.

8. Attempt to right-click and drag the flathead block from the palette into the drawing area.

As you can see, this is not an option. The Tool Palettes window does not give you the option of inserting blocks using the Insert dialog box, similar to right-clicking and inserting them from DesignCenter.

Adding Commands to a Tool Palette

AutoCAD makes it easy to add commands to a tool palette.

1. Right-click in the Fasteners tool palette and pick **Customize...** from the menu.

This displays the Customize dialog box.

2. With the Customize dialog box open, click and drag the **Polyline** button from the Draw toolbar to the Fasteners tool palette.

The Polyline button becomes part of the palette. Notice that a small black triangle appears next to it. The button is now a flyout button—a set of Draw commands are nested under this single button.

3. Click the black triangle next to the Polyline button on the Fasteners tool palette.

As you can see, the buttons for eight commands from the Draw toolbar are displayed on the flyout button.

4. Pick the **Close** button in the Customize dialog box.

5. Display the **Dimension** toolbar.

6. Right-click in the Fasteners tool palette and pick **Customize...** from the menu once again.

7. Click and drag the **Linear Dimension** button from the Dimension toolbar to the Fasteners tool palette.

8. Drag the **Erase** button from the docked Modify toolbar to the Fasteners tool palette.

9. Close the Customize dialog box.

10. Try each of the commands that you added to the Fasteners tool palette.

You can also arrange the order of tools on a tool palette and rearrange the order of tabs in the Tool Palettes window.

11. On the Fasteners tool palette, click and drag the **Erase** button up to the top of the palette.

The Erase button is now the first button on the palette.

12. Right-click on the Fasteners tab and pick **Move Up** or **Move Down** to rearrange the order of tabs in the Tool Palettes window.

13. Right-click in an open area of the Fasteners tool palette, select **Delete Palette** from the menu, and pick **OK** to confirm that you want to delete the tool palette.

AutoCAD deletes the tool palette.

14. Close the Tool Palettes window and save your work.

Blocks and Drawing Files

It is possible to insert a drawing (DWG) file as if it were a block. This can be useful when you want to use part or all of another drawing in the current drawing. Also, it's possible to create a drawing file from a block. This can be especially helpful when you need to transport a block to another computer.

Inserting a Drawing File

Draw

1. Begin a new drawing from scratch using **Imperial** units.

2. Pick the **Insert Block** button from the docked Draw toolbar and pick the **Browse...** button from the dialog box.

3. Find and open the folder with your name.

4. Open the file named **toler.dwg**.

Toler appears in the Name box.

5. Uncheck **Specify On-screen** under Insertion point, Scale, and Rotation.

6. Pick the **Explode** check box and pick **OK**.

Toler.dwg appears.

7. Pick one of the objects from the toler.dwg drawing.

This is possible because we selected the Explode check box to make the components of the block selectable individually.

8. Close the current drawing without saving.

The drawing named blks.dwg (containing the flathead screws) is now the current drawing.

Creating a Drawing File from a Block

WBLOCK, short for Write BLOCK, writes (saves) objects or a block to a new drawing file.

1. Enter the WBLOCK command.

AutoCAD displays the Write Block dialog box.

2. Under Source, pick the **Block** radio button.

3. Click in the box at the right of Block.

4. Pick **flathead**.

The Base point and Objects areas are grayed out because both base point and objects are a part of the block definition.

5. Review the information listed under Destination, but do not change any of it.

Notice that the proposed file name is flathead.

6. At the right, pick the button containing the three dots, select the folder with your name, and pick **Save**.

7. Pick the **OK** button to close the Write Block dialog box.

AutoCAD creates a new file named flathead.dwg.

Standard

8. Pick the **Open** button from the docked Standard toolbar.

9. Find flathead.dwg and preview it, but do not open it.

10. Pick the **Cancel** button.

Copying and Pasting Objects

AutoCAD's Windows-standard *copy and paste* feature provides an alternative to using the WBLOCK and INSERT commands. Copying and pasting can be a faster approach when you want to transfer a block or set of objects to another drawing on the same computer. Note that this approach does not work when you need to transfer a block or set of objects to another computer.

Standard

1. From the Standard toolbar, pick the **Copy to Clipboard** button.

This enters the COPYCLIP command.

2. Pick a couple of objects and press **ENTER**.

This copies the objects to the Windows Clipboard.

> **NOTE:** ──────────────────────────────
>
> You can select the object(s) first and then pick the Copy button.

3. Save your work.

4. Create a new drawing from scratch using **Imperial** units.

Standard

5. From the Standard toolbar, pick the **Paste** button.

This enters the PASTECLIP command.

6. Move the crosshairs and notice that the objects are attached to it.

7. Pick an insertion point.

As you can see, this is a fast and simple way of copying objects from drawing to drawing.

8. Exit AutoCAD without saving the current drawing.

Review Questions

1. Briefly describe the purpose of blocks.

2. Explain how the INSERT command is used.

3. How can you list all defined blocks contained within a drawing file?

4. A block can be inserted with or without selecting the Explode check box. Describe the difference between the two.

5. Describe the function of the WBLOCK command.

6. When would WBLOCK be useful?

7. How can you rename blocks?

8. Why would you want to purge unused blocks from a drawing file?

9. What is an advantage to using the Tool Palettes window to organize and insert blocks?

10. When copying an object from one drawing to another, why might you prefer the copy and paste method over the WBLOCK and INSERT approach?

Challenge Your Thinking

1. An electrical contractor using AutoCAD needs many electrical symbols in his drawings. He has decided to create blocks of the symbols to save time. Describe at least two ways the contractor can make the blocks easily available for all his AutoCAD drawings. Which method would you use? Why?

2. If you were to copy a block from one drawing and paste it into another, would AutoCAD recognize the pasted object as a block? Explain.

Applying AutoCAD Skills

Work the following problems to practice the commands and skills you learned in this chapter.

1. Open range.dwg, which you created in the "Applying AutoCAD Skills" section of Chapter 17. Make a block of the range and reinsert it into the drawing at a scale factor of .25, with a rotation of 90°. Save the drawing as range2.dwg.

2. Begin a new drawing named livroom.dwg. Draw the furniture representations shown in Fig. 32-4 and store each as a block. Then draw the living room outline. Don't worry about exact sizes or locations, and omit the text. Insert each piece of furniture into the living room at the appropriate size and rotation angle. Feel free to create additional furniture and to use each piece of furniture more than once. This file is also available on the Instructor Resource CD as livroom.dwg.

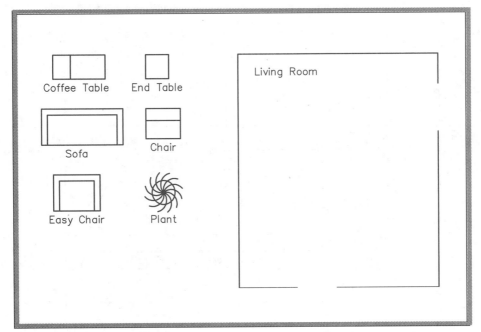

Fig. 32-4

3. After creating the blocks in problem 2, write two of them (of your choice) to disk using WBLOCK. Store them in the folder with your name.

4. Copy the block of the easy chair and paste it into another drawing.

5. Explode the PLANT block and erase every fourth arc contained in it. Then store the plant again as a block.

6. Rename two of the furniture blocks.

7. Purge unused objects.

8. Create a new drawing named revplate.dwg. Create the border, title block, and revisions box according to the dimensions shown in Fig. 32-5. (Do not include the dimensions.) Produce a block of the revisions box using the insertion point indicated. Then insert the block in the upper right corner of the drawing. This drawing is available as revplate.dwg on the Instructor Resource CD.

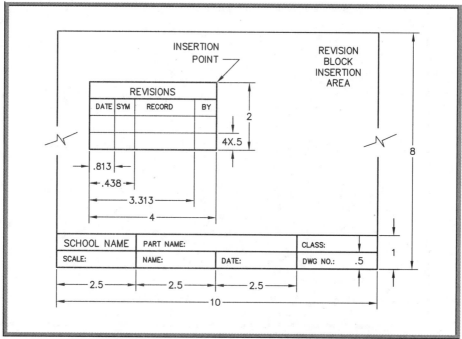

Fig. 32-5

Problem 8 courtesy of Mark Schwendau, Kishwaukee College

Using Problem-Solving Skills

Complete the following problems using problem-solving skills and your knowledge of AutoCAD.

1. Your company wants to make the revisions box you created in problem 8 (on the previous page) available to all the designers in the company. Open revplate.dwg from your named folder, create a block of the revisions box, and save it as a DWG file. Be sure to give the file a descriptive name so that the designers will know what is in the file.

2. Draw the electric circuit shown in Fig. 32-6 as follows: Draw the resistor using the mesh shown; then save it as a block. Draw the circuit, inserting the blocks where appropriate. Grid and snap are handy for drawing the resistor and circuit. Finish the circuit by inserting small donuts at the connection points. Add the text. The letter omega (Ω), which is used to represent the resistance in ohms, can be found under the text style GREEKC (character W).

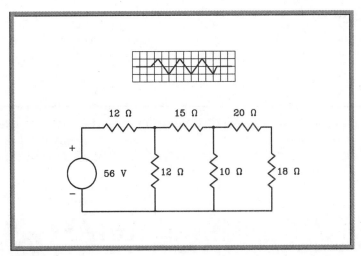

Fig. 32-6

Chapter 33   Dynamic Blocks

Objectives:

- **Create a dynamic block**
- **Assign actions to parameters**
- **Insert a dynamic block in a drawing**
- **Create a set of constraining values for a dynamic block**
- **Use a constrained dynamic block in a drawing**

Key Terms

constrain   parameters

dynamic blocks   value set

In the last chapter, you were introduced to blocks. By using blocks, AutoCAD users avoid drawing the same object twice. This can be extremely useful, because many objects in architecture, manufacturing, and engineering are used repeatedly and need to be drawn again and again in many drawings.

AutoCAD allows you to take advantage of the convenience of blocks even further with dynamic blocks. A *dynamic block* is a block with certain characteristics or *parameters*, such as size or location, that can be changed after you insert the block into a drawing. With dynamic blocks, you first define the parameters that you want to be changeable using AutoCAD's block editor. Once the block has been created, you only need the one dynamic block to draw a range of sizes or lengths of a part. For example, you could create a dynamic block for drawing a commonly used fastener, such as a hex nut, and vary the size of the nut when you insert the block into the drawing.

Creating a Dynamic Block

Let's create a dynamic block for a simple shape: a bed. We'll start with the shape of a twin bed and add parameters and actions that will allow its length and width to be stretched into a bed of any size.

1. Start AutoCAD and start a new drawing from scratch using **Imperial** units.

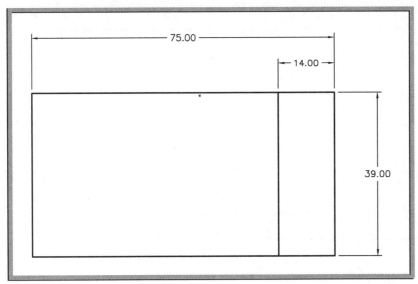

Fig. 33-1

2. Using the **LIMITS** command, set the drawing area to **120 × 100** and **ZOOM All**.

3. Create the top view of a twin-size bed using the dimensions shown in Fig. 33-1. Do not include the dimensions in your drawing. These are the standard dimensions (in inches) for a twin bed.

4. Name the drawing **bed.dwg**.

Draw

5. Pick the **Make Block** button from the docked Draw toolbar or enter the **BLOCK** command to display the Block Definition dialog box.

6. Name the block **Bed**, pick the lower left corner of the bed for the base point, and select the entire bed to be included in the block.

7. Near the bottom of the dialog box, check the **Open in block editor** box.

8. Pick **OK** to close the dialog box and open the new block in the block editor.

Assigning Parameters to the Block

There are two ways to work with blocks in the block editor. You can use either the Block Authoring Palettes window, which appears on the left side of the block editor, or the buttons on the toolbar near the top of the block editor. The following steps use both methods.

1. In the Block Authoring Palettes window, select the **Parameters** tab if it is not already selected.

2. From the available options, pick **Linear Parameter**.

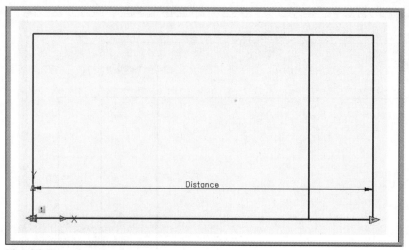

Fig. 33-2

3. Pick the bottom left corner of the bed as the start point and the bottom right corner of the bed as the endpoint. Be sure to snap accurately to these two points.

4. Indicate a point inside the bed for the label location, as shown in Fig. 33-2.

Notice that the parameter is automatically labeled Distance. Also notice the yellow icon with an exclamation mark near the lower left corner of the bed. This symbol means that the parameter is not currently associated with any action. You will set up the actions later. First, let's create a second parameter.

5. From the Block Authoring Palettes window, pick **Linear Parameter** again. This time, pick the lower left corner as the start point and the upper left corner for the endpoint. Place the label inside the bed, as shown in Fig. 33-3.

The second parameter is automatically named Distance1. However, you can rename parameters to make them easier to identify.

6. Select the first **Distance** parameter you created, right-click to display the shortcut menu, and pick Rename Label.

7. Highlight the word Distance, enter **Length** for the new name, and press **ENTER**.

8. Select the **Distance1** parameter and rename it **Width**.

NOTE:

You can also name parameters at the time you create them. Before you specify the parameter, press N to enter the Name option.

Chapter 33: Dynamic Blocks

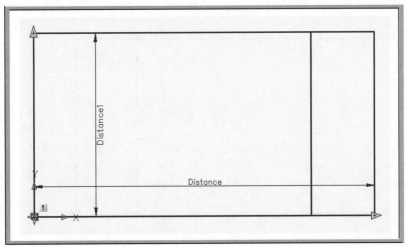

Fig. 33-3

Because we will stretch only one end of each parameter, let's change the grip display.

9. Pick the **Length** parameter, right-click, and select **Grip Display**.

10. Pick a Grip Display of **1**.

Notice that the closed blue arrow on the left endpoint disappears.

11. Change the Grip Display of the **Width** parameter to **1**.

12. Save your work.

Assigning Actions to the Parameters

The two parameters that you have established will allow you to adjust both the length and the width of the dynamic block named bed. Next, we will set up the actions to be associated with each parameter.

1. Locate the button on the Block Editor toolbar that looks like a thunderbolt, located near the middle of the toolbar.

This is the Action button.

2. Pick the **Action** button to begin to define an action.

Block Editor

> **NOTE:**
>
> You can also define an action using the Actions tab of the Block Authoring Palettes window.

3. Select the **Length** parameter.

4. From the Enter action type list, select **sTretch**.

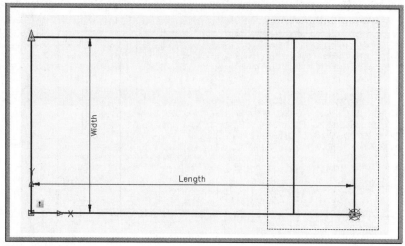

Fig. 33-4

Notice that a red symbol appears at the lower right corner of the text.

5. Pick the point in the lower right corner of the bed as the parameter point to associate with the action.

6. In reply to Specify first corner of stretch frame, use a standard crossing window. Place the window as shown in Fig. 33-4. Be sure to include the interior line that represents the bedcovers in the crossing window.

7. At the Select Objects prompt, use another crossing window to select roughly the same area. Press **ENTER** when you are finished.

A thunderbolt labeled Stretch becomes attached to the crosshairs.

8. Move the crosshairs outside the lower right corner of the bed and click to place the label.

Now let's rename the label for our new action.

9. Pick the **Stretch** action, right-click to display the shortcut menu, and pick **Rename**.

10. Rename the action **Stretch Length**.

Next, we will define the action needed to change the width of the bed.

11. Pick the **Action** button again and select the **Width** parameter.

12. From the Enter action type list, select **sTretch**.

13. Pick the endpoint at the upper left corner as the parameter point to associate with the action.

14. For the stretch frame, place a crossing window across the upper half of the bed, as shown in Fig. 33-5.

Block Editor

Chapter 33: Dynamic Blocks

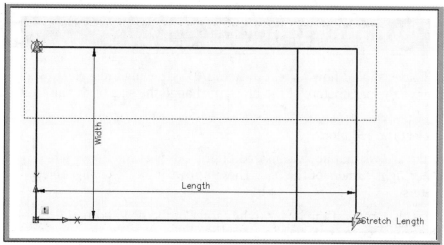

Fig. 33-5

15. At the Select objects prompt, use another crossing window to select roughly the same area. Press **ENTER** when you are finished.

16. Place the thunderbolt near the upper left corner of the bed.

17. Rename the action to **Stretch Width**.

The Length and Width parameters are now completely defined, so the alert icons for the parameters have disappeared. Your block should now look similar to the one in Fig. 33-6.

18. Pick the **Close Block Editor** button in the Block Editor toolbar to return to the AutoCAD window. Pick **Yes** when prompted to save changes to the Bed block.

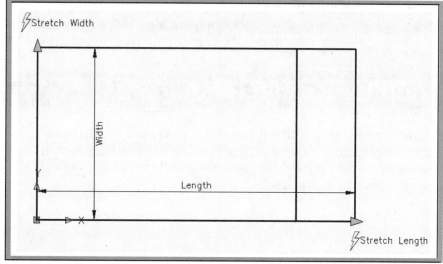

Fig. 33-6

Using Dynamic Blocks

The Bed block should now appear in AutoCAD's drawing area. Next, we will change the parameters of the block to change the size of the inserted block.

1. Select the **Bed** block.

Notice the arrows that appear instead of grip boxes at the upper left and lower right corners of the bed. These arrows indicate changeable parameters.

2. Pick the arrow at the lower right corner of the bed and notice that it becomes active (red) just as a grip would.

The Length parameter appears above the bed, and an edit box appears to show the current length of the bed, which is 75 inches.

3. Move the crosshairs and notice that the length of the bed changes dynamically.

4. At the keyboard, enter a new length of **80** and press **ENTER** to change the length of the bed to 80 inches.

The bed updates automatically.

This is the standard length for a queen, king, or extra-long twin bed. Let's change the Width dimension to create a queen bed, which has a standard width of 60 inches.

5. Select the bed again.

6. Pick the arrow at the upper left corner of the bed.

7. Enter a new width of **60**.

Once again, the Bed block automatically updates when you press ENTER.

Constraining Parameters

It may have occurred to you that it is possible to change the dynamic Bed block to any length and width using the dynamic block defined in the previous procedure. In real life, beds are usually made to standard dimensions, as shown in Fig. 33-7.

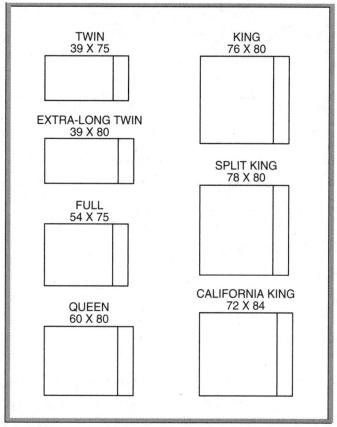

TWIN
39 X 75

KING
76 X 80

EXTRA-LONG TWIN
39 X 80

SPLIT KING
78 X 80

FULL
54 X 75

CALIFORNIA KING
72 X 84

QUEEN
60 X 80

Fig. 33-7

Creating a Value Set

AutoCAD makes it easy to *constrain*, or limit, the choices that can be used for a given parameter. You can do this by creating a *value set* for the parameter. For example, the three standard lengths for a bed are 75, 80, and 84 inches. The following steps constrain the Length parameter of the Bed block using a value set that contains these three values.

1. Double-click the block to display the **Edit Block Definition** dialog box.

2. Select **Bed** and pick **OK** to enter the block editor.

3. Select the **Length** parameter.

4. From the Tools pull-down menu, select **Properties**.

This displays the properties of the linear parameter in the Properties window.

5. Near the bottom of the Properties window, find the Value Set section and click to activate the box next to **Dist type**. (You may need to scroll down to find this section.)

6. Pick the down arrow and then pick **List** from the options that appear.

Notice that the next row, Dist value list, now shows 75.0000. This is the value you used when you originally created the block, and it becomes the default value.

7. Pick to activate the **Dist value list** row and pick the three dots to the right of the edit box.

This displays the Add Distance Value dialog box.

8. In the Distances to add edit box, enter **80** and pick the **Add** button.

The new value of 80.0000 appears in the window below the edit box.

9. With the cursor still active in the Distances to add edit box, enter **84** and pick the **Add** button to add 84.0000 to the list.

10. Pick **OK** to close the dialog box.

11. Press **ESC** to deselect the Length parameter.

The standard widths for beds are 39, 54, 60, 72, 76, and 78. Let's constrain the Width parameter to these values.

12. With the Properties window still open, pick to select the **Width** parameter.

13. In the Dist type row in the Properties window, change the value to **List**.

14. Click to activate the **Dist value list** row and pick the button with three dots to display the Add Distance Value dialog box.

Once again, the default value appears in the window.

15. Add the other standard width values: **54, 60, 72, 76**, and **78**. Pick **OK** when you are finished.

16. Close the Properties window and save the file. If AutoCAD displays a message asking whether you want to save the drawing and update the existing block definition in the drawing, pick **Yes**.

17. Close the block editor.

Using a Constrained Block

Constrained blocks can be altered on-screen in the same way as unconstrained dynamic blocks. The only difference is that, when you pick one of the parameter arrows, small tick marks appear to mark the valid values for the parameter.

1. Select the Bed block to display the parameter arrows.

2. Pick the lower right parameter arrow so that it turns red.

Just below the arrow, notice the faint tick marks on either side of the current bed length. See Fig. 33-8.

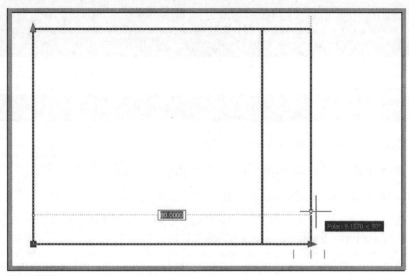

Fig. 33-8

You may recall that the bed is currently configured with queen dimensions: 60 × 80 inches. The 80 inches is one of the three standard values. The two ticks mark the other two values allowed for the Length parameter: 75 and 84. Let's change the Bed block to represent a full bed (54 × 75 inches).

3. Move the cursor to the left and notice that a tentative line appears at the 75-inch tick mark.

4. Pick any point while that line is present to reset the length of the bed to 75 inches.

5. Pick the upper left parameter arrow to activate the Width parameter and notice the tick marks representing allowed values.

The current value of 60 shows up in the edit box for reference.

6. Move the cursor down until the edit box shows a value of 54 and pick any point to change the width of the bed to 54.

The block is now configured to represent a full-size bed.

If you enter an undefined value into the edit box for a constrained parameter, AutoCAD automatically chooses the defined value that is closest to the value you entered. For example, if you entered a value of 63 for the Width parameter, the width of the bed would change to 60, which is the closest value defined in the value set.

7. Save your work and exit AutoCAD.

Chapter 33
Review & Activities

Review Questions

1. Define *dynamic block* in your own words.

2. Briefly describe the process of creating a dynamic block.

3. How can you rename a parameter or an action?

4. What is a value set? How would you create a value set to constrain a parameter for a dynamic block?

5. Why is careful selection of a dynamic block's base point important?

Challenge Your Thinking

1. One dynamic block action is visibility, which allows you to make a block invisible. This is somewhat similar to freezing a layer to make it invisible. When would each method of hiding drawing elements be preferable?

2. Review the Parameter Sets tab of the Block Authoring Palettes. Which parameter set would we have used to create the adjustments of the bed's length?

Applying AutoCAD Skills

Work the following problems to practice the commands and skills you learned in this chapter.

1. Open the bed.dwg file and save it as hotelplan.dwg. Create a hotel floor plan of your choosing, making a variety of rooms and room sizes. Place beds in each of the rooms. In about half of the rooms, insert two twin beds. In most of the others, insert a single king bed. In the remaining rooms, install two queen beds. Save your work.

2. Open the file named flathead.dwg that you created in Chapter 32. Create a block named flathead with the screw head's center as the base point. In the block editor, create a polar parameter with the center of screw head as the base point. Specify a Grip Display of 1. Assign a scale action to the parameter. Your dynamic block should look similar to Fig. 33-9. Save the dynamic block and close the

Chapter 33: Dynamic Blocks

block editor. Insert several of the blocks into the drawing area. Scale the blocks to a variety of sizes. Use the keyboard to enter numeric values for the diameter on several, then scale the others by dragging the grip with the cursor. Your finished drawing should look similar to Fig. 33-10. Save your work as dynflathead.dwg.

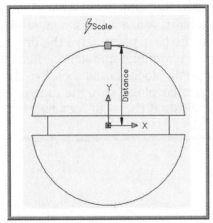

Fig. 33-9

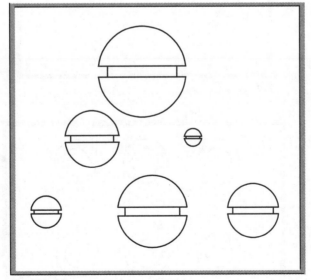

Fig. 33-10

Using Problem-Solving Skills

Complete the following activities using problem-solving skills and your knowledge of AutoCAD.

1. Open the chair.dwg file, which you created in Chapter 16. Delete all but one of the chairs. Make a block named chair, using the center of the circles as the base point. Save the drawing as dynchair.dwg. Create a rotation parameter and action. Rename the parameter Chair Angle. Your dynamic block should look like the one in Fig. 33-11. Insert six of the chair blocks into the drawing. Use dynamic rotation to rotate the individual chairs to the angles specified in Fig. 33-12.

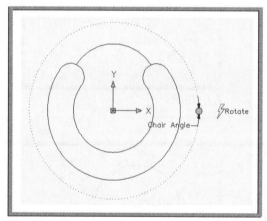

Fig. 33-11

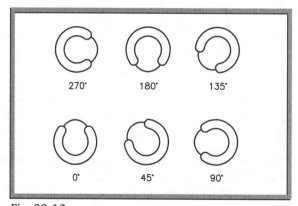

Fig. 33-12

2. Create the hex bolt shown in Fig. 33-13. Approximate the size and shape of the hex head. Then make a dynamic block of the bolt, creating a parameter and action to stretch the length of the bolt's shank. Create a value set of the following shank lengths: 1.00, 1.25, 1.50, 1.75, 2.00, 2.50, and 3.00 inches. Place seven blocks in the drawing, and use the constraints of the value set to create one bolt of each length. Your finished drawing should look similar to Fig. 33-14. Save the file as dynbolt.dwg.

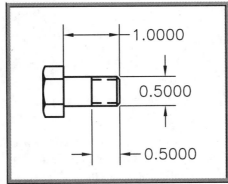

Fig. 33-13

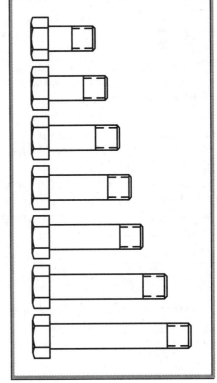

Fig. 33-14

Careers

Using AutoCAD

Tim Fuller Photography

Graphic Designer

Think about the last CD you purchased. Can you remember what the cover looks like? Does it include fancy artwork or wording? Now, consider the breakfast cereal you may have eaten this morning. Can you picture the designs on the box? Almost every item available for purchase today is enclosed in some sort of packaging. To entice consumers to select their product, companies want the product's exterior to be attractive. They hire graphic designers to create an appealing image to represent their product.

While graphic designers do not typically create plans used for manufacturing or building products, as most drafters do, their duties are similar. Graphic designers create visual images for product packaging, company logos, posters, banners, and advertisements. They are responsible for producing the images and text that represent a particular product or service.

Using Technology

While some graphic designers may create images using freehand drawing methods, computer software is becoming increasingly more important in this field. Programs such as Adobe Illustrator® and Adobe Photoshop® are used to create and alter images. Desktop publishing programs, such as Microsoft Publisher®, Adobe PageMaker®, and Quark XPress®, help graphic designers position and align the text and images or lay out a design.

Conveying Concepts

Graphic designers work closely with clients to figure out a way to communicate an idea or concept to the targeted audience. An understanding of what the client wants is critical. Graphic designers must also understand who will be using the product (or reading the book, poster, or billboard). They target their designs for the intended customer group according to published demographic surveys.

Graphic designers use fonts, colors, shapes, and visual effects (shading, blurring, distortion, etc.) to create an effective image for a product, service, or company. Ultimately, the artwork will help motivate consumers to purchase a product or use a service.

Career Activities

- If you were considering becoming a graphic designer, what type of experience do you think would be beneficial to your goals?
- Compare and contrast the skills required for a graphic designer and those required for a drafter.

Chapter 34 Symbol Libraries

Objectives:

- **Create a library of symbols and details**
- **Insert symbols and details using a symbol library**
- **Insert layers, dimension styles, and other content from drawings using DesignCenter**

Key Term

symbol library

Figure 34-1 shows a collection of electrical substation schematic symbols in an AutoCAD drawing file. Each of the symbols was stored as a single block and given a block name. (In this particular case, numbers were used for block names rather than words.) The crosses, which show the blocks' insertion base points, and the numbers were drawn on a separate layer and frozen when the blocks were created. They are not part of the blocks; they are used for reference only. A drawing file such as this that contains a series of blocks for use in other drawings is known as a *symbol library*.

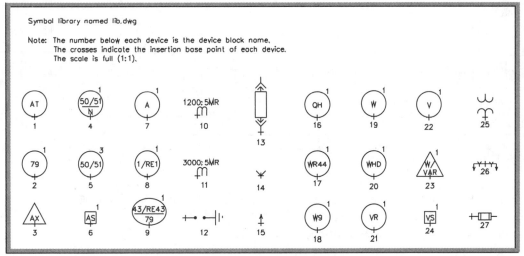

Fig. 34-1

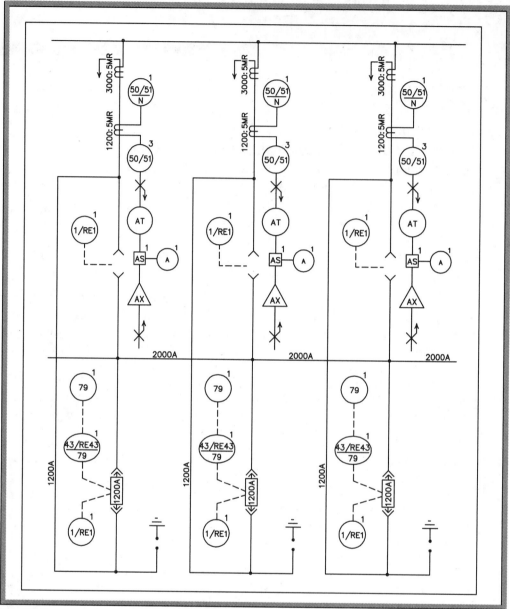

Fig. 34-2

After the symbols were developed and stored in a drawing file, DesignCenter was used to insert the symbols into a new drawing for creation of the electrical schematic shown in Fig. 34-2.

In this example, the blocks were then inserted into their proper locations, and lines were used to connect them. As a result, about 80% of the work was complete before the drawing was started. This is the primary advantage of grouping blocks in symbol libraries.

Creating a Library

Let's step through a simple version of the procedures just described. Keep in mind that a symbol library may be nothing more than a drawing file that contains a collection of blocks for use in other drawings.

1. Start AutoCAD and start a new drawing using the **tmp1.dwt** template file.

2. Using the **LIMITS** command, set the drawing area to **24′ × 14′** and **ZOOM All**.

3. Create the schematic representations of tools shown in Fig. 34-3. Set snap at **3″**. Construct each tool on layer **0**. Omit the text.

NOTE:

A block can be made up of objects from different layers, with different colors and linetypes. The layer, color, and linetype information of each object is preserved in the block. When the block is inserted, each object is drawn on its original layer, with its original color and linetype, no matter what the current drawing layer and object linetype are.

A block created on layer 0 and inserted onto another layer inherits the color and linetype of the layer on which it is inserted and resides on this layer. Therefore, it is important to create blocks on layer 0 in most cases. Other options exist, but they can cause confusion. Creating blocks on layer 0 is recommended if you want the blocks to take on the characteristics of the layer on which they are inserted.

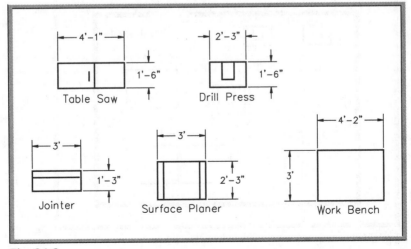

Fig. 34-3

4. Save your work in a file named **lib1.dwg**.

5. Create a block from each of the tools. Use the following information:
   - Use the names shown in Fig. 34-3.
   - Pick the lower left corner of each tool for the insertion base point.
   - Under Objects, pick the **Convert to block** radio button.
   - For Block unit, use inches.
   - The block names adequately describe the blocks, so do not add a description for each block.
   - Uncheck the Open in block editor box.

6. Save your work and close lib1.dwg.

Draw

Using a Symbol Library

We're going to use the new lib1.dwg file to create the workshop drawing shown in Fig. 34-4. With DesignCenter, we can review and insert blocks efficiently.

1. Start a new drawing using the **Quick Setup** wizard and the following information:
   - Use **Architectural** units.
   - Make the drawing area **22′ × 17′**.
   - Set snap at **6″** and grid at **1′**.
   - Be sure to **ZOOM All**.

2. Create a new layer named **Objects**, assign color red to it, and make it the current layer.

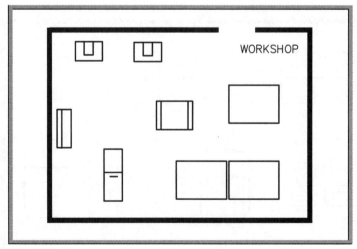

Fig. 34-4

3. Using **PLINE**, create the outline of the workshop as shown in Fig. 34-4. Make the starting and ending width **4″** and make it nearly as large as possible.

4. Save your work in a file named **workshop.dwg**.

5. Pick the **DesignCenter** button from the docked Standard toolbar.

DesignCenter appears. Notice the row of buttons along the top of the DesignCenter window.

6. Pick the **Load** button located in the upper left corner.

The Load dialog box appears.

7. Find the folder with your name and open it unless it is already open.

8. Find and open the drawing file named **lib1.dwg**.

This displays several types of lib1.dwg content in DesignCenter. As you can see, one of them is Blocks.

9. Pick the **Open Drawings** tab.

10. Double-click the **Blocks** icon.

The icons of the five blocks you created appear.

Inserting Blocks

1. Right-click and drag the jointer into the drawing area and release the right button.

2. Pick **Insert Block...** from the shortcut menu.

The Insert dialog box appears.

3. Under Insertion point and Rotation, check **Specify On-screen**, but uncheck this option under Scale.

4. Uncheck the Explode check box and pick the **OK** button.

5. Insert the block in the position shown in Fig. 34-4.

NOTE: ───────────────────────────────

You may need to ZOOM Extents to see the entire workshop drawing.

6. Insert the remaining blocks and save your work.

NOTE:

If you do not need to rotate the block when you insert it, you can save steps by left-clicking and dragging the block into position in the drawing.

You can insert symbols and details that are stored as blocks from any drawing file. AutoCAD provides several good examples.

1. Pick the **Load** button in DesignCenter to display the Load dialog box.

2. Find and open AutoCAD's **Sample** folder, and then open the **DesignCenter** folder.

3. Find the file named **House Designer.dwg** and open it.

4. Double-click **Blocks** in DesignCenter.

5. Insert a **36″** right-swing door into the doorway of the workshop. (You may need to edit the size of the opening.)

6. Experiment with the blocks in some of the other drawing files located in the DesignCenter folder, such as **Home - Space Planner.dwg**.

7. Erase any blocks that are not appropriate for this drawing, and save your work.

Inserting Other Content

As you may have noticed, DesignCenter permits you to drag other content into the current drawing. Examples include layouts, text styles, layers, and dimension styles.

1. Pick the **Load** button in DesignCenter and open the **lib1.dwg** file.

2. Double-click **Dimstyles**.

The Preferred and Standard dimension styles appear because they are stored in lib1.dwg.

3. Left-click and drag **Preferred** into the drawing area.

AutoCAD adds the Preferred dimension style to the current drawing.

4. Review the dimension styles in the docked Styles toolbar.

As you can see, Preferred was indeed added to the current drawing.

5. Pick the **Up** button (the button that has a folder with an Up arrow in it) once in DesignCenter and double-click **Layers**.

This displays the layers contained in lib1.dwg.

Chapter 34: Symbol Libraries

6. Click and drag **Center** to the drawing area.

7. Check the list of layers in the current drawing.

Center is now among them. So, you see, it is very easy to add content such as blocks, dimension styles, layers, and text styles from another drawing to the current drawing.

Other Features

The buttons along the top of the DesignCenter window offer several Windows-standard features. The Tree View Toggle button displays and hides the tree view.

1. Pick the **Tree View Toggle** button (fourth button from the right).

When this button is selected, it splits the window, with the left side consisting of an Explorer-style tree view of your computer's files, folders, and devices.

2. Pick the **Tree View Toggle** button again and again until the display of the tree view is present.

The Favorites button displays the contents of the Favorites folder. You can select a drawing or folder or another type of content and then choose Add to Favorites. AutoCAD creates a shortcut to that item, which is then added to the AutoCAD Favorites folder. The original file or folder does not actually move. Note that all the shortcuts you create using DesignCenter are stored in the AutoCAD Favorites folder.

The Search button displays the Search dialog box. You can enter search criteria and locate drawings, blocks, and nongraphic objects within drawings. The Preview button (third selection from the right) displays a preview of the selected item in a pane below the content area if a preview image has been saved with the selected item. If not, the Preview area is empty.

The Description button displays a text description of the selected item in a pane below the content area. When creating a new block, you are given the opportunity to enter a description; this is where you can view this information.

3. Save your work, close DesignCenter, and exit AutoCAD.

Chapter 34
Review & Activities

Review Questions

1. What is the primary purpose of creating a library of symbols and details?

2. When you create a library, on what layer should you create and store the blocks? Why?

3. What types of content does DesignCenter display?

4. From DesignCenter, how do you insert a dimension style into the current drawing?

5. Is it possible to use DesignCenter to insert content from other AutoCAD drawings into the current drawing? Explain.

Challenge Your Thinking

1. Identify an application for creating and using a library of symbols and details. Identify the symbols the library should include. Discuss your idea with others and make changes and additions according to their suggestions.

2. Why are blocks, rather than groups, used in symbol libraries? Can you think of any applications for which you could use groups as well as blocks in a symbol library? Explain.

3. Explore reasons it might be useful to add shortcuts to the Favorites folder that appears in DesignCenter. How might adding shortcuts to libraries of symbols and details be especially helpful?

Applying AutoCAD Skills

Work the following problems to practice the commands and skills you learned in this chapter.

1. Based on steps described in this chapter, create an entirely new symbol library specific to your area of interest. For example, if you practice architectural drawing, create a library of doors and windows. First create and/or specify a drawing template. If you completed the first "Challenge Your Thinking" item, you may choose to create the symbol library you planned.

2. After you have completed the library of symbols and details, begin a new drawing and insert the blocks. Create a drawing using the symbols and details.

3. The logic circuit for an adder is shown in Fig. 34-5. Draw the circuit by first constructing the inverter and the AND gate as blocks. Use the DONUT command for the circuit connections.

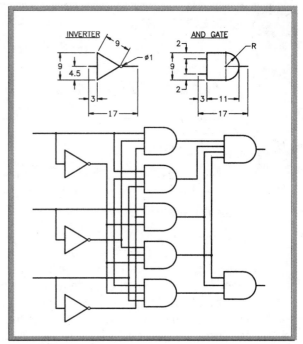

Fig. 34-5

Using Problem-Solving Skills

Complete the following activities using problem-solving skills and your knowledge of AutoCAD.

1. As a building contractor, you want to make a customized library from AutoCAD's existing libraries to suit your own needs. Open DesignCenter, select the Sample folder, and then pick the DesignCenter folder. Find out what blocks are most suitable for a building contractor. Browse through AutoCAD's drawing libraries and select the blocks that are most appropriate. Adjust the scale so that they fit on the screen. Save the drawing as contractor.dwg.

2. As a drafter for an electronics firm, you need a library of electronic symbols. Consult reference books to determine the correct schematic symbols for a PNP transistor, NPN transistor, resistor, capacitor, diode, coil, and transformer. Draw the symbols and make blocks of them to create an electronics symbol library. Save it as electsymbol.dwg.

Problem 3 courtesy of Gary J. Hordemann, Gonzaga University

Chapter 35 Attributes

Objectives:

- Create fixed and variable attributes
- Store attributes in blocks
- Edit individual attributes

Key Terms

attributes
attribute tag
attribute values

fixed attributes
variable attributes

Attributes consist of text information stored within blocks. The information describes certain characteristics of a block, such as size, material, model number, cost, etc., depending on the nature of the block. The advantage of adding attribute information to a drawing is that it can be extracted to form a report such as a bill of materials. The attribute information can be made visible, but in most cases, you do not want the information to appear on the drawing. Therefore, it usually remains invisible, even when plotting.

The electrical schematic shown in Fig. 35-1 contains attribute information, even though you cannot see it. (The numbers you see in the components are not the attributes.)

Figure 35-2 shows a zoomed view of one of the components after the attribute information has been made visible. In this example, the attribute information is displayed near the top of the component.

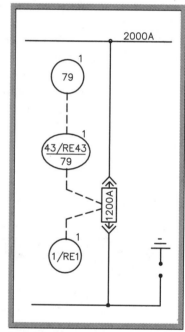

Fig. 35-1

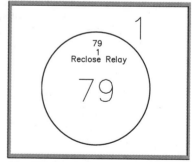

Fig. 35-2

DESCRIPTION	DEVICE	QUANTITY/UNIT
Recloser Cut-out Switch	43/RE43/79	1
Reclose Relay	79	1
Lightning Arrester	--	3
Breaker Control Switch	1/RE1	1
1200 Amp Circuit Breaker	52	1

Fig. 35-3

All of the attributes contained in this schematic were compiled into a file and placed into a program for report generation. The report (bill of materials) in Fig. 35-3 was generated directly from the electrical schematic drawing.

Fixed Attributes

Attributes can be fixed or variable. *Fixed attributes* are those whose values you define when you first create the block. You will learn about variable attributes later in this chapter.

Creating Attributes

Attributes are created, or defined, using the ATTDEF command.

1. Start AutoCAD and open the drawing named **lib1.dwg**.

It should look similar to the one in Fig. 35-4, but without the text.

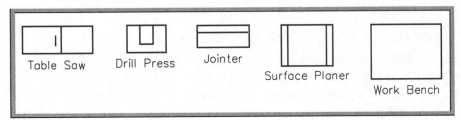

Fig. 35-4

NOTE:

If you do not have lib1.dwg on file, create it using the steps outlined in Chapter 34.

Let's assign information (*attribute values*) to each of the tools so that we can later generate a bill of materials. We'll design the attributes so that the report will contain a brief description of the component, its model, and the cost.

2. Zoom in on the table saw. It should fill most of the screen.

3. From the **Draw** pull-down menu, select **Block** and **Define Attributes…**, or enter **ATTDEF** (short for "attribute definition") at the keyboard.

This enters the ATTDEF command and displays the Attribute Definition dialog box, as shown in Fig. 35-5.

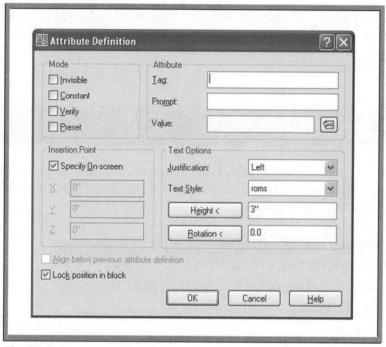

Fig. 35-5

4. In the Mode area, check **Invisible** and **Constant**.

Invisible specifies that attribute values are not displayed when you insert the block. Constant gives attributes a fixed value for all insertions of that block. You will learn more about these two items when you insert blocks that contain attributes.

5. In the Attribute area, type the word **description** (in upper- or lowercase letters) in the box at the right of Tag.

The *attribute tag* identifies each occurrence of an attribute in the drawing. Once again, this will become clearer after you create and use attributes.

6. In the box at the right of Value, type **Table Saw** (exactly as you see it here).

Table Saw becomes the default attribute value.

7. Under Text Options, enter the following information:
   - Justification: **Left**
   - Text Style: **roms**
   - Height: **3″**
   - Rotation: **0.0**

8. Pick the **OK** button.

9. In reply to Specify start point, pick a point inside the table saw, near the top left.

The word DESCRIPTION appears on the table saw.

10. Press the spacebar to repeat the ATTDEF command.

11. This time, enter **model** for the tag and **1A2B** for the value.

12. Pick the **Align below previous attribute definition** check box in the lower left corner of the dialog box and pick **OK**.

The word MODEL appears below DESCRIPTION.

13. Repeat Steps 10 through 12, but enter **cost** for the tag and **$625.00** for the value. Be sure to pick the **Align below previous attribute definition** check box.

14. Save your work.

You are finished entering the table saw attributes. Let's assign attributes to the remaining tools.

15. Zooming and panning as necessary, assign attributes to the remaining tools. Use the information shown in Table 35-1.

16. **ZOOM All** and save your work.

Description	Model	Cost
Drill Press	7C-234	$590.00
Jointer	902-42A	$750.00
Surface Planer	789453	$2070.00
Work Bench	31-1982	$825.00

Table 35-1

HINT:

Be sure to pick the Align below previous attribute definition check box in the Attribute Definition dialog box when adding the second and third attributes to each tool. This will save time and will automatically align the second and third attributes under the first one.

Storing Attributes

Let's store the attributes in the blocks.

Modify

1. Pick the **Explode** button from the docked Modify toolbar and explode each of the tools.

Exploding the blocks permits you to redefine them using the same name.

Draw

2. Pick the **Make Block** button from the docked Draw toolbar.

3. Pick the down arrow at the right of Name and pick **Table Saw**.

4. Pick the **Pick point** button and pick the lower left corner of the table saw for the insertion base point.

5. Pick the **Select objects** button, select the table saw and its attributes, and press **ENTER**.

6. Pick the **OK** button.

7. When AutoCAD asks if you want to redefine Table Saw, pick **Yes**.

This redefines the Table Saw block, and the attribute information disappears.

8. Repeat Steps 2 through 7 to redefine each of the remaining tool blocks.

9. Save your work.

All of the tools in lib1.dwg have been redefined and now contain attributes. When you insert these tools into another drawing, the attributes will insert also.

INFOLINK

You will learn more about attribute extraction and bills of materials in Chapter 36.

Displaying Attributes

Let's display the attribute values using the ATTDISP (short for "attribute display") command.

1. Enter **ATTDISP** and specify **On**.

You should see the attribute values, similar to those shown in Fig. 35-6.

2. Reenter **ATTDISP** and enter **N** for Normal.

The attribute values should again be invisible.

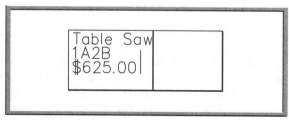

Fig. 35-6

Chapter 35: Attributes

Variable Attributes

Thus far, you have experienced the use of fixed attribute values. With *variable attributes*, you have the freedom of changing the attribute values as you insert the block. Let's step through the process.

1. Using the **tmp1.dwt** template file, begin a new drawing.

2. Zoom in on the lower left quarter of the display, make layer **0** the current layer, and set the snap resolution to **2"**.

3. Draw the architectural window symbol shown in Fig. 35-7. The dimensions not given are **2"** in length. Do not place dimensions on the drawing. (The symbol represents a double-hung window for use in architectural floor plans.)

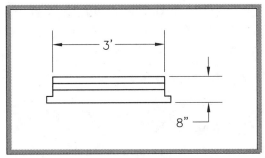

Fig. 35-7

4. Save your work in a file named **window.dwg**.

5. Enter the **ATTDEF** command and check **Invisible**, but Constant, Verify, and Preset should be unchecked.

6. For the tag, type the word **type**.

7. In the Prompt edit box, type **What type of window?**

8. For the value, type **Double-Hung**.

9. Under Text Options, enter the following information.
   - Justification: **Center**
   - Text Style: **roms**
   - Height: **3"**
   - Rotation: **0.0**

10. Pick the **OK** button and pick a point over the top center of the window, leaving space for two more attributes.

The word TYPE appears.

11. Redisplay the dialog box, and leave the attribute modes as they are.

12. Type **size** for the attribute tag, **What size?** for the attribute prompt, and **3′ × 4′** for the attribute value.

13. Pick the **Align below previous attribute definition** check box, and pick **OK**.

The word SIZE appears on the screen below the word TYPE.

14. Repeat Steps 11 through 13 using the following information.
    - Tag: **manufacturer**
    - Prompt: **What manufacturer?**
    - Value: **Andersen**

15. Store the window symbol and attributes as a block. Name it **DH** (short for "double-hung") and pick the lower left corner for the insertion base point.

The Edit Attributes dialog box appears.

16. We do not want to edit the attributes at this time, so pick the **Cancel** button and save your work.

17. Enter **ATTDISP** and **On**.

Did the correct attribute values appear?

Inserting Variable Attributes

1. Enter the **INSERT** command and select the block named **DH** in the Insert dialog box.

2. Uncheck **Explode** if it is checked.

3. Make sure **Specify On-screen** is checked under Insertion point (Scale and Rotation should be unchecked) and pick **OK**.

4. Pick an insertion point at any location.

5. In reply to What manufacturer?, press **ENTER** to accept Andersen.

6. In reply to What size?, enter **3′ × 5′**.

7. In reply to What type of window?, press **ENTER** to accept Double-Hung.

The block and attributes appear.

Editing Attributes

ATTEDIT, short for "attribute edit," allows you to edit attributes in the same way as you edited the size attribute of the double-hung window. Your drawing should currently look similar to the one in Fig. 35-8. The exact location of the attributes is not important.

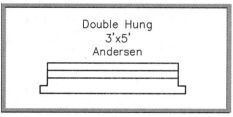

Fig. 35-8

1. Enter the **ATTEDIT** command and pick one of the two blocks.

This displays the Edit Attributes dialog box.

2. Change **Andersen** to **Pella** and pick **OK**.

Andersen changes to Pella.

3. Save your work and close the file named window.dwg.

The lib1.dwg file again becomes the current drawing.

Block Attribute Manager

AutoCAD's Block Attribute Manager permits you to edit the attribute definitions in blocks, remove attributes from blocks, and change the order in which you are prompted for attribute values when inserting a block.

1. Enter the **BATTMAN** command.

This displays the Block Attribute Manager dialog box as shown in Fig. 35-9.

2. Pick the down arrow located at the right of Block near the top of the dialog box and pick **Surface Planer**.

3. Select the line that includes **COST** and **$2070.00** and pick the **Edit...** button.

This displays the Edit Attribute dialog box.

4. Pick each of the tabs and review the information and options in each one and then pick the **Cancel** button.

5. Pick **Cancel** and exit AutoCAD without saving.

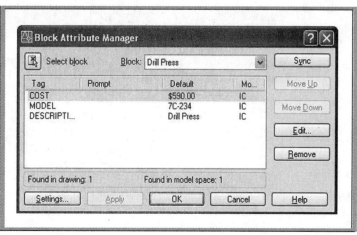

Fig. 35-9

Chapter 35
Review & Activities

Review Questions

1. Explain the purpose of creating and storing attributes.

2. Briefly define each of the following commands.
   ATTDEF
   ATTDISP
   ATTEDIT
   BATTMAN

3. What are attribute tags?

4. What are attribute values?

5. Explain the attribute modes Invisible and Constant.

Challenge Your Thinking

1. Discuss the advantages and disadvantages of using variable and fixed attribute values. Why might you sometimes prefer one over the other?

2. Brainstorm a list of applications for blocks with attributes. Do not limit your thinking to the applications described in this chapter. Compare your list with lists created by others in your group or class. Then make a master list that includes all the ideas.

Applying AutoCAD Skills

Work the following problems to practice the commands and skills you learned in this chapter.

1. Load the livroom.dwg drawing containing the furniture representations you created in Chapter 32. If this file is not available, create a similar drawing. Outline a simple plan for assigning attributes to each of the components in the drawing. Create the attributes and redefine each of the blocks so the attributes are stored within the blocks.

2. The small hardware shop in which you work is computerizing its fasteners inventory. The shop carries both U.S. and metric sizes. All of the available fasteners are saved as blocks in the DesignCenter folder in AutoCAD's Sample folder. Place all of the blocks from Fasteners – Metric.dwg and Fasteners – US.dwg into a drawing named hardware.dwg. Place them in an orderly and logical fashion. Then assign attributes to each fastener to describe it. Use attribute tags of Type and Size. For the individual attribute values, use the information given in the name of each block.

3. Look in a computer hardware catalog to find the sizes and basic shapes of several printers from different manufacturers. Draw the basic shape of at least five different printers. Create attributes to describe the printers in terms of type (inkjet, laser, etc.), brand, and cost. Save the drawing as printers.dwg.

4. Refer to schem.dwg and schem2.dwg on the Instructor Resource CD. Display and edit the attributes contained in the drawing files.

 ## Using Problem-Solving Skills

Complete the following problems using problem-solving skills and your knowledge of AutoCAD.

1. Attributes are used for many reasons besides creating bills of materials. For example, you can use attributes to track items in a collection of anything from model cars to unique bottles to action figures. For this problem, suppose you are a dealer in and collector of Avon bottles that have a transportation theme. You have been asked to participate in a display of various collections being sponsored by the local library. Display space is limited, so you need to make a layout to determine your specific needs. Create a library of schematic symbols for the following car-shaped bottles. The cars vary in size from 2″ × 5″ for the Corvette and Studebaker to 2½″ × 8″ for the Pierce Arrow and 3″ × 9″ for the Cord and Deusenberg. Assign the attributes as shown in Table 35-2 on page 488. Save the drawing as cars.dwg.

Bottle	Contents	Price
1937 Cord	Wild Country	$12
1951 Studebaker	Spicy	$8
1988 Corvette	Spicy	$8
Silver Deusenberg	Oland	$13
1933 Pierce Arrow	Deep Woods	$10

Table 35-2

2. You are setting up sprinklers for new landscaping for a residential plot. Create a new drawing with architectural units and a drawing area of 300′ × 200′. Draw the plot boundaries with corners at the following absolute coordinates:

45′,170′
235′,140′
210′,45′
35′,25′

The house will be a rectangle of 60′ × 32′. The northwest corner of the house is at absolute coordinates 90′,112′. Your client wants a sprinkler system with enough sprinkler heads to cover the entire property. Each sprinkler head can be set to turn a full 360° or any number of degrees greater than 30°. When set to turn 360°, each sprinkler can cover a radial area of approximately 30′. Decide on the number of sprinklers needed, their placement on the property, and the angle (number of degrees) that each should be set to cover. Then create the sprinkler heads, load them from your library, or load them from DesignCenter. Place them in the locations you have determined. On a separate layer, draw the range of each sprinkler to show the client that the sprinkler heads will indeed cover the entire property.

Finally, define attributes with the following tags: Location, Angle, and Direction. Assign these attributes to each sprinkler head in your drawing and give them the appropriate values according to the plan you have developed. Save the drawing as sprinklers.dwg.

Chapter 36

Bills of Materials

Objectives:

- **Extract attributes from blocks for report generation**
- **Create a bill of materials**

Key Terms

attribute extraction   template file
comma-delimited file   xref
nested block

After finishing the attribute assignment process (Chapter 35), you are
ready to create a report such as a bill of materials. The first step in this
process involves *attribute extraction* (gathering the attribute data and
placing it into an AutoCAD table or an electronic file that can be read by
a computer program).

1. Start AutoCAD and begin a new drawing using the **tmp1.dwt**
   template file.

2. Create the four walls of a workshop (fill the upper two-thirds of
   the drawing area) and insert each of the blocks on layer Objects as
   shown in Fig. 36-1 on page 490.

HINT:

Pick the DesignCenter button from the docked Standard toolbar. Find
the file named lib1.dwg located in the folder with your name and
insert the five blocks contained in this drawing file.

3. Save your work in a file named **extract.dwg**.

4. Enter **ATTDISP** and **On**.

Each of the tools contains attributes, as shown in Fig. 36-1.

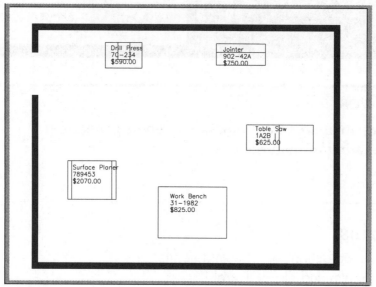

Fig. 36-1

Attribute Extraction

The EATTEXT command permits you to extract the attributes from the extract.dwg file using the Attribute Extraction wizard.

1. From the Tools pull-down menu, select **Attribute Extraction...** or enter the **EATTEXT** command.

This starts the Attribute Extraction wizard, as shown in Fig. 36-2. There are six pages in the Attribute Extraction wizard.

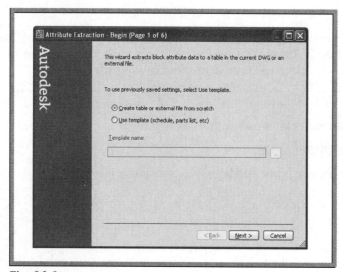

Fig. 36-2

2. Make sure the radio button for **Create table or external file from scratch** is selected and pick **OK**.

3. On page 2 of the wizard, select the **Current drawing** radio button, because we want to extract attributes from all blocks in the current drawing. The current drawing, extract.dwg, should be listed in the Drawing files area.

4. Pick the **Additional Settings...** button before continuing. Review the options available in the dialog box.

The Block settings area allows you to select different blocks to be included in the extraction. The extract.dwg file does not contain any external references (xrefs) or nested blocks, so it does not matter whether these three boxes are checked. An *xref* is a drawing that is attached (linked) to the current drawing. Any changes made to the attached drawing are reflected in the current drawing. A *nested block* is a block that contains one or more other blocks (*i.e.*, a block within a block).

The Count settings area allows you to extract blocks from the entire drawing or only blocks from the model space. Again, it does not matter which radio button is selected, because extract.dwg contains no blocks in layout space.

5. Pick **Cancel** to exit the Additional Settings dialog box.

6. Pick **Next** to go to page 3 of the wizard.

Selecting and Reviewing the Attributes

Page 3 of the Attribute Extraction wizard displays the block names and attribute data associated with each one, as shown in Fig. 36-3 on page 492. The Blocks area on the left lists all the blocks in the drawing. The check box above the list gives you the option to exclude blocks without attributes. The Properties for checked boxes area allows you to choose which properties (attributes) to include or exclude from the extraction. If you uncheck the Exclude general block properties check box above the list, you can also extract general properties such as layer, linetype, and lineweight.

1. We want to extract attribute information from all five blocks, so make sure that all of the check boxes are checked next to the blocks, but *do not* pick the Next button.

2. We want to extract only the cost, model, and description attributes, so uncheck the Name property on the right.

3. Pick the **Next** button.

Page 4 of the wizard presents the attributes in a table format.

4. Review the attributes. Pick the **Full Preview** button, review them again, and then switch back to the original view by closing the Full Preview window.

Fig. 36-3

We want to rearrange the columns of the table into a more logical order for a bill of materials.

5. Click in the **COST** column heading and drag it to the far right of the table.

6. Right-click on the **DESCRIPTION** column heading and select **Sort Ascending**.

The preview of your attribute table should now look similar to the table in Fig. 36-4. As you can see, you have many options for arranging the rows and columns of your table.

Fig. 36-4

Now focus your attention on the Extract attribute data to area. The default destination is an AutoCAD table. You also have the option of creating a comma-delimited (CSV) file, a Microsoft Access® database (MDB) file, a Microsoft Excel® (XLS), or a text extract (TXT) file. A *comma-delimited file* is a popular format for importing data into database and other programs.

7. Make sure the box next to AutoCAD table is checked; then pick **Next**.

Page 5 of the Attribute Extraction wizard allows you to select a style for the attribute table. You can modify an existing table style or create a new one.

8. Pick **Next** to accept the Standard table style.

Page 6 of the Attribute Extraction wizard is displayed.

Creating a Template File

Before completing the attribute extraction, AutoCAD gives you the opportunity to save a *template file* that contains the settings you have made using the wizard for later use.

1. Pick the **Save Template...** button, locate and open your named folder, and type **extract** for the name.

2. Pick the **Save** button. (A BLK file extension is assigned automatically.)

3. Pick the **Finish** button.

Placing the Attribute Table in a Drawing

When you pick Finish, the table attaches to the crosshairs so that you can place it in the drawing.

1. Pick a point below the workshop in response to Specify insertion point.

As you can see, our bill of materials is much too small for the drawing.

2. Type **SCALE**, select the table, and use a scale factor of **20**.

3. Move the table so it is centered below the walls of the workshop.

4. Save your work and exit AutoCAD.

Chapter 36
Review & Activities

Review Questions

1. Describe the purpose of AutoCAD's EATTEXT command.

2. Describe the process of creating a report or bill of materials from a drawing that contains blocks with attributes.

3. What is the purpose of the template (BLK) file?

4. Name the file types into which you can store the attributes using the Attribute Extraction wizard.

Challenge Your Thinking

1. Find out more about comma-delimited files. After you have extracted a CSV file, explore and list options for opening and printing its contents.

2. Research template (BLK) files. Write a short paragraph explaining what they do, what they contain, and why they are useful.

Applying AutoCAD Skills

Work the following problems to practice the commands and skills you learned in this chapter.

1. Open the livroom.dwg file from Chapter 32, which contains several pieces of furniture. Add attributes to the furniture and create blocks. Using the Attribute Extraction wizard, create a bill of materials and place it in the drawing.

2. Using the attribute data file created in problem 1, insert its contents into Microsoft Excel or another spreadsheet if you have access to one.

3. Open sprinklers.dwg from Chapter 35. Create a list of the sprinkler heads for the landscaping contractor to use as reference when the sprinkler system is installed.

4. Refer to schem.dwg and schem2.dwg on the Instructor Resource CD. Using the procedure outlined in this chapter, create a bill of materials from each of the two drawings.

Using Problem-Solving Skills

Complete the following activities using problem-solving skills and your knowledge of AutoCAD.

1. Open cars.dwg from Chapter 35. Using the schematic blocks, create a display to fit in a 2½′ × 3′ display space. Before you begin, you may want to consult antique car or Avon references if you are unfamiliar with any of the car models. Knowing what the cars look like will help you position them attractively in the display. Then extract the attributes and create a list to distribute to potential customers.

2. Open printers.dwg from "Applying AutoCAD Skills" problem 3 in Chapter 35. Extract the attributes from the printer blocks you created and create a table of the printer attributes. Display the table and all of the blocks in the drawing.

Computer Assembly Library

One rule of thumb for CAD drafters and designers is not to draw anything twice. Although that may sound strange at first, what it really means is to use CAD commands and functions to reproduce the same details again and again without having to redraw them. You are already familiar with the MIRROR, COPY, and ARRAY commands. To extend the capabilities of these and other commands, AutoCAD also permits you to block two or more objects so that you can manipulate them as one. These blocks of objects may also be assigned characteristics or attributes, which can be retrieved and listed later if desired.

Description

A business that assembles computers has decided that it wishes to automate its advertising and billing practices. To do so, the CAD drafter has been assigned the task of drawing basic symbols to represent the various parts of the computer. Figure P7-1 shows four possible configurations each for CPUs, CD-ROM drives, and hard disk drives as an example, and Fig. P7-2 shows a possible symbol library based on these configurations.

Using the examples shown in Figs. P7-1 and P7-2, create a symbol library that contains various components of a computer. Do not limit the library to just the three parts shown in Figs. P7-1 and P7-2. Do research if necessary to find out what other components should be included in the symbol library and the configurations for each component. After you have created the symbols, assign appropriate attributes to them, as shown in Fig. P7-1, block the individual symbols and save the symbol library.

Create a new drawing and use the symbol library to configure three computers—a low-cost model, a high-end "supermodel," and a middle-of-the-road model. Extract the attributes and create a list of materials needed to assemble the three computers.

CPUs	
300 MegaHertz	$300
350 MegaHertz	$350
400 MegaHertz	$400
500 MegaHertz	$450
Hard Drives	
4.2 Gigabyte	$125
6.4 Gigabyte	$175
8.0 Gigabyte	$225
12.0 Gigabyte	$300
CD-ROM Drives	
24x	$35
42x	$75
Read/write	$300
DVD-ROM	$350

Fig. P7-1

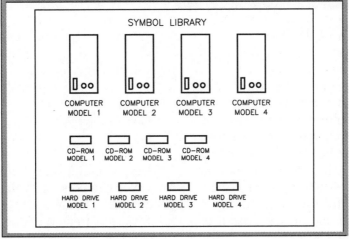

Fig. P7-2

Hints and Suggestions

1. Create all blocks on layer 0.

2. Create only one symbol for each type of component (CPU, hard disk drive, etc.). Assign attributes to it. Then assign parameters to create a dynamic block that can be used to represent all of the models in that category.

Summary Questions

Your instructor may direct you to answer these questions orally or in writing. You may wish to compare and exchange ideas with other students to help increase efficiency and productivity.

1. Distinguish between groups and blocks.

2. Describe the function of a symbol library.

3. What might be an advantage of maintaining a library of dynamic blocks?

4. What factors might determine the attributes assigned to a block?

5. Why is creating a bill of materials from a set of blocks a timesaving technique? Explain.

Careers Using AutoCAD

Electrical/Electronics Drafter

If you've ever taken an electronics class, you may have studied circuit structures and other means of directing electricity from one source to another. Maybe you have watched someone install a new light switch or ceiling fan at home, and you have taken a look at the wiring instructions. Because electricity is a potentially powerful source of energy, it is necessary to understand how to install or disconnect electrical items safely.

Understanding Systems

Drafters who specialize in electricity or electronics must understand the basics of circuitry and wiring systems. They use this knowledge, plus input from electrical engineers, to develop diagrams for the wiring and layout of electronic equipment. The detailed diagrams they create are used by individuals who build and repair electronic equipment.

Electricians also rely on diagrams created by electrical drafters to work with electrical equipment safely and properly. Electricians install wiring systems in buildings, vehicles, power plants, city streets, and anywhere else electricity is routed.

Obtaining Training

Electrical drafters should understand the basics of electricity and wiring. Some high schools offer electronics classes, and many college-level vocational programs provide training and certification for aspiring electrical drafters.

Tim Fuller Photography

Safety consciousness should be a concern for anyone interested in this field. Electricity benefits us in many ways, but it can be harmful if it is not treated with caution.

Electrical or electronics drafters who want to advance in their field often go back to school to become electrical or electronics engineers. In general, to become an electrical or electronics engineer, you must have at least a four-year engineering degree from an accredited university. Experience in the field is also helpful.

Career Activities

- Interview someone at a local electrical repair shop. Ask to look at examples of the kinds of diagrams the repair staff use in their work. Write a short report about your discoveries.
- Find out more about opportunities for electronics engineers and electrical engineers in your area.
- Do research to find out the difference between an electrical engineer and an electronics engineer. How do their jobs differ? Which sounds more interesting to you?

Chapter 37 Isometric Drawing

Objectives:

- Set up a drawing file for isometric drawing
- Create an isometric drawing
- Dimension an isometric drawing correctly

Key Terms

isometric drawing
isometric planes
pictorial representation

An *isometric drawing* is a type of two-dimensional drawing that gives a three-dimensional (3D) appearance. Isometric drawings are often used to produce a *pictorial representation* of a machined part, injection-molded part, architectural structure, or some other physical object. Pictorial representations are useful because they provide a realistic view of the object being drawn, compared to orthographic views. The drawing in Fig. 37-1 is an example of an isometric drawing created with AutoCAD.

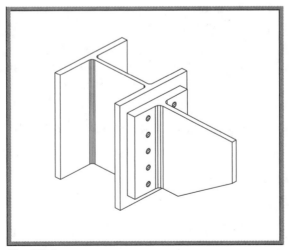

Fig. 37-1

Figure 37-1 courtesy of Gary J. Hordemann, Gonzaga University

Setting Up an Isometric Drawing

Because isometric drawings are two-dimensional, all of the lines in an isometric drawing lie in a single plane parallel to the computer screen. The drawing achieves the 3D effect by using three axes at 120° angles, as shown in Fig. 37-2. In AutoCAD, isometric drawing is accomplished by changing the SNAP style to Isometric mode.

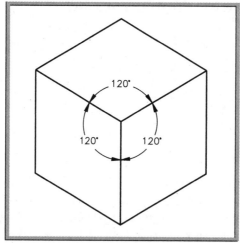

Fig. 37-2

Isometric Planes

As you can see from Fig. 37-2, the 120° angles produce left, right, and top imaginary drawing planes. These three planes are known as the *isometric planes*.

1. Start AutoCAD and begin a new drawing from scratch using **Imperial** units.

2. Set the drawing limits to **15,12**.

3. Set the grid at **1** unit and the snap resolution at **.5** unit.

4. Enter the **SNAP** command.

5. Enter **S** for Style.

6. Enter **I** for Isometric.

7. Enter **.5** for the vertical spacing. (This should be the default value.)

You should now be in Isometric drawing mode, with the crosshairs shifted to one of the three isometric planes.

8. Move the crosshairs and notice that it runs parallel to the isometric grid.

9. Pick **Drafting Settings...** from the **Tools** pull-down menu and pick the **Snap and Grid** tab in the Drafting Settings dialog box.

Notice in the lower right area of this dialog box that you can turn isometric snap on and off. Currently, it is on because you turned it on with the SNAP command.

10. Pick the **Cancel** button.

Toggling Planes

When you produce isometric drawings, you will often use more than one of the three isometric planes. To make your work easier, you can shift the crosshairs from one isometric plane to the next.

1. Press the **CTRL** and **E** keys and watch the crosshairs change.

2. Press **CTRL** and **E** again, and again.

As you can see, this toggles the crosshairs from one isometric plane to the next.

NOTE:

You can also toggle the crosshairs using the ISOPLANE command. Enter ISOPLANE and then enter L, T, or R to change to the left, top, and right planes, respectively. The F5 key also toggles among the three planes.

Creating an Isometric Drawing

Let's construct a simple isometric drawing.

1. Create a layer named **Objects**, assign a color of your choice to it, and make it the current layer.

2. Enter the **LINE** command and draw the aluminum block shown in Fig. 37-3. Make it **3 × 3 × 5** units in size.

3. Save your work in a file named **iso.dwg**.

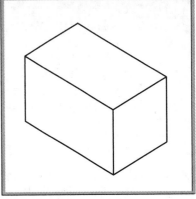

Fig. 37-3

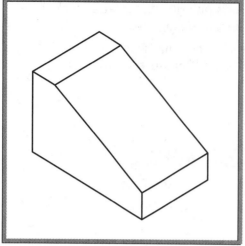

Fig. 37-4

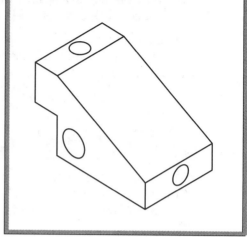

Fig. 37-5

4. Alter the block so that it looks similar to the one in Fig. 37-4, using the **LINE**, **BREAK**, and **ERASE** commands.

5. Further alter the block so that it looks similar to the one in Fig. 37-5. Use the **ELLIPSE** command to create the holes.

HINT:

Since the ellipses are to be drawn on the three isometric planes, toggle the crosshairs to the correct plane. Then choose the ELLIPSE command's Isocircle option. You will need to change the snap spacing to .25.

Dimensioning an Isometric Drawing

To dimension an isometric drawing properly, you must rotate the dimensions so that they show more clearly what is being dimensioned. To do this, you can use the Oblique option of the DIMEDIT command. Compare the two drawings shown in Fig. 37-6. The drawing at the right is the result of applying oblique to the drawing at the left.

Now let's dimension the aluminum block.

1. Create a new layer named **Dimensions**, assign a different color to it, and make it current.

2. In the lower left corner of the block, draw one horizontal and one vertical extension line as shown in Fig. 37-7.

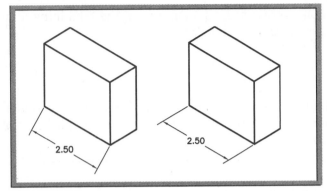

Fig. 37-6

3. Display the **Dimension** toolbar and use the **Aligned Dimension** button to create the dimensions as shown in Fig. 37-7.

Dimension

Your drawing should now look similar to the one in Fig. 37-7.

Notice that dimensions are not well-oriented. This is because you have not yet applied oblique to them.

Dimension

4. From the Dimension toolbar, pick the **Dimension Edit** button to enter the DIMEDIT command.

5. Enter **O** for Oblique.

6. Select the **3.000** dimension at the top of the block and press **ENTER**.

7. Enter **–30** for the obliquing angle.

The dimension rotates –30° to take its proper position.

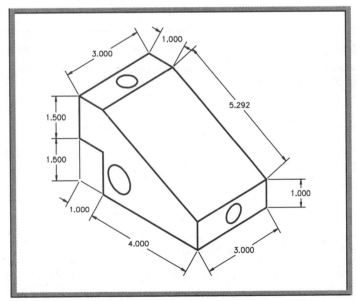

Fig. 37-7

8. Use DIMEDIT to correct the appearance of the remaining dimensions so that they match those in Fig. 37-8.

HINT:

Enter an obliquing angle of 30 or –30, depending on the dimension you are editing. Use grips if necessary to adjust the positions of the dimension lines.

9. Explode the **1.000** dimension at the top of the block and erase the right arrow and dimension line for a cleaner appearance.

When you finish, your drawing should look like the one in Fig. 37-8.

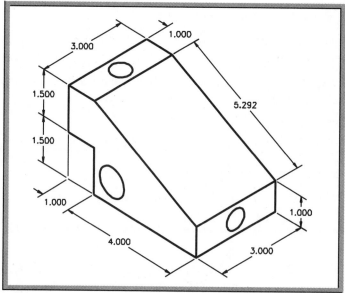

Fig. 37-8

NOTE:

In this drawing, we have not dimensioned the holes in the aluminum block. The completed drawing would include size and location dimensions for all three holes.

10. Return to AutoCAD's standard drawing format using the **SNAP** command or the **Drafting Settings** dialog box.

11. Close the Dimension toolbar, save your work, and exit AutoCAD.

Review Questions

1. Describe two methods of changing from AutoCAD's standard drawing format to isometric drawing.

2. Describe how to change the isometric crosshairs from one plane to another.

3. Explain how to create accurate isometric circles.

4. How do you correct the appearance of dimensions used on an isometric drawing?

5. How do you change from isometric drawing to AutoCAD's standard drawing format?

Challenge Your Thinking

1. Discuss the purpose(s) of isometric drawings. When and why are they used?

2. If you were to dimension the holes in the aluminum block created in this chapter, how would you go about it?

Applying AutoCAD Skills

Work the following problems to practice the commands and skills you learned in this chapter.

1-3. Create the machined parts shown in Figs. 37-9 through 37-11 using AutoCAD's isometric capability. Estimate their sizes.

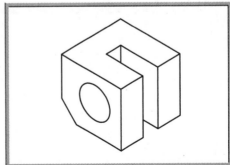

Fig. 37-9

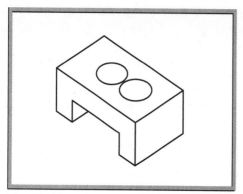

Fig. 37-10

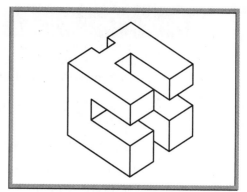

Fig. 37-11

4. Draw the isometric solid according to the dimensions shown in Fig. 37-12A. Then center the hole in the top surface of the part. Specify a diameter of 10.61. Your finished drawing should look like the one in Fig. 37-12B.

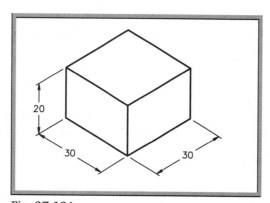

Fig. 37-12A

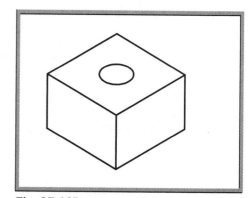

Fig. 37-12B

5. Create the cut-away view of the wall shown in Fig. 37-13.

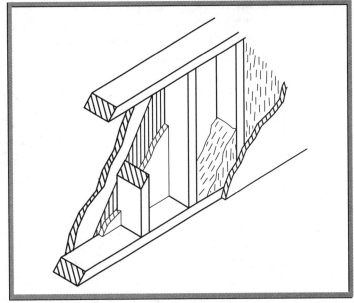

Fig. 37-13

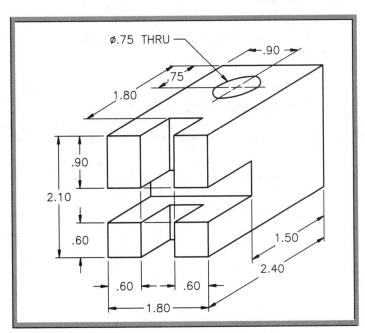

Fig. 37-14

6. A drawing of a guide block is shown in Fig. 37-14. Draw a full-scale isometric view using AutoCAD's SNAP and ELLIPSE commands. An appropriately spaced grid and snap will facilitate the drawing.

Problem 5 courtesy of Autodesk, Inc.
Problem 6 courtesy of Gary J. Hordemann, Gonzaga University

7. An exploded, sectioned orthographic assembly is shown in Fig. 37-15. Draw the assembly as a full-scale exploded isometric assembly. Use isometric ellipses to represent the threads as shown in the isometric pictorial.

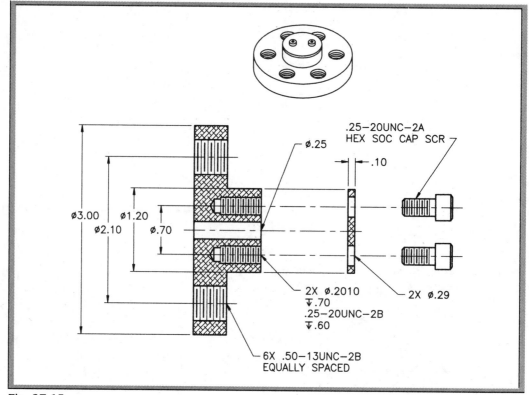

Fig. 37-15

Problem 7 courtesy of Gary J. Hordemann, Gonzaga University

Chapter 37: Isometric Drawing

8. Accurately draw an isometric representation of the orthographic views shown in Fig. 37-16. Draw and dimension the isometric according to the dimensions provided.

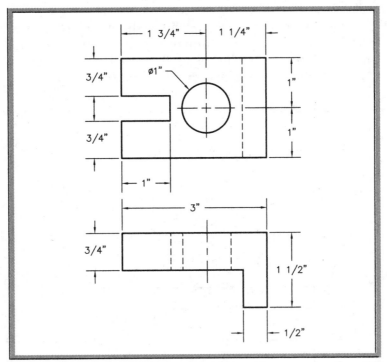

Fig. 37-16

Using Problem-Solving Skills

Complete the following problems using problem-solving skills and your knowledge of AutoCAD.

1. The design engineer for the quality assurance division of your company asked you for an isometric drawing of the dovetail block shown in Fig. 37-17, which is used in a milling machine setup. Draw the block, dimension it, and save it as ch37dove.dwg.

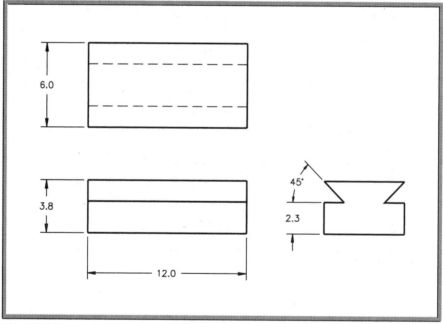

Fig. 37-17

2. The engineer also needs an isometric drawing of the slide shown in Fig. 37-18 for the same milling machine setup. This one, however, does not need to be dimensioned. Save the drawing as ch37slide.dwg.

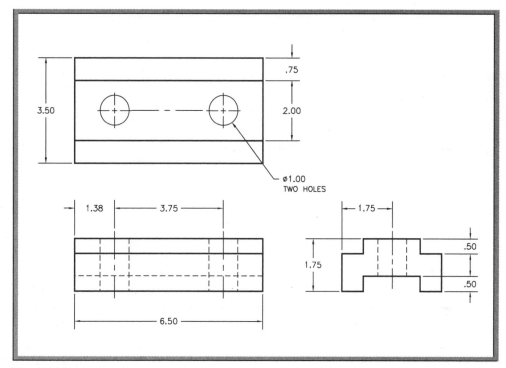

Fig. 37-18

Objectives:

- Create a basic 3D model
- Apply several 3D viewing options
- View and control a model interactively in 3D space

Key Terms

arcball   extrusion thickness   parallel projection
camera   facet shading   perspective projection
clipping plane   flat shading   roll
compass   Gouraud shading   Z axis
equator   hidden line removal

Three-dimensional (3D) drawing and modeling is becoming increasingly popular due, in large part, to improvements in hardware and software. AutoCAD's 3D capabilities allow you to produce realistic views of car instrument panels, snow skis, office buildings, and countless other products and structures. A key benefit of creating 3D drawings is to communicate better a proposed design to people and groups inside and outside a company. Also, the availability of 3D data speeds many downstream processes such as engineering analysis, digital and physical prototyping, machining, and mold-making. In the field of architecture, 3D modeling can help companies persuade customers to move ahead with a proposed construction project such as a shopping mall.

This chapter introduces AutoCAD's 3D capabilities with five easy-to-use commands. These commands permit you to create simple 3D models and view them from any point in space.

An example of a model generated in AutoCAD is shown on the left side of Fig. 38-1. The model on the right is the same object, viewed from the top.

The model consists of a section of steel rod and a sheet-metal shroud that fasten into an aircraft fuel control system. The model shows how the shroud protects a wiring harness that runs through the square notch in the shroud next to the rod. The shroud protects the wires from the heat of the rod. This prevents the plastic insulation from melting and causing an electrical short.

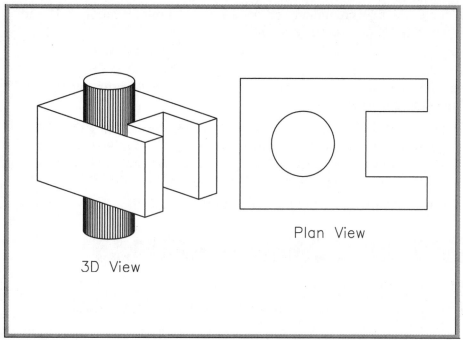

Fig. 38-1

Creating a Basic 3D Model

To work in three dimensions in AutoCAD, you will use a third axis on the Cartesian coordinate system. This axis, called the *Z axis*, shows the depth of an object, adding the third dimension. (In this context, the X axis shows the width of the object, and the Y axis shows the length or height.) The Z axis runs through the intersection of the X and Y axes at right angles to both of them.

Let's draw a steel rod and sheet-metal shroud similar to those shown in Fig. 38-1.

1. Start AutoCAD and create a new drawing using the following information.
    • Use the **Quick Setup** wizard.
    • Set the units to **Decimal**.
    • Set the area to **12″ × 9″**.
    • **ZOOM All**.

2. Create a **1″** grid and a **.5″** snap.

3. Create a new layer named **Objects**, assign the color magenta to it, and make it the current layer.

4. Enter the **ELEV** command and set the new default elevation at **1″** and the new default thickness at **3″**.

An elevation of 1″ means the base of the object will be located 1″ above a baseplane of 0 on the Z axis. The thickness of 3″ means the object will have a thickness on the Z axis of 3″ upward from the elevation plane. This is called the *extrusion thickness*.

5. Draw the top (plan) view of the sheet-metal shroud using the **LINE** command. Construct it as shown on the right in Fig. 38-1. Omit the rod at this time. The width and length of the shroud are 4″ × 6″, respectively. The square notch measures 2″ × 2″.

6. Save your work in a file named **3d.dwg**.

Viewing and Hiding

Let's view the object in 3D.

1. From the **View** pull-down menu, pick **3D Views** and **Viewpoint**.

HINT:

Instead, you may choose to type the VPOINT command and press ENTER twice. You may find this to be faster than picking items from the pull-down menu.

2. Move the pointing device and watch what happens.

3. Place the crosshairs inside the globe representation, also referred to as the *compass*, as shown in Fig. 38-2 and pick that approximate point.

A 3D model of the object appears on the screen.

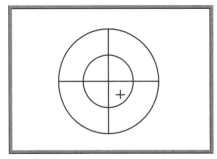

Fig. 38-2

Chapter 38: The Third Dimension

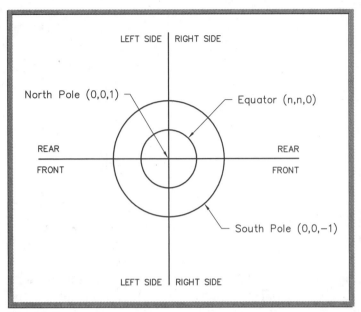

LEFT SIDE | RIGHT SIDE

North Pole (0,0,1)

Equator (n,n,0)

REAR

REAR

FRONT

FRONT

South Pole (0,0,−1)

LEFT SIDE | RIGHT SIDE

Fig. 38-3

Study the globe representation in Fig. 38-3 carefully. The placement of the crosshairs on the globe indicates the exact position of the viewpoint. Placing the crosshairs inside the inner ring (called the *equator*) results in viewing the object from above. Placing the crosshairs outside the inner ring results in a viewpoint below the object.

If the crosshairs is on the right side of the vertical line, the viewpoint will be on the right side of the object. Similarly, if the crosshairs is in front of the horizontal line, the viewpoint will be in front of the object.

4. Enter **Z** (for ZOOM) and **.9x**.

The object should look somewhat like the one in Fig. 38-4.

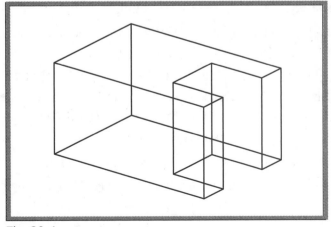

Fig. 38-4

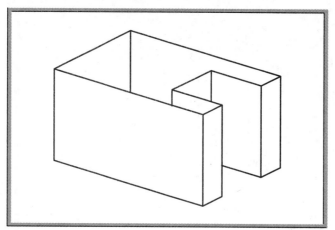

Fig. 38-5

The HIDE command removes edges that would be hidden if you were viewing a real, solid 3D object. Hiding these edges produces a more realistic view of the object. This is called *hidden line removal*.

5. From the **View** pull-down menu, select **Hide**, or enter the **HIDE** command at the keyboard.

The object should now look similar to the one in Fig. 38-5.

6. Return to the plan view of the object by selecting the **View** pull-down menu and then selecting **3D Views** and **Top**.

7. **ZOOM All** to restore the original plan view of the model.

Changing Elevation and Thickness

Let's add the steel rod to the 3D model at a new elevation and thickness.

1. Enter the **ELEV** command and set the elevation at **−1"** and the thickness at **6"**.

2. Draw a **2"**-diameter circle in the center of the model, as shown in Fig. 38-1.

Visualize how the cylinder will appear in relation to the existing object.

3. Enter **VPOINT** and press **ENTER** a second time. Place the crosshairs in approximately the same location as before and pick a point.

Does the model appear as you had visualized it?

4. Remove the hidden lines by entering the **HIDE** command.

The model should now look similar to the one in Fig. 38-1 on page 513.

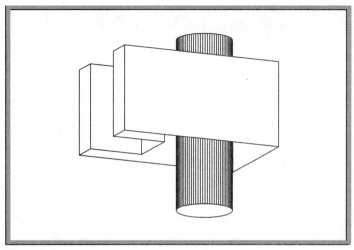

Fig. 38-6

5. Experiment with **VPOINT** to obtain viewpoints from different points in space, and create a 3D view of the model that is similar to the one in Fig. 38-6.

HINT: —————————————————————————

Enter VPOINT and press ENTER a second time. Select a point in the globe. Press the spacebar (or ENTER) to reissue the VPOINT command, and press ENTER to display the globe. Repeating this is a quick way to view the object from several different viewpoints.

6. Save your work.

Viewing Options

AutoCAD offers several options for viewing models in 3D space. One method, which you've practiced, is to use the VPOINT command and globe. Another is to use the 3D Views option in the Views pull-down menu. A third is to select buttons from a toolbar. Fourth, you can make selections from a dialog box. Fifth is 3D Orbit, an option that is both visual and easy to use. In the future, the method you choose should be based on your personal preference. One method is only better than another if it's faster for you and if it accomplishes what you are trying to achieve.

Using the Predefined Viewpoint Buttons

AutoCAD offers several predefined viewpoint buttons in its View toolbar.

1. Display the View toolbar.

2. Rest the pointer on each of the buttons to view its name.

The dark blue side of the button's icon shows the view that will display when the button is picked.

3. Pick each of the buttons except for the first and last ones and enter **HIDE** after each one.

As you can see, this is a fast way of displaying preset views of a model. The SW Isometric View button displays a southwest isometric view of the model, as if the model were oriented on a map. Likewise, SE stands for southeast, NE stands for northeast, and NW stands for northwest.

Using a Dialog Box

1. Display the top view of the model and **ZOOM All**.

2. Enter **VP** at the keyboard.

This displays the Viewpoint Presets dialog box, as shown in Fig. 38-7.

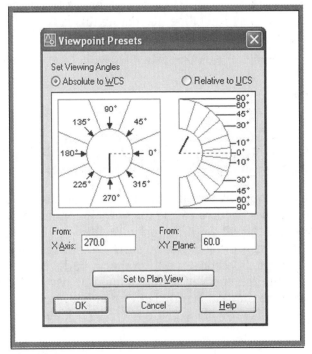

Fig. 38-7

Chapter 38: The Third Dimension

For now, ignore the Absolute to WCS and Relative to UCS radio buttons at the top of the dialog box.

Focus on the half-circle located at the right. It allows you to set the viewpoint height.

3. Pick a point to move the dial (as shown in Fig. 38-7) so that you view the object from above at a **60°** angle and pick the **OK** button.

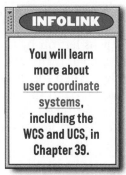

INFOLINK

You will learn more about <u>user coordinate systems</u>, including the WCS and UCS, in Chapter 39.

HINT:

On the semicircle on the right side of the dialog box, pick a point as indicated by the location of the dial in Fig. 38-7.

You should now be viewing the object from above at a 60° angle.

4. Enter **HIDE**.

5. Display the Viewpoint Presets dialog box again.

The full circle (located in the left half of the dialog box) allows you to set the viewpoint rotation.

6. Pick a point to view the object at a **135°** angle and pick **OK**.

7. Remove hidden lines.

8. Display the dialog box again.

The two edit boxes enable you to enter values for the rotation and height.

9. Pick the **Set to Plan View** button and pick **OK**.

10. Enter **ZOOM** and **.9x**.

11. Experiment further with the Viewpoint Presets dialog box.

12. Display the top view and enter **ZOOM** and **1**.

3D Orbit

3D Orbit is a tool for manipulating the view of 3D models by clicking and dragging. It's especially useful because you can use it to manipulate 3D models while they are shaded or when hidden lines are removed. 3D Orbit's right-click shortcut menu offers options for removing hidden lines, rotating objects, panning, zooming, shading, perspective viewing, and clipping planes. You will explore these options in the following steps.

1. Enter **3DO**.

This activates an interactive 3D Orbit view in the current viewport. 3D Orbit displays an object called an *arcball*, as well as a grid of lines (unless the grid was left off by the last person to use AutoCAD on your computer). The arcball is the green circle divided into four quadrants by smaller circles. When the 3DORBIT command is active, the target of the view stays stationary and the point of view (also called the *camera*) moves around the target. The target point is the center of the arcball, not the center of the object(s) you are viewing.

2. Position the cursor inside the arcball and click and drag upward to rotate the model.

3. With the 3DORBIT command active, right-click to display a shortcut menu.

4. In the menu, rest the pointer on **Shading Modes** and then select **Hidden**.

This removes the hidden lines from the model.

5. With the cursor inside the arcball, click and drag.

As you can see, the model remains in hidden-line mode.

A red, green, and blue icon made up of the X, Y, and Z axes appears in the lower left area (unless the last person to use AutoCAD on your computer turned it off). The icon serves as a visual aid to help you understand the direction from which you are viewing the model. If you do not see it now, that's okay.

Rotating Objects in 3D Space

When you move the cursor over different parts of the arcball, the cursor icon changes. When you click and drag, the appearance of the cursor indicates the rotation of the view.

1. Study the cursor; then move it outside the arcball and then back inside it.

The cursor changes to a small sphere encircled by two lines. When the cursor is a small sphere, as shown in Fig. 38-8A, you can manipulate the view freely.

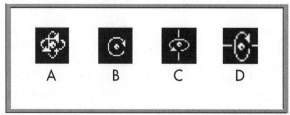

Fig. 38-8

2. With the cursor inside the arcball, click and drag up and down, back and forth.

3. Move the cursor outside the arcball.

The cursor changes to a circular arrow around a small sphere as shown in Fig. 38-8B.

When you click and drag outside the arcball, AutoCAD moves the view around an axis at the center of the arcball, perpendicular to the screen. This action is called a *roll*.

4. With the cursor outside the arcball, click and drag.

5. Move the cursor over one of the smaller circles on the left or right side of the arcball.

The cursor becomes a horizontal ellipse around a small sphere as shown in Fig. 38-8C.

When you click and drag from either of these points, AutoCAD rotates the view around the Y axis that extends through the center of the arcball.

6. With the cursor over one of the smaller circles at the left or right, click and drag to the left and right.

7. Move the cursor over one of the smaller circles on the top or bottom of the arcball.

The cursor becomes a vertical ellipse around a small sphere as shown in Fig. 38-8D.

When you click and drag from either of these two points, AutoCAD rotates the view around the horizontal or X axis that extends through the center of the arcball.

8. With the cursor over one of the smaller circles at the top or bottom, click and drag up and down.

Panning and Zooming

1. With the **3DORBIT** command active, right-click to display the shortcut menu and select **Pan**.

This causes AutoCAD to change to the pan mode within the 3DORBIT command.

2. Click and drag to adjust the view of the model.

3. Right-click and pick **Zoom**.

4. Zoom in and out on the model.

5. Display the shortcut menu and select **Orbit**.

The arcball returns.

Consider the two buildings shown in Fig. 38-9. Both are shown from the same point in space. However, the one at the top is shown in *parallel projection*. This is the standard projection that AutoCAD uses when displaying 3D models, and this is how you are currently viewing the 3D model. The lines are parallel and do not converge toward one or more vanishing points. The building at the bottom is a *perspective projection*. Parts of the building that are farther away appear smaller, the same as in a real-life photograph.

Fig. 38-9

1. With 3DORBIT still active, right-click to display the shortcut menu.

2. Pick **Projection** and then pick **Perspective**.

You are now viewing the model in perspective projection.

3. Rotate the model.

4. If necessary, zoom out.

You now get a more realistic view of the model.

Shading the Object

1. Display the shortcut menu, pick **Shading Modes**, and pick **Flat Shaded**.

AutoCAD flat-shades the model. *Flat shading* is a very basic method of shading a model using a single shade of color. You will learn more about flat shading (also called *facet shading*) later in this book.

2. Rotate the model.

The model remains shaded, which is another benefit of using 3D Orbit to rotate models.

3. From the shortcut menu, select **Shading Modes** and **Gouraud Shaded**.

Gouraud shading is a method of smooth shading that results in a more realistic model. The cylinder now has a smoother appearance.

4. Rotate the model.

The last two shading modes, Flat Shaded, Edges On and Gouraud Shaded, Edges On, shade the model and display the edges that represent the model.

5. Select **Flat Shaded, Edges On**; then select **Gouraud Shaded, Edges On**.

Notice the difference in the appearance of the model.

Controlling Visual Aids

AutoCAD provides several visual aids to help orient you in 3D space as you work with objects in 3D Orbit.

1. From the shortcut menu, pick **Visual Aids** and **Compass**.

This displays a 3D sphere within the arcball composed of three ellipses representing the X, Y, and Z axes.

2. Rotate the model and pay attention to the 3D sphere.

3. From the shortcut menu, turn off **Compass**.

The same shortcut menu controls the grid, which represents the 0 elevation baseplane discussed earlier in this chapter. It also controls the display of the UCS icon, which is the red, green, and blue icon that shows the current position of the X, Y, and Z axes.

4. From the shortcut menu, turn off **Grid** and **UCS Icon**, and rotate the model.

Setting Clipping Planes

A *clipping plane* is a plane that slices through one or more 3D objects, causing part of the object on one side of the plane to be omitted. It is useful in viewing the interior of a part or assembly. You can use clipping planes to achieve views of an object that are similar to sectional views.

1. Display the shortcut menu, select **More**, and pick **Adjust Clipping Planes**.

The Adjust Clipping Planes window appears.

2. In the Adjust Clipping Planes window, depress the **Front Clipping On/Off** button. (It may already be depressed.)

3. With the cursor inside this window, click and drag up and down.

This moves the clipping plane up and down and provides a corresponding view of the model in the drawing area.

4. In the Adjust Clipping Planes window, depress the **Back Clipping On/Off** button, as well as the **Adjust Back Clipping** button.

5. With the cursor inside this window, click and slowly drag up and down to adjust the location of the back clipping plane.

6. Pick the **Create Slice** button, click, and move the cursor up and down.

With the front and back clipping planes on, you can adjust their location at the same time.

7. Close the Adjust Clipping Planes window.

The last two options on the shortcut menu help you control the current view of the object. Reset View resets the view back to how it was when you first started 3D Orbit. The Preset Views option displays a list of predefined views such as Top, Front, and NW Isometric.

8. Display the shortcut menu and pick **Reset View**.

9. Change the projection to parallel and the shading to wireframe.

10. Press **ESC** or **ENTER** to exit 3D Orbit.

11. Close the View toolbar, save your work, and exit AutoCAD.

Review Questions

1. Describe the purpose of the VPOINT command.

2. The extrusion thickness of an object is specified with which command?

3. Briefly explain the process of creating objects (within the same model) at different elevations and thicknesses.

4. When you are using the VPOINT command and the small crosshairs is in the exact center of the globe, what is the location of the viewpoint in relation to the object?

5. With what command are hidden lines removed from 3D objects?

6. Sketch the VPOINT globe shown in Fig. 38-10 and indicate where you must position the small crosshairs to view an object from the rear and underneath.

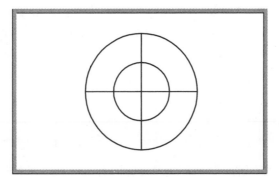

Fig. 38-10

7. Why is the 3D Orbit feature especially useful, compared to other 3D viewing options?

8. What is the difference between parallel projection and perspective projection?

9. What is a clipping plane?

Challenge Your Thinking

1. Match the VPOINT globe representations in Fig. 38-11 with the objects. The first one has been completed to give you a starting point. Write your answers on a separate sheet of paper.

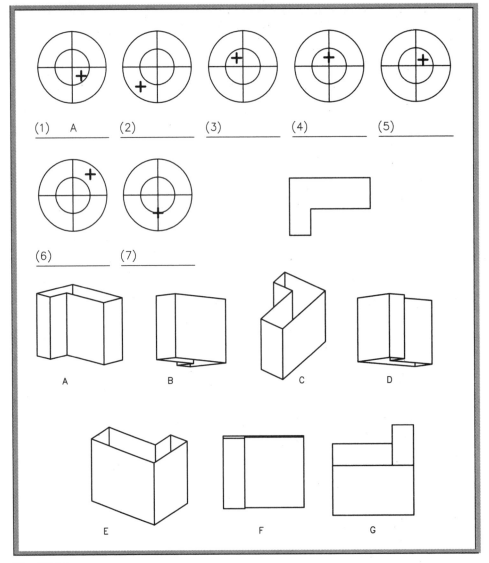

(1) ___A___ (2) _____ (3) _____ (4) _____ (5) _____

(6) _____ (7) _____

Fig. 38-11

Chapter 38: The Third Dimension

2. Of what benefit are front and back clipping planes? Describe at least one situation in which clipping planes could save drawing time or reveal important information about a design.

3. Enter the *x*, *y*, and *z* coordinate values for the alphabetically identified points in Fig. 38-12. To get you started, point A has coordinate values of 0,0,0. Write your answers on a separate sheet of paper.

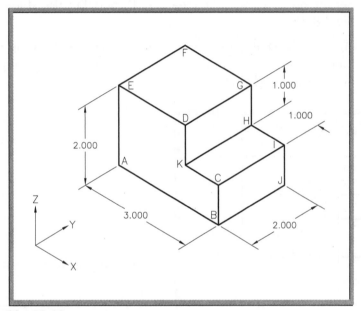

Fig. 38-12

Applying AutoCAD Skills

Work the following problems to practice the commands and skills you learned in this chapter. Draw the following objects and generate a 3D model of each.

1. Set the elevation at 1" to create a 3D model of the guide shown in Fig. 38-13.

2. To draw the shaft shown in Fig. 38-14, set the elevation for the inner cylinder at 0. Set the snap resolution before picking the center point of the first circle.

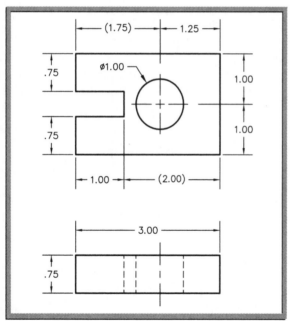

Fig. 38-13

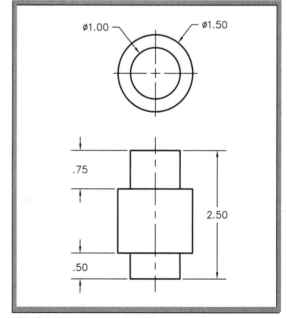

Fig. 38-14

3. Draw the gear blank and hub shown in Fig. 38-15 as a basic 3D model. View the model using each of the buttons on the View toolbar. Are all the views practical for this model? Which ones are not? Enter the HIDE command and view the model again from each position. Is the HIDE command practical for all views? Why?

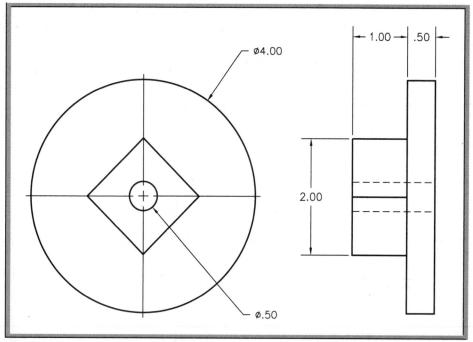

Fig. 38-15

4. Open the file named Welding Fixture Module.dwg located in AutoCAD's Sample folder. Using Save As..., save the file in your folder with a new name of Welding Fixture Module38.dwg. Use 3D Orbit to rotate, zoom, and pan while the model is shaded. As you are manipulating the model, AutoCAD may reduce the shaded mode of the model to wireframe, depending on the speed of your computer. AutoCAD uses this method to maintain speed of the rotation, zoom, or pan, rather than require you to wait for the model to shade. Use each of the shading modes in 3D Orbit. Produce front and back clipping planes through the model to view the interior of the geometry.

Using Problem-Solving Skills

Complete the following activities using problem-solving skills and your knowledge of AutoCAD.

1. The gear manufacturer for whom you supply the gear blank (see "Applying AutoCAD Skills" problem 3) wants to see the blank in a realistic form. Provide a three-dimensional view of the gear blank flat-shaded with the edges on. Save and plot the drawing. Did it plot the way you expected?

2. Use your knowledge of AutoCAD to create the shaft, gear, and cam shown in Fig. 38-16 as closely as possible.

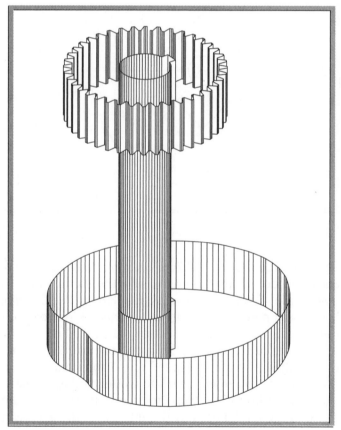

Fig. 38-16

AutoCAD drawing in problem 2 courtesy of Autodesk, Inc.

Chapter 39 — User Coordinate Systems

Objectives:

- **Create 3D faces**
- **Apply X/Y/Z point filters**
- **Create user coordinate systems to simplify work in 3D drawings**
- **Move among UCSs as necessary to work on a 3D model**
- **Generate a plan view of a 3D model**
- **Delete unused UCSs**

Key Terms

point filtering
3D face
UCS icon

user coordinate systems (UCSs)
X/Y/Z point filtering

X/Y/Z point filters allow you to enter coordinates using both the pointing device and the keyboard. With the 3DFACE command, you can create surfaces in 3D space using *x*, *y*, and *z* coordinates. For instance, you can create inclined and oblique surfaces, such as a roof on a building.

Creating 3D Faces

The 3DFACE command creates a three-dimensional object similar in many respects to a two-dimensional solid object. The 3DFACE prompt sequence is identical to that of the SOLID command. However, unlike the SOLID command, the 3DFACE command allows you to enter points in a natural clockwise or counterclockwise order to create a normal *3D face*, or planar surface. *Z* coordinates are specified for the corner points of a 3D face, forming a section of a plane in space.

1. Start AutoCAD and create a new drawing as follows.
   - Use the **Quick Setup** wizard.
   - Set the units to **Architectural**.
   - Set the area to **120′ × 90′**.
   - **ZOOM All**.

2. Create a **10'** grid and a **5'** snap.

3. Create a new layer named **Objects**, assign the color red to it, and make it the current layer.

4. Enter the **ELEV** command and set the new default elevation at **10'** and the new default thickness at **30'**.

5. Draw the outline of the building shown in Fig. 39-1 using the **LINE** command. The width and length of the building are **40'** and **60'**, respectively. The square notch measures **20'** × **20'**.

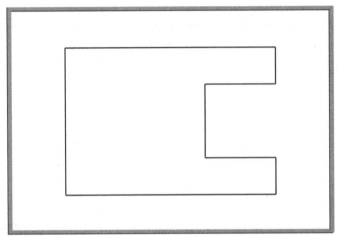

Fig. 39-1

6. Save your work in a file named **3d2.dwg**.

7. With the **VPOINT** command, view the outline of the building from above, front, and left.

8. Remove hidden lines using the **HIDE** command.

9. Undo the last two steps so that you again see the top view.

10. Display the Surfaces toolbar.

Surfaces

11. From the Surfaces toolbar, pick the **3D Face** button.

This enters the 3DFACE command.

X/Y/Z Point Filtering

Point filtering is the process of using the pointing device to enter any two of the 3D coordinates that define a point and then entering the third coordinate from the keyboard. This method allows you to specify points in 3D space without constantly changing your point of view to make the points accessible.

Chapter 39: User Coordinate Systems

1. Type **.xy** and press **ENTER**.

2. In reply to of, approximate the x, y position of point 1 in Fig. 39-2 and pick that point. (Point 1 is about 5′ from the corner of the object.)

This is the start of one side of the roof.

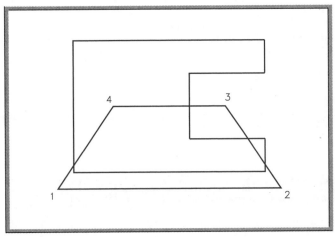

Fig. 39-2

As shown at the AutoCAD prompt, the first point also requires a z coordinate.

3. Enter **35′** for the z coordinate.

This method of entering 3D points is referred to as *X/Y/Z point filtering*.

4. In reply to Second point, enter **.xy** and pick point 2, as shown in Fig. 39-2.

5. Enter **35′** for the z coordinate.

6. Repeat these steps for points 3 and 4, but enter **70′** for the z coordinate of these points. (Be sure to enter **.xy** before picking the point.)

7. Press **ENTER** to terminate the 3DFACE command.

Mirroring a 3D Face

The MIRROR command is not limited to two dimensions. When you mirror an object such as the 3D face you just created, the object is mirrored in 3D space. Let's use the MIRROR command to create the other half of the roof.

1. Enter the **MIRROR** command, select the 3D face object you just created, and press **ENTER**.

2. Place the mirror line so that the two sections of the roof meet along the shorter horizontal edge of the 3D face. Do not delete the "source object."

Now let's view the object in 3D.

3. Select a viewpoint that is to the right, in front of, and above the building.

4. Remove hidden lines.

Your model should look similar to the one in Fig. 39-3.

5. Save your work.

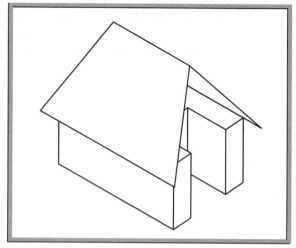

Fig. 39-3

Snapping to Points in 3D Space

A third way to create a 3D face is to snap to existing points at different x,y,z coordinates. Let's use this method to finish the roof so that it looks similar to the one in Fig. 39-4.

Surfaces

1. With the current 3D view still on the screen, enter the **3DFACE** command and snap to point 1, as shown in Fig. 39-4.

2. Snap to point 2.

3. Snap to point 3.

4. Snap to point 1 to complete the surface boundary.

5. Press **ENTER** to terminate the 3DFACE command.

6. Remove hidden lines.

7. View the object from above, front, and left side.

8. Create the remaining portion of the roof using the **3DFACE** command and object snap, and then remove hidden lines.

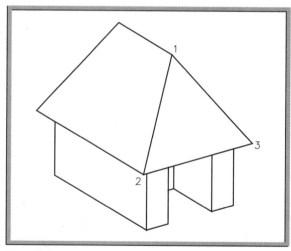

Fig. 39-4

HINT:

To complete Step 8, you may choose to return to the plan view and use the MIRROR command to complete the roof.

9. Generate a view similar to the one in Fig. 39-5.

10. Close the Surfaces toolbar.

11. Save your work.

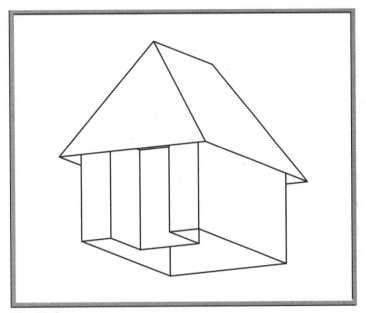

Fig. 39-5

Creating UCSs

AutoCAD's default coordinate system is the world coordinate system (WCS). In this system, the X axis is horizontal on the screen, the Y axis is vertical, and the Z axis is perpendicular to the XY plane (the plane defined by the X and Y axes).

You may create drawings in the world coordinate system, or you may define your own coordinate systems, called *user coordinate systems (UCSs)*. The advantage of a UCS is that its origin is not fixed. You can place it anywhere within the world coordinate system. You can also rotate or tilt the axes of a UCS in relation to the axes of the WCS. This is a useful feature when you're creating a 3D model.

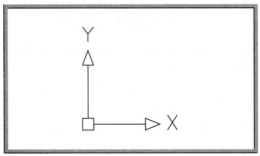

Fig. 39-6

The *UCS icon*, located in the lower left corner of the drawing area, shows the orientation of the axes in the current coordinate system. When you are viewing the plan view of the world coordinate system, the Z axis of the UCS icon is not visible, as shown in Fig. 39-6.

Consider the 3D model shown in Fig. 39-8. A UCS was defined to match the inclined plane of the roof as indicated by the grid of points. Once a UCS such as this is established and made current, all new objects lie on this plane.

The UCS command enables you to create a new current UCS.

1. Using **Save As...**, create a new file named **3d3.dwg**.

2. Alter the plan view so that it resembles the illustration in Fig. 39-7.

3. Obtain a view similar to the one shown in Fig. 39-8.

4. Display the UCS toolbar.

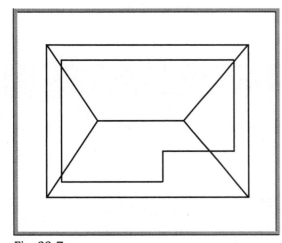

Fig. 39-7

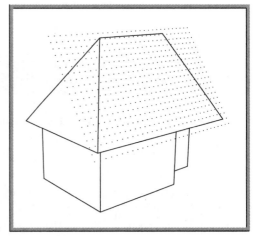

Fig. 39-8

Using Three Points

One of the most common and useful ways of defining a UCS in AutoCAD is to select three points that lie on the axes. The first point becomes the origin, and the other two define the X and Y axes.

UCS

1. From the UCS toolbar, pick the **3 Point** button.

AutoCAD enters the UCS command and, because you chose the 3 Point button, AutoCAD enters the 3point option automatically.

2. In reply to Specify new origin point, snap to point 1 (shown in Fig. 39-9).

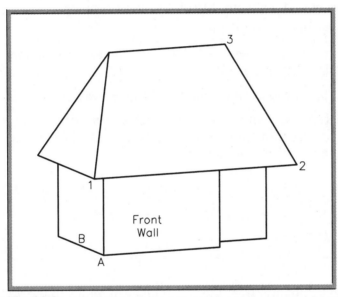

Fig. 39-9

3. In reply to the next prompt, snap to point 2.

This defines the positive X direction from the first point.

4. In reply to the next prompt, snap to point 3. (Point 3 lies in the new XY plane and has a positive *y* coordinate.)

Note that the drawing does not change. However, the crosshairs, grid (if it is on), and coordinate system icon shift to reflect the new UCS.

5. With the coordinate display on, notice the *x* and *y* values in the status bar as you move the crosshairs to each of the three points you selected.

UCS

6. From the UCS toolbar, pick the **UCS** button.

This enters the UCS command.

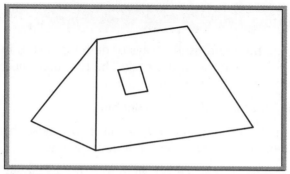

Fig. 39-10

7. Enter the **Save** option and enter **frtroof** for the UCS name.

This names and saves the current UCS.

Suppose you want to construct the line of intersection between the roof and a chimney passing through the roof.

8. Change both the elevation value and the thickness value to **0**.

9. With the **LINE** command, draw the line of intersection where a chimney would pass through (as shown in Fig. 39-10) by creating a rectangle at any location on the roof.

 HINT: ─────────────────────────────

Turn on snap.

UCS

10. Pick the **World UCS** button from the UCS toolbar to return to the WCS.

11. To prove that the rectangle lies on the same plane as the roof, view the 3D model from different points in space.

 HINT: ─────────────────────────────

The viewpoint **4,–.1,1** illustrates it well.

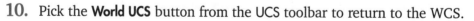
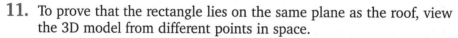
Aligning with an Object

Another convenient way to create a UCS is to align it with a planar object. AutoCAD allows you to specify any object that lies entirely within one plane (such as a 3D face) to define the UCS. Let's create a second UCS, this time using the Object option of the UCS command.

Chapter 39: User Coordinate Systems

1. View the model from an orientation similar to the one used before (as shown in Fig. 39-8).

2. Pick the **Object** button from the UCS toolbar.

UCS

The Object option of the UCS command allows you to align the UCS with a planar object, such as a 3D face, on the screen.

3. Select the bottom edge of the roof. (Use the same section of the roof as before.)

The crosshairs shifts to reflect the orientation of the new UCS. Notice that the Object option creates a UCS that is similar to the one named frtroof.

Using a Z Axis Vector

You are likely to use the 3 Point and Object buttons for most applications, at least at first. However, other UCS creation options are available. With the Z Axis Vector button, you specify the positive direction of the Z axis. AutoCAD then determines the directions of the X and Y axes using an arbitrary but consistent method.

UCS

1. From the UCS toolbar, pick the **Z Axis Vector** button. (Turn off ortho.)

2. Snap to point A (see Fig. 39-9) in reply to Specify new origin point.

3. In reply to the next prompt, pick any point on line B using the **Nearest** object snap.

The new UCS is on the same vertical plane as the front wall of the building. In relation to the wall, notice the positive X, Y, and Z directions that make up the UCS, and notice the UCS icon.

Using Other Options

The View button creates a new UCS perpendicular to the current viewing direction; that is, parallel to the screen. This is helpful when you want to annotate (add notes to) the 3D model.

UCS

1. Pick the **View** button.

2. Using the **DTEXT** command, place your name near the model.

The Z Axis Rotate option creates a new UCS by rotating the current UCS a specified number of degrees around the Z axis.

UCS

3. Pick the **Z** button to enter the Z Axis Rotate option.

4. Enter **30** for the rotation angle.

The current UCS rotates 30° about the Z axis. The X and Y buttons rotate the current UCS about the X and Y axes.

The Origin option allows you to create a new UCS by moving the origin of the UCS, leaving the orientation of the axes unchanged.

UCS

5. Pick the **Origin UCS** button.

6. Pick the lower right corner of the roof (point 2; see Fig. 39-9).

Notice how the UCS icon moves.

Managing UCSs

The purpose of creating user coordinate systems is to make it easier to work with 3D models. When you have created several UCSs, you need to be able to move from one to another quickly. You may also want to delete unused UCSs. The UCS command and toolbars allow you to manage UCSs efficiently.

Changing the Current UCS

AutoCAD provides several methods to change from one UCS to another. Examples include the Previous and World options. The UCS II toolbar also allows you to switch from one UCS to another quickly.

UCS

1. From the UCS toolbar, pick the **UCS Previous** button.

This restores the previous UCS. AutoCAD saves the last ten user coordinate systems in a stack, so you can step back through these systems with repeated UCS Previous operations.

UCS

2. Pick the **World UCS** button.

This restores the world coordinate system.

> **NOTE:** ——————————————————————
>
> If you want to place a point in the WCS while you are in another UCS, do so by entering an asterisk prior to the coordinates (*e.g.*, *7.5,4). During line construction, you can also place relative coordinates (*e.g.*, @*5,3) and polar coordinates (*e.g.*, @*3.5<90) in the WCS regardless of the current UCS.

3. Display the UCS II toolbar and pick the down arrow.

This displays all named UCSs, as well as predefined orthographic UCSs, such as Top, Bottom, and Front. Notice also that the list contains the Previous and World options.

4. Pick the **frtroof** UCS from the list.

This restores the frtroof UCS.

5. Pick **Back** from the list.

This restores the Back UCS. This UCS would be suitable for editing the back side of the building.

NOTE: ───

You can also restore named UCSs at the AutoCAD prompt. Enter the UCS command and press R (for Restore). AutoCAD asks for the name of the UCS to restore.

Generating the Plan View

The PLAN command enables you to generate the plan view of any user coordinate system, including the WCS.

1. Enter the **PLAN** command.

2. Press **ENTER** to generate the plan view of the current UCS.

3. Enter **PLAN** and **W** for World.

This displays the plan view of the world coordinate system.

4. Pick the **World UCS** button from the UCS toolbar.

5. Erase your name and **ZOOM All**.

UCS

NOTE: ───

If you want to display the plan view of previously saved user coordinate systems, enter the UCS option of the PLAN command.

Deleting a UCS

The UCS Del option, which does not have a corresponding button in the UCS toolbar, allows you to delete one or more saved coordinate systems. You may want to delete a UCS that you have defined incorrectly.

1. Enter the **UCS** command and press **D** (for Del).

2. Type **frtroof** and press **ENTER**.

AutoCAD deletes the definition of the frtroof UCS from the drawing database.

3. Pick the **Undo** button to undo the deletion of the UCS.

4. Close the UCS and UCS II toolbars, save your work, and exit AutoCAD.

Chapter 39
Review & Activities

1. Describe a 3D face object.

2. What basic AutoCAD command is the 3DFACE command most like?

3. Describe the X/Y/Z point filtering process.

4. What is the primary benefit of using X/Y/Z point filters?

5. What is the purpose and benefit of using AutoCAD's user coordinate systems?

6. Describe the UCS command's 3point option.

7. What purpose does the UCS icon serve?

8. Briefly describe the UCS command's View option.

9. Which command permits you to quickly see the plan view of the current UCS?

Challenge Your Thinking

1. Experiment with xlines and mlines in three dimensions. Can you use the X/Y/Z point filters to place these objects? In what ways do they behave differently in 3D space?

2. Explain the difference between a view and a UCS, and describe how the two can be used together to create complex 3D models.

3. Using AutoCAD's on-line help, research the UCSFOLLOW system variable. What is the purpose of this variable? Why might it be useful when you are working with several different UCSs?

Chapter 39
Review & Activities

Applying AutoCAD Skills

Work the following problems to practice the commands and skills you learned in this chapter.

1. Create a new drawing with a drawing area of 24′ × 18′. Set the thickness to 8′, and draw a 16′ × 12′ rectangle to represent a room. View the room from the SE Isometric viewpoint. Use the 3Point option of the UCS command to create a new UCS that will allow you to create a door on the right end wall of the room. Using the new UCS, draw a door that is 3′ wide and 6′-6″ high. Center the door on the right wall of the room. Save the drawing as prb39-1.dwg.

2. With the UCS 3point option, create a UCS using one of the walls that make up the building in this chapter. Save the UCS using a name of your choice. With this as the current UCS, add a door and window to the building. Save this drawing as prb39-2.dwg.

3. Obtain a 3D view of the building from this chapter. With the UCS View option, create a UCS. Create a border and title block for the 3D model. Include your name, the date, the file name, etc., in the title block. Save your work as prb39-3.dwg.

Using Problem-Solving Skills

Complete the following activities using problem-solving skills and your knowledge of AutoCAD.

1. Your proposed lampshade design needs to be drawn in three dimensions before your supervisor will approve the design. Draw a 3D model of the lampshade shown in Fig. 39-11 on page 544; omit the hanger. Use the appropriate method(s) for creating the 3D faces. Save the drawing as prb39-lampshade.dwg. Plot it in the view of your choice.

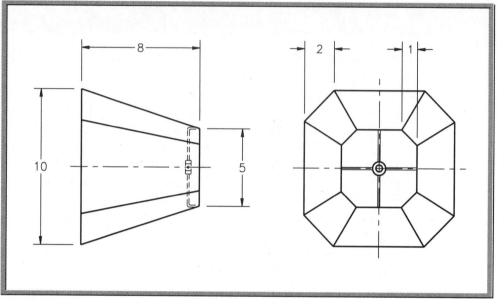

Fig. 39-11

2. Draw the 3D model of the die head shown in Fig. 39-12. Complete the four angled surfaces with the 3DFACE command and X/Y/Z point filtering. Save the drawing as prb39-diehead.dwg.

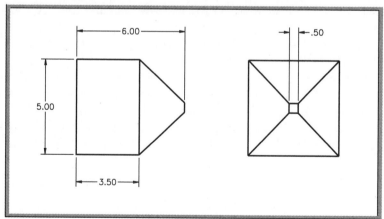

Fig. 39-12

3. Make a 3D drawing of the track shown in Fig. 39-13. Your supervisor wants further design work on the 60° angular face. Set up a UCS on the face in preparation for the redesign. Save the drawing as prb39-track.dwg.

4. Your job as a heating contractor requires that you show the exact location of baseboard hot water heating elements. Open prb39-1.dwg, which you created in "Applying AutoCAD Skills" problem 1. Locate two holes, one for input and the other for output, in each of the remaining three walls. Each hole is 2" in diameter, and the center of each hole is 4" up from the floor and 1' in from the adjoining wall. Make a UCS appropriate for each wall before drawing the respective holes. Save the drawing as prb39-baseboard.dwg.

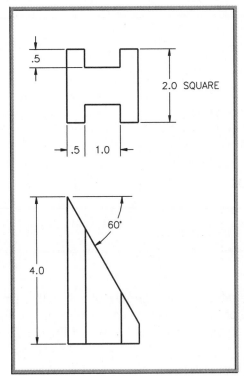

Fig. 39-13

Part 8 Project

3D Modeling

Sometimes, to model an object accurately, you need an alternative to orthographic projections. Two such alternatives are isometric views and 3D models. AutoCAD provides special options to facilitate such drawings.

Description
This project includes two problems:

1. Draw an isometric version of a die (one of two dice), as shown in Fig. P8-1. Include the spots, indicating values of one, two, and three on the three faces showing. Center the spots accurately, and make the length of each side 2″ long. Dimension the length, height, and depth of the die.

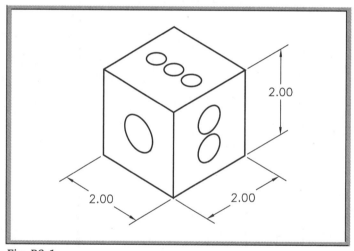

Fig. P8-1

2. Create a 3D model of a die the size of a 3″ cube. Draw the spots (numbered one through six) on each of the six faces, symmetrically arranged upon each surface, as shown in Fig. P8-2.

Hints and Suggestions

1. Remember that circles appear elliptical in isometric drawings. Use the Isocircle option with the correct isoplane orientation.

2. Use the DIVIDE command to place the spots. Remember to set PDMODE to a value that allows you to see the divisions, and use REGEN to make the divisions appear.

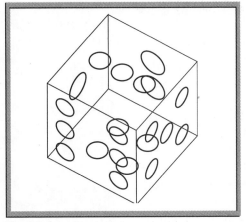

Fig. P8-2

3. Use the HIDE command before viewing and plotting the final 3D model.

4. You may wish to use the BOX primitive to create the 3D model; this command allows you to create a perfect cube of any size.

5. Create two viewports of the 3D model to show all six surfaces. You should only need to create one model.

Summary Questions

Your instructor may direct you to answer these questions orally or in writing. You may wish to compare and exchange ideas with other students to help increase efficiency and productivity.

1. What effect does toggling the isometric plane have upon the use of the ortho feature?

2. Distinguish between thickness and elevation.

3. Of what value is the 3D globe (compass) when you work on 3D models?

4. Describe one possible advantage of reorienting the UCS icon.

Careers Using AutoCAD

Tim Fuller Photography

Interior Designer

A variety of related careers fall under the umbrella of "drafting." While many drafters and designers create plans to manufacture or build products, others work to create attractive surroundings for daily living. Try to think of an old, run-down building in your area that has recently been restored. Although architectural designers certainly played a role in ensuring that the building was foundationally and structurally secure, other design professionals were responsible for making the inside usable and aesthetically appealing. These individuals, better known as interior designers, are responsible for designing and supervising interior construction and restoration efforts.

Details of Design

Interior designers cover a lot of territory when planning a comfortable indoor atmosphere. They create drawings and plans specifying the types and quantities of materials necessary for construction, the sort of lighting systems that will best compliment a room, and how to furnish an area creatively. Interior designers may be employed by design firms or may work as independent consultants. They become involved with building and restoration projects for homes, offices, and even film and television sets.

Being Versatile

Interior designers wear many hats; that is, they are responsible for various aspects of the design process. They should be able to estimate the costs involved with construction and restoration, such as specific costs of materials and labor. Leadership skills are also important, as interior designers need to be able to convey their ideas to those involved with creating or restoring an area. Interior designers must also be able to think creatively in order to design an appropriate look for various indoor areas.

Career Activities

- How is an interior designer's job similar to that of an architectural drafter? What are the differences?
- Find out about courses of study available for interior designers in your area. What are the prerequisites for these courses?

Chapter 40 Solid Primitives

Objectives:

- Explain the difference between solid models and surface models
- Create solid primitives, including a cylinder, torus, cone, wedge, box, and sphere

Key Terms

solid models
solid primitives
surface models

Solid modeling is becoming increasingly popular. One reason is that designs created as solid models more accurately reflect a physical prototype. The appearance of a solid model is similar to that of a surface model, but the two are different. *Solid models* are closed volumes like a real physical object, and they contain physical and material properties. *Surface models* can represent open or closed volumes. An example of a surface is the roof for the building that you created in Chapter 39. The roof is not a closed volume because it does not have any thickness.

AutoCAD permits you to generate predefined *solid primitives* (basic shapes you can use to build more complex solid models) by specifying the necessary dimensions. Examples include boxes, cones, cylinders, and spheres, as shown in Fig. 40-1.

Predefined and user-defined primitives make up the basic building blocks for creating moderately complex solid models, as will be illustrated in the following chapters.

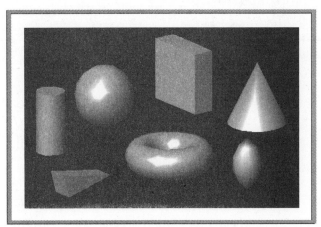

Fig. 40-1

Creating a Cylinder

The CYLINDER command allows you to create a solid cylinder primitive of any size.

1. Start AutoCAD and use the following information to create a new drawing.
   - Use the **Quick Setup** wizard.
   - Use **Decimal** units.
   - Create a drawing area of **11 × 8.5**.
   - **ZOOM All**.

2. Create a new layer named **Objects**, assign the color red to it, and make it the current layer.

3. Set grid to **1** and snap to **.25**.

4. Save your work in a template file named **solid.dwt**. For the template description, enter **For solid models**, and pick **OK**.

5. Close solid.dwt and use it to create a new drawing file named **primit.dwg**.

6. Turn off the grid.

7. Display the Solids toolbar and pick the **Cylinder** button from it.

Solids

This enters the CYLINDER command.

NOTE:

The Elliptical option enables you to create an elliptical cylinder using prompts similar to those used in the ELLIPSE command.

8. Pick a point at any location on the screen.

9. Enter **.5** for the radius and **2** for the height of the cylinder.

10. Select a viewpoint that is to the left, in front of, and above the cylinder.

The cylinder should appear similar to the one in Fig. 40-2.

11. Gouraud-shade the cylinder.

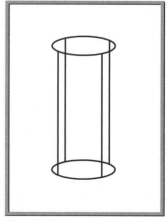

Fig. 40-2

Creating a Torus

The TORUS command lets you create donut-shaped solid primitives.

1. Enter **ZOOM** and **.5x**. Pan if necessary to make space for the torus.

Solids

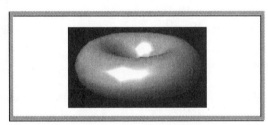

2. Pick the **Torus** button from the Solids toolbar.

This enters the TORUS command.

3. Pick a center point for the torus (anywhere) and enter **1** for the radius of the torus.

The radius of the torus specifies the distance from the center of the torus to the center of the tube.

4. Enter **.5** for the radius of the tube.

A torus appears as shown in Fig. 40-3.

5. Save your work.

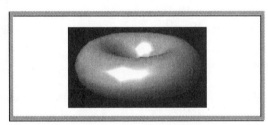

Fig. 40-3

NOTE:

If you enter a negative number for the radius of the torus, a football-shaped solid primitive appears as shown in Fig. 40-4. Try it.

Fig. 40-4

Creating a Cone

The CONE command allows you to create a cone-shaped solid primitive.

Solids

1. From the Solids toolbar, pick the **Cone** button.

This enters the CONE command.

NOTE:

The function of the Elliptical option is similar to that of the CYLINDER command.

2. Pick a center point at any location on the screen.

3. Enter **1** for the radius and **2** for the height of the cone.

A cone appears, as shown in Fig. 40-5.

4. Enter **ZOOM** and **.7x** to create more space on the screen.

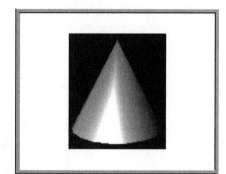

Fig. 40-5

Creating a Wedge

It is possible to create a wedge using the WEDGE command.

Solids

1. From the Solids toolbar, pick the **Wedge** button.

This enters the WEDGE command.

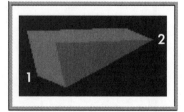

Fig. 40-6

2. Pick two corner points as shown in Fig. 40-6.

3. Enter **.75** for the height.

A wedge-shaped solid primitive appears.

     Chapter 40: Solid Primitives

Creating a Box

The BOX command enables you to create a solid box.

1. From the Solids toolbar, pick the **Box** button.

This enters the BOX command.

Solids

 NOTE: ────────────────────────────────

> The Center option permits you to create a box by specifying a center point.

2. Pick a point to represent one of the corners of the box.

3. Pick the diagonally opposite corner of the base rectangle as shown in Fig. 40-7. You may pick the two points in either order.

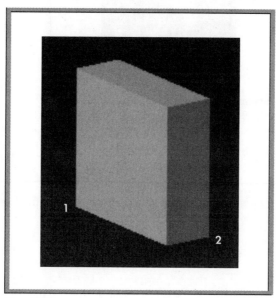

Fig. 40-7

4. Enter **2** for the height.

A box appears.

 NOTE: ────────────────────────────────

> To create a solid cube, you can select the BOX command's Cube option and enter a length for one of its sides.

Creating a Sphere

The SPHERE command creates a solid sphere primitive.

Solids

1. From the Solids toolbar, pick the **Sphere** button.

This enters the SPHERE command.

2. Pick a point for the center of the sphere.

3. Enter **1** for the radius of the sphere.

The sphere appears as shown in Fig. 40-8.

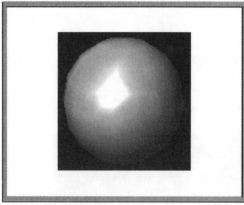

Fig. 40-8

4. Save your work and exit AutoCAD.

Review Questions

1. Describe the function of each of the following commands.
   CYLINDER
   TORUS
   CONE
   WEDGE
   BOX
   SPHERE

2. With what command can you create a football-shaped primitive?

3. With what command can you create a solid cube?

4. What is the purpose of the Elliptical option of the CYLINDER and CONE commands?

Challenge Your Thinking

1. If you could slice open a solid primitive, would it be hollow or solid inside?

2. Experiment with the CYLINDER and CONE commands. Is it possible to create a solid primitive cylinder or cone at other than a 90° (right) angle to its base? Explain.

Applying AutoCAD Skills

Work the following problems to practice the commands and skills you learned in this chapter.

1. Create a solid shaft that measures .5125″ in diameter by 10″ in length. Rotate the shaft so that its length is parallel to the WCS.

2. Create a solid ball bearing that measures .3625″ in diameter. Use a polar array to copy the bearing 16 times using a 1″ radius. View the bearings from above, and smooth-shade them as shown in Fig. 40-9.

3. Create an inflated bicycle inner-tube that measures 24″ in diameter. Make the tube diameter 2″.

4. Create a football-shaped primitive of any size on top of a 6″ cube.

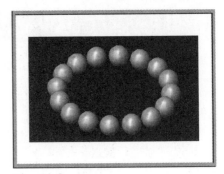

Fig. 40-9

 ## Using Problem-Solving Skills

Complete the following activities using problem-solving skills and your knowledge of AutoCAD.

1. Your company's graphics department is designing ornaments and decorations for various holidays. They need forms for the art decals. Create the solid primitives for them, as shown in Fig. 40-10. Shade the primitives with Gouraud shading. Save the drawing as ornament.dwg.

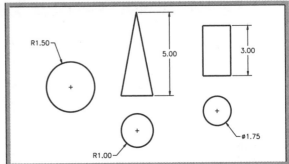

Fig. 40-10

2. You need forms around which to design a laptop computer carrying case. Create two solid boxes: one that measures 13″ × 10″ × 2″ to simulate the computer, and another that measures 5″ × 6″ × 1″ to simulate a plug-in module. Shade the boxes as appropriate. Save the drawing as case.dwg.

Chapter 40: Solid Primitives

Objectives:

- Revolve a 2D polyline to create a 3D solid model
- Adjust the quality, accuracy, and appearance of a solid model
- Extrude a polyline to produce a solid model

Key Terms

revolve
tessellation

Many solid models are produced by first creating a 2D object. For example, it's possible to *revolve* a polyline (turn it around an axis) to create a solid volume. That's how the model of a steel shaft shown in Fig. 41-1 was created. In this chapter, we will do this. We will also extrude a polyline to create a solid.

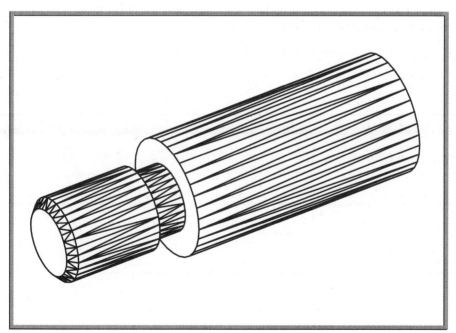

Fig. 41-1

Solid Revolutions

We will use the REVOLVE command to create a solid model of the shaft shown in Fig. 41-1.

Defining the Shape

1. Start AutoCAD and use the **solid.dwt** template file to create a new drawing named **shaft.dwg**.

2. Create the top half of the shaft using a polyline as shown in Fig. 41-2. Make sure the polyline is closed, place it in the center of the screen, and omit the dimensions.

HINT:

Begin at point 1, work counterclockwise, and take advantage of polar tracking. Be sure that the polar tracking increment angle is set at 45°, which will help you create the 45° chamfer.

3. Using one of the 3D viewing options, select a viewpoint that is to the left, above, and in front of the polyline.

You should now see a view similar to the one in Fig. 41-3.

4. Enter **ZOOM** and **.7x**.

5. Save your work.

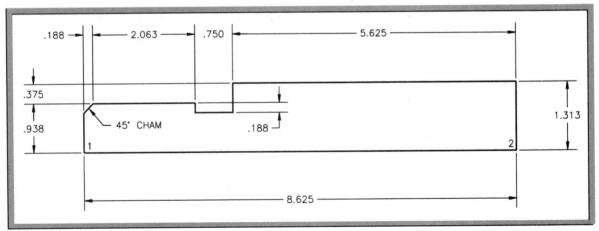

Fig. 41-2

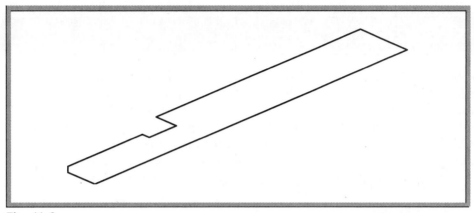

Fig. 41-3

Revolving the Polyline

Solids

1. Display the Solids toolbar and pick the **Revolve** button.

This enters the REVOLVE command.

2. Select the polyline and press **ENTER**.

3. Pick point 1 shown in Fig. 41-2 in reply to the default Specify start point for axis of revolution.

4. Pick point 2 in reply to Specify endpoint of axis.

5. Press **ENTER** to select the default **360**.

The 3D solid model of the shaft generates on the screen, as shown in Fig. 41-4.

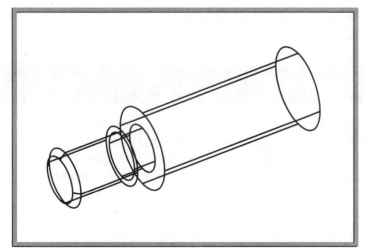

Fig. 41-4

Tessellation

ISOLINES is a system variable that controls the number of tessellation lines used in solid objects. *Tessellation* is the creation of the lines used to describe a curved surface. The default value of 4 often produces a crude-looking object, but the solid model generates more quickly than when ISOLINES is set at a higher value of up to 2047. A higher value, however, improves the quality and accuracy of the curved areas on the solid object.

1. Enter **ISOLINES** and a new value of **20**.
2. Enter **REGEN**.

As you can see, increasing the value of ISOLINES improves the appearance of the solid model. However, it increases the time required by the Boolean subtraction, union, and intersection operations. You will have the opportunity to practice these operations in the following chapters.

The solid model looks like a 3D surface model—not a solid model. However, it is a solid model. Let's prove it.

3. Select the model and display its properties.

Notice that the object is a 3D solid.

4. Close the window and turn off the grid.
5. Remove hidden lines.
6. Enter **ISOLINES**, enter **4**, and enter **REGEN**.
7. Set **ISOLINES** at **10** and enter **REGEN**.
8. Set **ISOLINES** back to **20** and enter **REGEN**.
9. Save your work and close the drawing file.

Solid Extrusions

The solid model of an aluminum casting shown in Fig. 41-5 was created using the EXTRUDE command. This command extrudes and creates a closed volume from a planar 3D face, closed polyline, polygon, circle, ellipse, closed spline, donut, or region. It is not possible to extrude objects contained within a block or a polyline that has a crossing or self-intersecting segment.

1. Begin a new drawing using the **solid.dwt** template file.

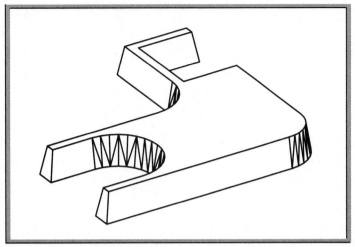

Fig. 41-5

2. Using the **PLINE** command's **Arc** and **Line** options, approximate the size and shape of the casting shown in Fig. 41-6. The casting's overall size is **6.25″** × **6.25″**. Pick the points in the order shown. Do not place the numbers on the drawing.

3. Save your work in a file named **extrude.dwg**.

4. Select a viewpoint that is to the left, above, and in front of the polyline.

5. Pick the **Extrude** button from the Solids toolbar.

Solids

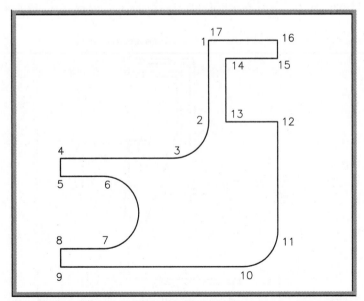

Fig. 41-6

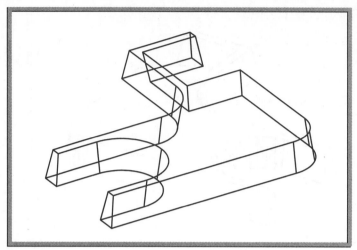

Fig. 41-7

6. Select the polyline and press **ENTER**.

7. Enter **.75** in reply to Specify height of extrusion.

8. Enter **5** (for degrees) in reply to Specify angle of taper for extrusion.

A solid primitive appears, as shown in Fig. 41-7.

9. Remove hidden lines.

Let's attempt to place a slotted hole in the solid model.

10. Enter the **PLAN** command to see the plan view of the current UCS.

11. Using **PLINE**, place a slotted hole in the object as shown in Fig. 41-8.

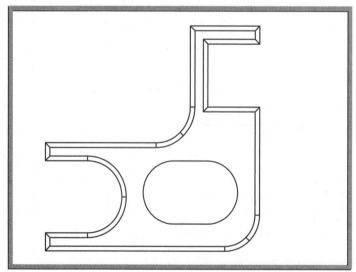

Fig. 41-8

12. Next, view the object in 3D space as before.

Note that the "hole" is still two-dimensional.

13. Pick the **Extrude** button, select the new polyline, and press **ENTER**. Enter **.75** for the height and **0** for the taper angle.

Solids

The object should now look similar to the one in Fig. 41-9.

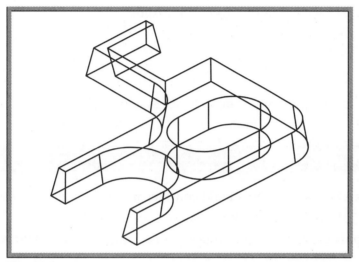

Fig. 41-9

14. Enter the **SHADE** command.

The slotted hole is filled in because it is solid, not hollow. Creation of a hole requires a Boolean subtraction operation. In other words, the solid object that represents the slotted hole must be subtracted from the larger solid object.

15. Erase the larger model so that only the slotted hole is present.

As you can see, it is indeed a solid object and not a hole.

16. Undo to restore the larger model.

17. Move the crosshairs over the model until the "slotted hole" highlights. Select it while it is highlighted and erase it.

18. Close the Solids toolbar, as well as any others you may have opened.

19. Save your work and exit AutoCAD.

> **INFOLINK**
>
> You will learn more about <u>Boolean subtraction</u> with solid models in Chapter 42.

Chapter 41
Review & Activities

Review Questions

1. Describe each of the following commands.
   REVOLVE
   EXTRUDE

2. What is tessellation?

3. Describe the purpose of the ISOLINES system variable.

Challenge Your Thinking

1. Discuss the factors you should consider when setting the ISOLINES system variable. Be specific.

2. Is it possible to extrude a wide polyline (one that has a defined width)? Explain.

Applying AutoCAD Skills

Work the following problems to practice the commands and skills you learned in this chapter.

1. Design a frame to fit a picture that measures 11″ × 14″. Create the frame with polylines and extrude it to a thickness of ³/₄″. Save the drawing as ch41frame.dwg.

2. Construct a solid bowling pin as shown on the left in Fig. 41-10. Then increase the number of tessellation lines and enter REGEN to create a higher resolution version of the bowling pin, as shown on the right.

3. Identify a simple object in the room. Create a profile of the object and then extrude it to create a solid model. Note that only certain objects—those that can be extruded—can be used for this problem.

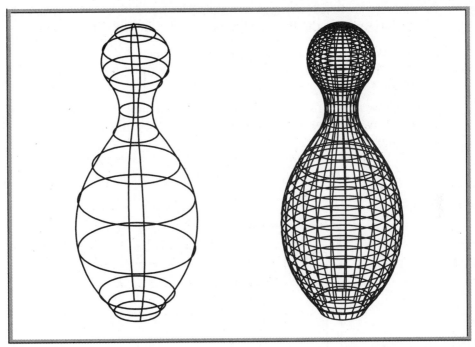

Fig. 41-10

4. Use the following three ways to create a solid cylinder with a radius of .4 and a height of 2 (Fig. 41-11). Set ISOLINES to 8 before you begin.

  a. Draw a rectangle measuring .2 × 2 and revolve it 360°.
  b. Draw a circle with a radius of .4 and extrude it to a height of 2.
  c. Use the CYLINDER command to create a solid cylinder with a radius of .4 and a height of 2.

Now use the AREA command to determine the surface area of all three objects. Calculate the exact answer and compare it to the area AutoCAD calculated for each of the three cylinders.

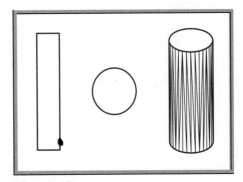

Fig. 41-11

5. Change ISOLINES to 1 and enter REGEN. Use AREA to determine the surface area of the three cylinders created in problem 4 again.

6. Set ISOLINES back to 8 and repeat the process for a right prism measuring .8 × .8 × 2. Use the following three ways to create the prism shown in Fig. 41-12.

   a. Draw a rectangle measuring .8 × 2. Extrude it to a height of .8.
   b. Draw a square measuring .8 on each side and extrude it to a height of 2.
   c. Use the BOX command to insert a box measuring .8 × .8 × 2.

   Use the AREA command to find and compare the surface area of the three objects.

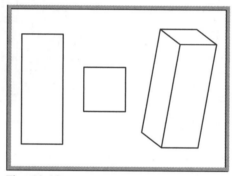

Fig. 41-12

Using Problem-Solving Skills

Complete the following activities using problem-solving skills and your knowledge of AutoCAD.

1. Create a solid model of the boat trailer roller. Define the shape of the roller with a polyline as shown in Fig. 41-13, revolve it, and view it from the SW Isometric viewpoint. Save the drawing as ch41roller.dwg.

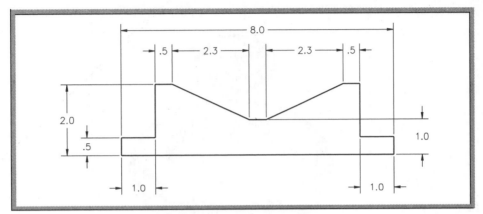

Fig. 41-13

2. The boat trailer manufacturer requires a variation of the boat trailer roller for use on a smaller trailer. Open the drawing and save it as ch41roller2.dwg. Reduce its size by one-third. Change the value of the ISOLINES system variable to define the roller more clearly. Shade the roller as appropriate. Save the drawing. Your drawing should look similar to the one in Fig. 41-14.

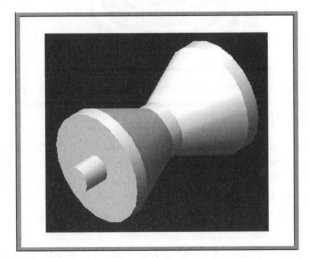

Fig. 41-14

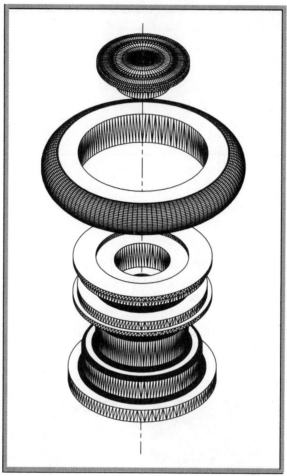

Fig. 41-15

3. The bumper assembly for a bumper pool table is shown in Fig. 41-15. Use the dimensions of the three parts (shown in Fig. 41-16), and use the REVOLVE command to model the parts as solids. Create each part as a separate drawing.

   Note that the bumper support has a square hole. Later you will find out how to create such a hole. For now, assume the hole to be circular.

   After you have finished each of the three parts, place it into an isometric view. You will find this much easier to do if you first use the ROTATE3D command to make the plan view of the world coordinate system coincide with the top view of the object. Remove hidden lines; then produce a smooth-shaded rendering.

Problem 3 courtesy of Gary J. Hordemann, Gonzaga University

Chapter 41: Basic Solid Modeling

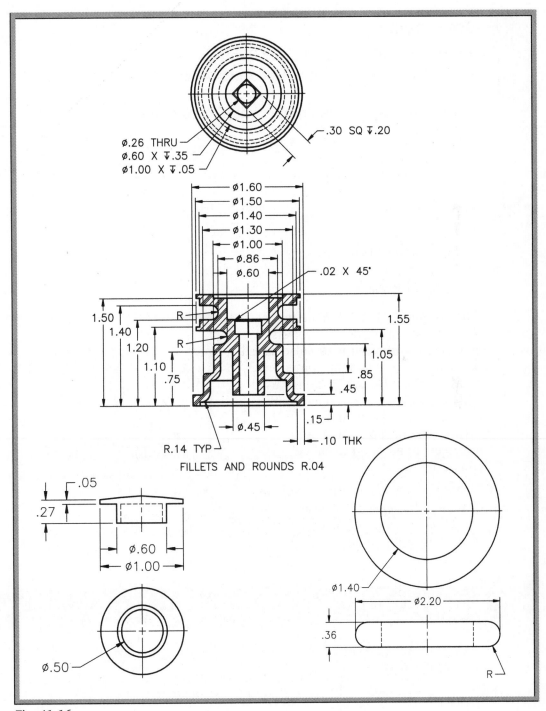

Fig. 41-16

Boolean Operations

Objectives:

- Prepare solid primitives for Boolean operations
- Use the Boolean subtraction operation to remove portions of a solid model
- Use the Boolean union operation to combine composite models

Key Terms

Boolean logic
Boolean mathematics
Boolean operations
composite solids

Much of AutoCAD solid modeling is based on the principles of *Boolean mathematics*. Boolean math, also called *Boolean logic*, is a system created by mathematician George Boole for use with logic formulas and operations. AutoCAD *Boolean operations*, such as union, subtraction, and intersection, are used to create composite solids.

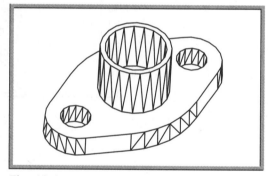

Fig. 42-1

Composite solids are composed of two or more solid objects. Using Boolean operations, let's create a composite solid of the steel support shown in Fig. 42-1.

Preparing the Base Primitive

1. Start AutoCAD and begin a new drawing using the **solid.dwt** template file.

2. Using the **PLINE Arc** and **Line** options, create the shape shown in Fig. 42-2 using the information below.

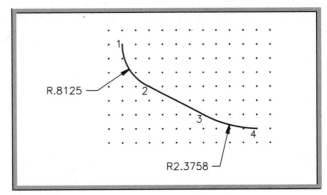
Fig. 42-2

- The grid of dots, representing the snap grid, are spaced **.25″** apart. Use the grid to produce the polyline accurately.
- Use the four points in the order shown. Each point falls on the snap grid.
- In the PLINE command, use the Arc's **Radius** option.

You have just completed one quarter of the object shown in Fig. 42-3.

3. Mirror the shape to create one half of the object, and mirror that to create the entire object.

4. Use the **PEDIT Join** option to join the four polylines into one.

5. Select a viewpoint that is to the left, in front of, and above the object.

6. Set **ISOLINES** to **8** and save your work in a file named **compos.dwg**.

7. Display the Solids and Solids Editing toolbars.

Solids

8. From the Solids toolbar, pick the **Extrude** button.

This enters the EXTRUDE command.

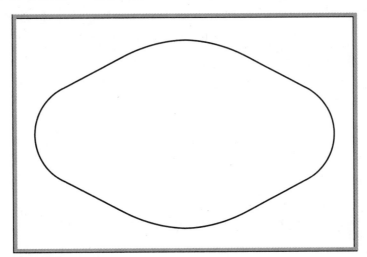

Fig. 42-3

9. Select the polyline and press **ENTER**.

10. Enter **.5** for the height, and specify a taper angle of **3** degrees.

A view similar to the one in Fig. 42-4 appears.

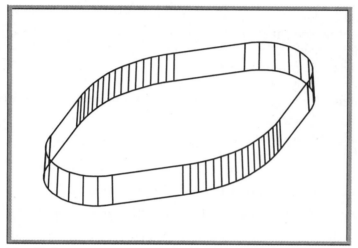

Fig. 42-4

11. Enter the **PLAN** command and press **W** for World; **ZOOM All**.

The plan view allows you to see the taper, as shown in Fig. 42-5, but without the two holes and the dimensions.

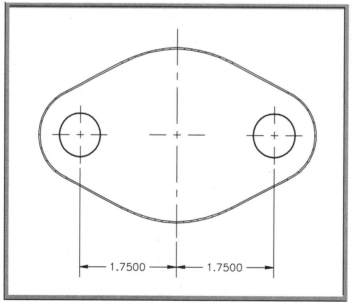

Fig. 42-5

12. Pick the **Cylinder** button from the Solids toolbar and place two cylinders as shown in Fig. 42-5 to represent the holes. Make their radius **.375** and their height **.5**.

HINT:

Create one of them and use the COPY or MIRROR command to create the second.

You have just created two solid cylinders within the original extruded solid.

13. Select a viewpoint to the left, in front of, and above the object.

Subtracting the Holes

The SUBTRACT command performs a Boolean operation that creates a composite solid by subtracting one solid object from another.

1. From the Solids Editing toolbar, pick the **Subtract** button.

This enters the SUBTRACT command.

2. Select the extruded solid (but not the cylinders) as the solid to subtract from, and press **ENTER**.

3. Select the two cylinders as the solids to subtract, and press **ENTER**.

AutoCAD subtracts the volume of the cylinders from the volume of the extruded solid.

4. Remove hidden lines.

The composite solid should look similar to the one in Fig. 42-6.

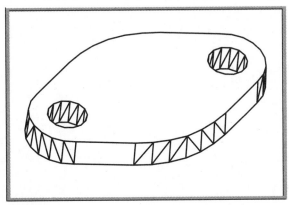

Fig. 42-6

Adding the Support Cylinder

The UNION command permits you to join two solid objects to form a new composite solid.

1. Create a plan view of the WCS and **ZOOM All**.

2. Place a solid cylinder at the center of the model. Make the diameter **1.75"**, and make it **1.75"** in height.

HINT:

Use the snap grid to snap to the center of the model.

3. Using the same center point, place a second cylinder inside the first. Specify a diameter of **1.5"** and a height of **1.75"**.

4. Specify a viewpoint in 3D space as you did before.

5. Subtract the smaller cylinder from the larger cylinder.

Solids Editing

The model is currently made up of two separate solid objects. Let's use the UNION command to join them to form a single composite solid.

6. From the Solids Editing toolbar, pick the **Union** button.

Solids Editing

This enters the UNION command.

7. Select the two solid objects and press **ENTER**.

The two objects become a single composite solid.

8. Enter **HIDE**.

Your model should look similar to the one in Fig. 42-1 (page 570).

9. Close the toolbars, save your work, and exit AutoCAD.

Review Questions

1. Describe the procedure used to create a hole in a solid object.

2. What is a composite solid?

3. Explain the purpose of the following solid modeling commands.
   SUBTRACT
   UNION

Challenge Your Thinking

1–2. Describe how you would create the solid models shown in Figs. 42-7 and 42-8. Be specific.

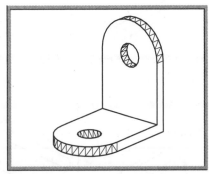

Fig. 42-7

Fig. 42-8

Applying AutoCAD Skills

Work the following problems to practice the commands and skills you learned in this chapter.

1–2. Apply the commands you learned in this chapter to create the composite solids shown in Figs. 42-9 and 42-10.

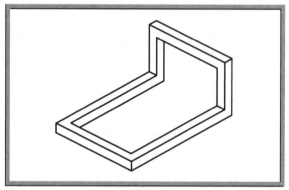

Fig. 42-9

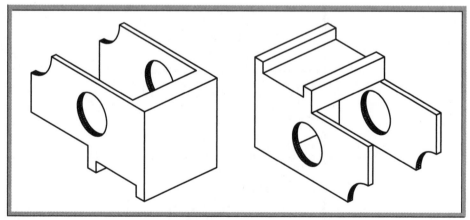

Fig. 42-10

3. Identify a machine part, such as a gear on a shaft or a bracket. Sketch the part in your mind or on paper. Using AutoCAD's solid primitives and Boolean operations, shape the part by adding and subtracting material until it is complete.

4. Draw a 3D model of the height gage shown in Fig. 42-11 using the **EXTRUDE** command and Boolean operations. Can you use a box or a wedge to create the V-groove? Why or why not? Save the drawing as ch42htgage.dwg.

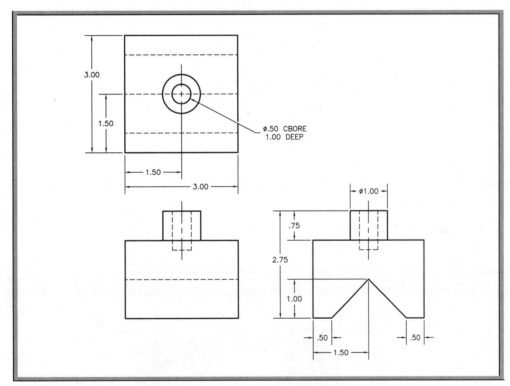

Fig. 42-11

5. Your customer wants to see a 3D view of the base plate shown in Fig. 42-12A. Create the base, extrude it, and then use Boolean operations to create the 3D model. Shade the base plate with flat shading edges on, as shown in Fig. 42-12B. Save the drawing as ch42baseplate.dwg.

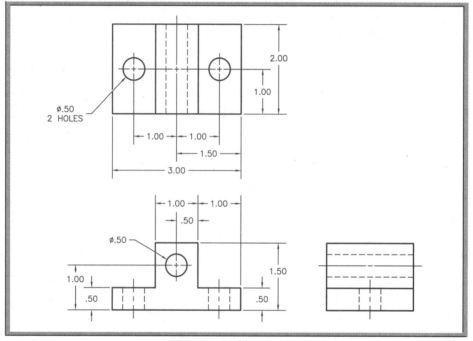

Fig. 42-12A   Figure 42-12A prepared by Gary J. Hordemann, Gonzaga University

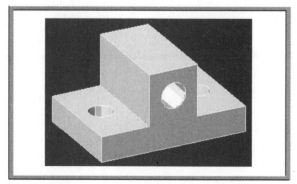

Fig. 42-12B

Using Problem-Solving Skills

Complete the following activities using problem-solving skills and your knowledge of AutoCAD.

1. Create the fluid coupling end cap shown in Fig. 42-13A using the REVOLVE command. Begin by creating a layer named Endcap, and set the color to cyan. Set ISOLINES to 25. The object's profile is shown in Fig. 42-13B.

   After you have finished drawing the end cap, produce an isometric view. You will find this much easier to do if you first use the ROTATE3D command to make the plan view of the world coordinate system coincide with the top view of the end cap. Remove hidden lines and then smooth-shade the end cap.

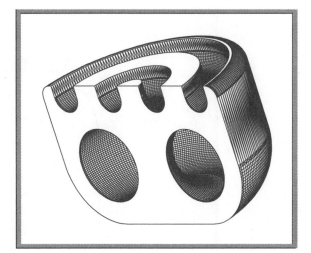

Fig. 42-13A

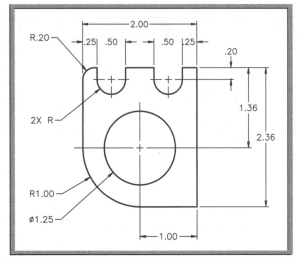

Fig. 42-13B

Problem 1 courtesy of Gary J. Hordemann, Gonzaga University

2. Create the shaft clamp shown in Fig. 42-14 using the SUBTRACT and EXTRUDE commands. Create the .40 diameter hole by inserting and subtracting a cylinder. Remove the hidden lines and smooth-shade the clamp.

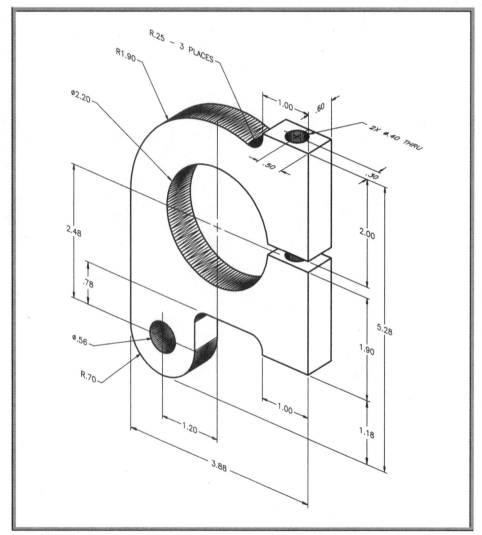

Fig. 42-14

Problem 2 courtesy of Gary J. Hordemann, Gonzaga University

Chapter 42: Boolean Operations

3. Create a solid model of the adjustable pulley half shown in Fig. 42-15. It will help if you first change the view to the right side view and the UCS to the current view. If you allow for the counterbored hole when drawing the profile, you will not have to bring in and subtract cylinders. After revolving the profile, go to the front view and insert and subtract the cylinders for the two smaller holes.

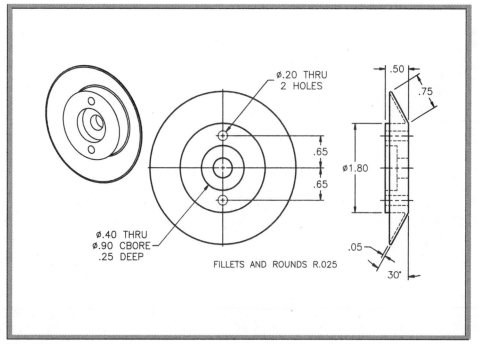

Fig. 42-15

Chapter 42: Boolean Operations

Chapter 43 | Adjusting Solid Models

Objectives:

- Apply bevels and rounded edges to a composite solid
- Create a composite solid by intersection
- Shell a solid to create a hollow object
- Check for interference between adjacent objects in 3D space

Key Terms

interference
interference solids
intersection
shelling

AutoCAD provides commands and options for modifying primitives and solid composites. You are already familiar with the CHAMFER and FILLET commands; in this chapter, you will see how they can be used to modify 3D objects.

Using a method called *intersection*, you can create a solid from the overlapping portion of two intersecting solid objects. Another method, called *shelling*, enables you to remove material from an object to create a shell or hollow object. *Interference* occurs when the two objects overlap or collide in 3D space. The INTERFERE command permits you to determine whether two solids or objects overlap.

In this chapter, you will create the model of a table and bowl shown in Fig. 43-1. You will create the basic table, round the top edges, bevel the legs, and place the bowl on top of it. Using the INTERFERE command, you will ensure that the bowl is positioned correctly on top of the table and does not collide with it.

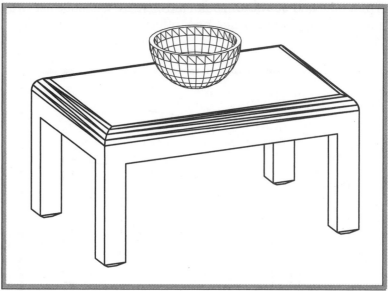

Fig. 43-1

Beveling and Rounding Solid Edges

Let's create the table shown in Fig. 43-1 using the methods we've used in previous chapters. Then we will apply the CHAMFER and FILLET commands to make the table look more like the one in the previous illustration.

1. Start AutoCAD and begin a new drawing using the **solid.dwt** template file. Name the file **table.dwg**.

2. Display the Solids and Solids Editing toolbars.

3. Pick the **Box** button from the **Solids** toolbar and create a solid box **6** units in the X direction and **4** units in the Y direction. Make it **3** units in height and place it in the center of the screen.

Solids

NOTE:

In this file, 1 unit = 1'.

4. Select a viewpoint that is to the left, in front of, and above the object.

5. Enter **ZOOM** and **.8x**.

6. Using the **BOX** and **SUBTRACT** commands, create and subtract two rectangular boxes from the first one to form the table shown in Fig. 43-2.

Solids

Solids Editing

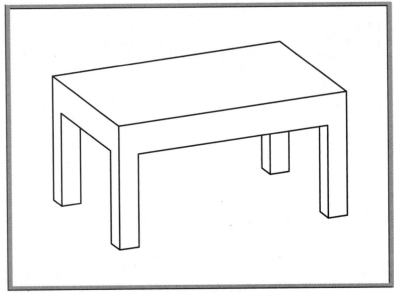

Fig. 43-2

HINT:

Create one box 5 units by 4 units by 2.25 units tall and another box 6 units by 3 units by 2.25 units tall. Subtract both from the first box. It may be helpful to switch to the plan view.

7. Enter **HIDE**.

The solid model should look similar to the one in Fig. 43-2.

8. Save your work.

9. Enter **REGEN**.

Beveling the Table Legs

Modify

1. Enter the **CHAMFER** command and select the bottom of one of the four table legs. If AutoCAD selects the side of the table rather than the bottom of the leg, type **N** for Next, and press **ENTER**. When AutoCAD selects the bottom, press **ENTER**.

2. Enter a base surface chamfer distance of **.15**.

3. At the Specify other surface chamfer distance prompt, enter **.1**.

4. At the next prompt, pick all four edges that make up the bottom of the leg and press **ENTER**.

AutoCAD chamfers the edges.

5. Repeat Steps 1 through 4 to chamfer the remaining three legs.

The bottoms of the legs should now look similar to those in Fig. 43-3.

Fig. 43-3

Rounding the Table Top

Modify

6. Enter **FILLET**, select any edge of the table top, and enter **.3** for the radius.

7. Enter **C** for Chain, select the remaining three edges of the table top, and press **ENTER**.

8. Enter **HIDE** and save your work.

The model should look very similar to the one in Fig. 43-3.

9. Enter **REGEN**.

Boolean Intersection

You have learned that you can create composite solids by adding and subtracting solid objects. You can also create them by overlapping two or more solid objects and calculating the solid volume that is common (intersecting) to each of the objects. AutoCAD's INTERSECT command performs this Boolean operation.

Solids

1. Enter the **BOX** command and create a 2-unit cube at any location on the floor (WCS) under the table. (The floor and WCS are on the same plane.) Be sure snap is on.

HINT:

You may want to return to the plan view to perform this step. After picking the first corner of the box, select the Cube option and enter 2 in reply to Specify length.

2. View the objects in 3D as you did before. If necessary, move the box away from the table or select a viewpoint so the two objects do not overlap.

3. Enter the **UCS** command, **N** for New, and enter the **3point** option.

4. Using the **Endpoint** object snap, pick any one of the top corners of the cube.

5. To specify the positive X direction, snap to an adjacent top corner of the cube.

6. For the positive Y direction, snap to the other adjacent top corner of the cube.

The UCS and UCS icon change to reflect the new UCS.

7. Enter **PLAN** to review the plan view of the current UCS.

8. Pick the **Sphere** button from the Solids toolbar.

Solids

9. Place the center point of the sphere at the center of the top of the cube and enter **1** for the radius.

A solid sphere appears.

10. Change to the **SW Isometric** viewpoint.

You should see a view that is similar to the one in Fig. 43-4.

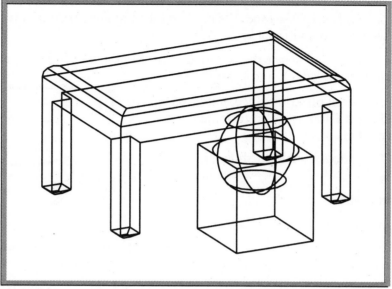

Fig. 43-4

Solids Editing

11. From the Solids Editing toolbar, pick the **Intersect** button.

12. Select the cube and sphere and press **ENTER**.

A half-sphere (hemisphere) appears. AutoCAD has calculated the solid volume that is common (intersecting) with each of the primitives.

13. Save your work.

Shelling a Solid Object

We want to make the hemisphere into a bowl. The fastest way to do this is to shell it out.

Solids Editing

1. Pick the **Shell** button from the Solids Editing toolbar.

This enters the Shell option of the SOLIDEDIT command.

2. Select the hemisphere.

Next, we need to remove the top face from the shelling. If we don't, AutoCAD will hollow out the hemisphere, but it will not have an opening.

3. In reply to Remove faces, pick the edge of the hemisphere.

The top face and the main body face share the same edge, so AutoCAD displays 2 faces found, 2 removed. We don't want to remove the main body face, so we need to add it back.

4. Enter **A** for Add, select the rounded part of the hemisphere (do not select the top edge), and press **ENTER**.

5. For the shell offset distance, enter **.05** and press **ENTER** twice to terminate the command.

AutoCAD produces a shell, as shown in Fig. 43-5.

6. Enter **HIDE**.

7. Save your work.

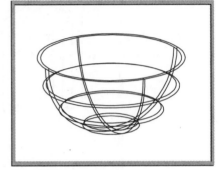

Fig. 43-5

Positioning Adjacent Objects

Let's move the bowl to a more logical position—on top of the table.

1. Produce a front view, as shown in Fig. 43-6.

HINT:

From the View pull-down menu, pick 3D Views and Front.

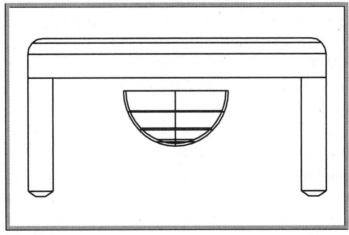

Fig. 43-6

Approximating the Location

2. Enter **ZOOM** and **.5x** and turn off object snap tracking if it is on.

3. Move the bowl upward and over, if necessary, so that it appears to rest on the tabletop.

4. Change to the WCS and produce a plan view.

5. If the bowl is not resting at the center of the table, move it to the approximate center.

6. Produce a viewpoint to the left, in front of, and above the tabletop.

7. Create a layer named **Bowl**, make it green, and move the bowl to this layer.

8. Turn off the grid, Gouraud-shade the table and bowl, and save your work.

Testing for Interference

The INTERFERE command enables you to check for interference (overlap) between two or more solid objects. It is particularly useful when fitting together solid objects into an assembly.

1. Produce a front view.

Solids

2. From the Solids toolbar, pick the **Interference** button.

This enters the INTERFERE command.

3. Pick either the table or the bowl and press **ENTER**.

4. Pick the other composite solid and press **ENTER**.

The message Solids do not interfere should appear. If you don't receive this message, the two interfere with one another. If the solids don't interfere with each other, proceed to Step 6.

5. If they do interfere, press **ESC** to cancel.

6. Move the bowl downward **.25** unit.

HINT:

Enter the MOVE command, select the bowl, and press ENTER. For the base point, pick any point on or near the bowl. For the second point of displacement, move downward a short distance with polar tracking on, and enter .25.

7. Enter the **INTERFERE** command.

8. Pick one of the two composite solids and press **ENTER**; pick the other and press **ENTER**.

AutoCAD displays the message Interfering pairs: 1. This is because the two interfere with one another.

When solids interfere with one another, you can create *interference solids* (solids that consist of only the interfering part of the two objects). Interference solids make it easier to see which parts of the solids overlap.

9. Enter **Yes** in reply to Create interference solids?

10. Using **MOVE Last**, move the new solid up and away from the bowl and table.

11. Produce a viewpoint similar to before.

The new solid should look similar to the one in Fig. 43-7.

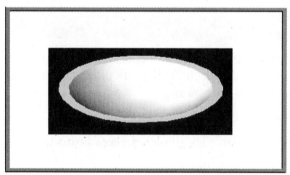

Fig. 43-7

12. Erase the new solid and move the bowl back to its previous location.

13. Close the toolbars you opened.

14. Save your work and exit AutoCAD.

Review Questions

1. How do the CHAMFER and FILLET commands affect the appearance of solid objects containing outside corners?

2. Explain how the INTERSECT command is used to form a composite solid.

3. Explain why shelling is useful.

4. Describe the purpose of the INTERFERE command.

5. How might the INTERFERE command be useful when designing parts of an assembly?

Challenge Your Thinking

1. How do the CHAMFER and FILLET commands affect the appearance of solid objects containing inside corners?

2. What factors should you consider when you are planning to create a solid model using interference solids? Explain.

Applying AutoCAD Skills

Work the following problems to practice the commands and skills you learned in this chapter.

1–2. Apply the commands you learned in this chapter to create the solid composites of a metal shroud shown in Figs. 43-8 and 43-9. Approximate all sizes. Use the INTERSECT command to complete the shroud in Fig. 43-8. Then modify it to complete the one in Fig. 43-9. (Increase the value of ISOLINES before you begin.)

Fig. 43-8

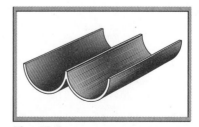

Fig. 43-9

3. Open ch42base plate.dwg. Chamfer the two edges where the upper horizontal and vertical surfaces meet. Use a .25″ chamfer at 45°. Create fillets with a radius of .15 where the upper vertical surfaces meet the plate's horizontal surfaces. Create fillets with a radius of .15 at the two outer corners. Refer to Fig. 43-10 for specific locations.

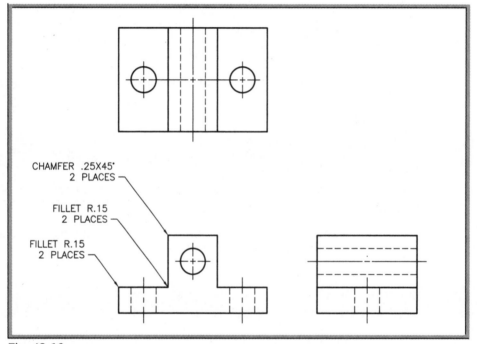

Fig. 43-10

Figure 43-10 prepared by Gary J. Hordemann, Gonzaga University

Using Problem-Solving Skills

Complete the following activities using problem-solving skills and your knowledge of AutoCAD.

1. Using the SUBTRACT and UNION commands, model the plate shown in Fig. 43-11A. The dimensions are given in the orthographic projections in Fig. 43-11B.

Use the FILLET command to create a fillet where the boss joins the main part of the plate.

After you have finished the model, produce an isometric view. (First rotate it using the ROTATE3D command.) Then remove hidden lines. The drawing should look like the one in Fig. 43-11A.

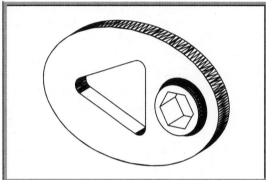

Fig. 43-11A

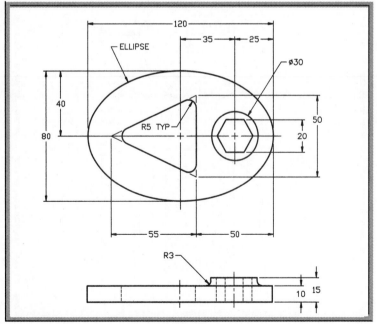

Fig. 43-11B

Problem 1 courtesy of Gary J. Hordemann, Gonzaga University

2. Using the SUBTRACT and EXTRUDE commands, model the tube bundle support shown in Fig. 43-12A. The dimensions are given in the orthographic projections shown in Fig. 43-12B.

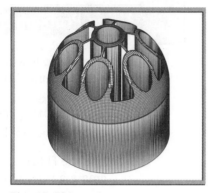

Fig. 43-12A

The rounded top can be produced in several ways. One way is to use the REVOLVE command to create a piece to be removed from the original extrusion (using the SUBTRACT command). A second way is to use the REVOLVE command to create a rounded piece (without holes); then use the INTERSECT command to obtain the common geometry. Try both ways. Obtain an isometric view, and then remove the hidden lines.

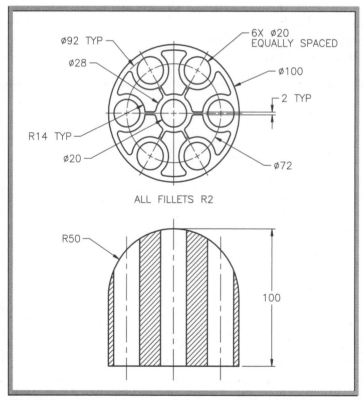

Fig. 43-12B

Problem 2 courtesy of Gary J. Hordemann, Gonzaga University

3. Three orthographic views of a T-swivel support are shown in Fig. 43-13. Model the piece as a solid. Use the FILLET command to fillet all of the indicated edges.

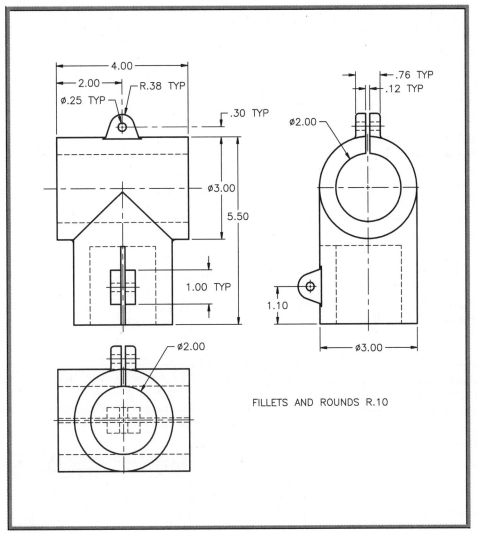

Fig. 43-13

Problem 3 courtesy of Gary J. Hordemann, Gonzaga University

Careers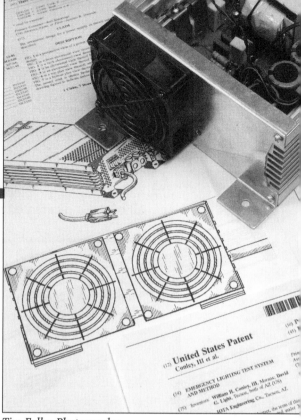
Using
AutoCAD

Patent Drafter

Have you ever had a brilliant idea for an invention? Something so simple, perhaps, that you couldn't believe someone else didn't think of it first? Well, then, you have a lot of work to do to make sure that you really are the first one to come up with your idea. Conducting research on previously patented inventions is necessary to avoid "stealing" someone else's idea.

The process inventors go through to obtain a patent is complicated and time-consuming, but if their ideas are truly unique, the feeling of accomplishment is worth the effort. Inventors don't have to do all this work by themselves, however. Many people are available to help make the research and approval process easier. Attorneys and patent drafters are examples of people who play key roles in securing a patent for an invention.

Drawing the Idea

Patent drafters produce drawings of both existing products and new product ideas. Drawings of existing products include carefully labeled components; this is necessary to identify parts that may have already been patented.

When designing drawings of new product ideas, drafters use information provided by the inventor (such as rough sketches and notes) to create a picture of what the new product will look like. Drafters will often create multiple drawings of the same product from various angles; this helps avoid cluttering a single drawing with excessive details.

Tim Fuller Photography

It Takes Teamwork

A patent drafter is just one of several individuals who helps an idea become an actual product. The inventor's attorney is responsible for keeping the process of obtaining a patent moving. It is the inventor's attorney who typically selects and hires a patent drafter. The drafter must then work closely with the attorney—to ensure his or her work is consistent with the written description of the item—and the inventor—who provides the overall concept of the product.

 Career Activities

- Research the patent process on the Internet. How would you go about applying for a patent?
- Make a list of at least three inventions that you can't imagine living without. Then visit the library and try to find out who invented them, when they were patented, and how they have changed since their first production.

Chapter 44   Downstream Benefits

Objectives:

- **Create a full section from a solid model**
- **Produce a profile view from a solid model**

Key Terms

cutting plane
full section
profile
tangential edges

Solid modeling provides many benefits that are "downstream" from the solid model. Examples include mass properties generation, detail drafting, finite element analysis, and the fabrication of physical parts.

In some cases, solid models can now be used directly to guide production processes. In others, the solid model is used to create production drawings such as sections and profiles. (A *profile* is an outline or contour, such as a side view, of an object.) In this chapter, you will create full section and profile views of a pulley.

1. Start AutoCAD and begin a new drawing using the **solid.dwt** template file. Name the file **pulley.dwg**.

2. Set snap to **.125** and grid to **.25**.

3. Set **ISOLINES** to **30**, and display the Solids toolbar.

4. Create the polyline shown in Fig. 44-1 on page 598 using the following information.

   - The dots of the grid are spaced **.125** apart. The bold dots are spaced **.25** apart.
   - All polyline endpoints fall on the grid, so use it to produce the object accurately.
   - Produce the polyline in the top half of the drawing area.
   - The axis of revolution is an imaginary line of arbitrary length.

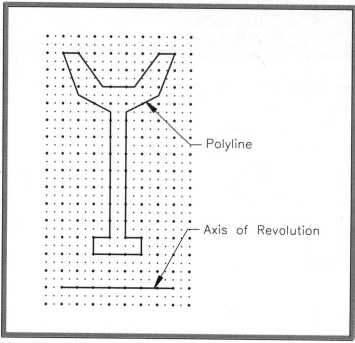

Fig. 44-1

5. Using the **REVOLVE** command, revolve the polyline to produce the pulley.

The pulley should appear, as shown on the left in Fig. 44-2.

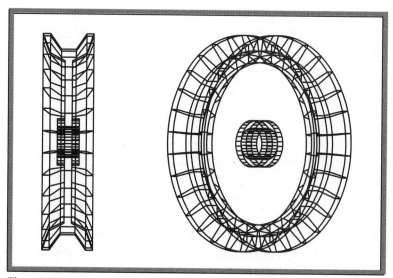

Fig. 44-2

6. Center the pulley in the drawing area and save your work.

7. Select a viewpoint that is to the left and in front of the model, and enter **ZOOM .8x**.

Your view of the model should be similar to the one on the right in Fig. 44-2.

8. Erase the line used for the axis of revolution. If necessary, zoom in.

Creating a Full Section

A *full section* of a part is a view that slices all the way through the part, showing the interior of the part as it appears at that slice, or cross section. The SECTION command helps you create a full cross section of a solid model, such as the pulley shown in Fig. 44-6 on page 601.

Defining the Cutting Plane

To create a full section, you must first define a *cutting plane*—the plane through the part from which the sectional view will be taken.

1. Enter the **PLAN** command and press **ENTER** to accept the Current UCS default setting.

A plan view of the pulley appears, as shown in Fig. 44-3. You will not see the box around the pulley.

2. **ZOOM All**.

The pulley should lie entirely within the drawing area. (The grid reflects the drawing area.)

3. Using the **PLINE** command, draw a box around the pulley as shown in Fig. 44-3. The exact size of the box is not important.

4. Create a view that is above, to the left, and in front of the pulley; turn off the grid.

The rectangular polyline passes through the center of the pulley, as shown in Fig. 44-4 on page 600.

5. Enter **HIDE**.

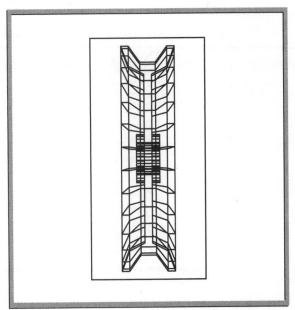

Fig. 44-3

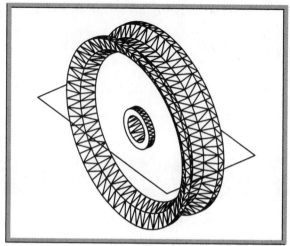

Fig. 44-4

Extracting the Section

1. Enter **PLAN** to obtain the plan view of the current UCS and enter **ZOOM All**.

2. Make layer **0** the current layer.

3. Pick the **Section** button from the Solids toolbar.

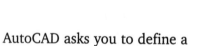
Solids

This enters the SECTION command.

4. Pick the pulley and press **ENTER**.

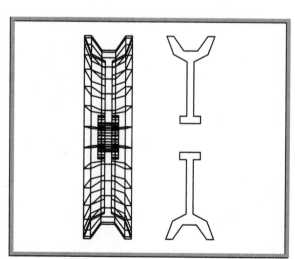

Fig. 44-5

AutoCAD asks you to define a plane.

5. Enter **O** (for object) and pick the rectangular polyline.

AutoCAD creates a cross section on the same plane as the rectangular polyline.

6. Enter **MOVE Last** and move the cross section to the right of the pulley.

7. Erase the rectangular polyline object and save your work.

The drawing should look similar to the one in Fig. 44-5.

The full section is not complete until you hatch it and add the missing lines.

1. Hatch the cross section using the **ANSI31** hatch pattern.

HINT:

Use the BHATCH command. In the dialog box, pick the Select Objects button, pick the object that you want to hatch, press ENTER, and then pick the OK button.

2. Add lines to the section, as shown in Fig. 44-6, to complete the sectional view.

3. Make **Objects** the current layer, and freeze layer **0**.

4. Produce a left view of the pulley, **ZOOM .7x**, and save your work.

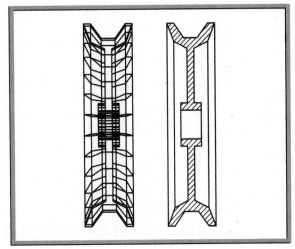

Fig. 44-6

Creating a Profile

The SOLPROF command creates 2D profile objects from a solid model. You must be in a paper space layout to use this command.

1. Pick the **Layout 1** tab located in the lower left portion of the drawing area.

2. Right-click the **Layout1** tab, select **Rename** from the shortcut menu, and name the layout **Profile**.

A single viewport of the pulley appears in the layout.

3. In the status bar, pick the **PAPER** button to change it to **MODEL**.

4. Enter **ZOOM .5x** and move the pulley to the right half of the layout.

5. Pick the **Profile** button from the Solids toolbar.

Solids

This enters the SOLPROF command.

6. Select the pulley and press **ENTER**.

7. Enter **Yes** in reply to Display hidden profile lines on separate layer?

This means that AutoCAD will place hidden lines on a separate layer, allowing you to control the display of these lines.

8. Enter **Yes** in reply to Project profile lines onto a plane?

This projects the visible and hidden profile lines onto a single plane.

9. Enter **Yes** in reply to Delete tangential edges?

This deletes transition lines, called *tangential edges*, that occur when a curved face meets a flat face.

AutoCAD creates profile objects on top of the pulley.

10. With polar tracking or ortho on, move the pulley away and to the left of the profile objects. (The pulley is red, and the profile objects are black.)

11. Select the profile objects and display properties information on them.

AutoCAD created a block for the visible profile lines and a block for the hidden profile lines. Also, AutoCAD created new layers to hold the profile objects.

12. Assign the **Hidden** linetype to the new layer beginning with PH.

AutoCAD stores hidden profile lines on this layer and visible lines on the layer that begins with PV.

13. Pick the **Model** tab and **ZOOM All**.

The new profile should look similar to the one in Fig. 44-7.

14. Close the floating toolbar, save your work, and exit AutoCAD.

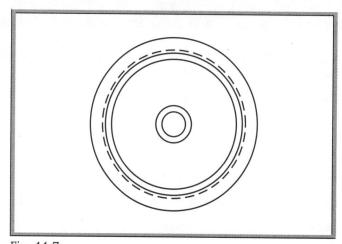

Fig. 44-7

Review Questions

1. Name two downstream benefits of solid modeling.

2. Explain the purpose of the SECTION command.

3. Explain how to define the cutting plane when using the SECTION command.

4. What kind of objects does the SOLPROF command produce?

5. What is the purpose of the layers that SOLPROF creates?

Challenge Your Thinking

1. Discuss how you would orient and print together the full section and profile views that you created in this chapter.

2. Investigate current manufacturing practices. Then explain why solid models can sometimes be used directly to drive production processes, but at other times, the solid model is used merely to create 3D views for visualization and production drawings such as sections and profiles.

Applying AutoCAD Skills

Work the following problems to practice the commands and skills you learned in this chapter.

1. Select any one of the solid models you created in any of the previous chapters, or create a new one on your own, and create a full section of the model.

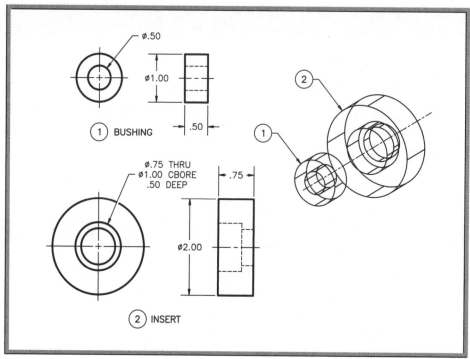

Fig. 44-8

2. Draw a solid model of the bushing and insert shown in Fig. 44-8. Then create a full sectional view to completely describe them. Remember to change your hatch for the insert. Save the drawing as ch44bushing.dwg.

3. You are given a file containing the solid model and sectional view of the bushing and insert (problem 2 above). The manufacturing department needs a set of 2D working drawings. Create a 2D profile, and add any other views you think are necessary to describe the bushing and insert fully. Save the drawing as ch44bushing2.dwg.

Figure 44-8 prepared by Gary J. Hordemann, Gonzaga University

Using Problem-Solving Skills

Complete the following activities using problem-solving skills and your knowledge of AutoCAD.

1. Fig. 44-9 shows a bumper assembly for a bumper pool table. The dimensions for the bumper support, bumper ring, and cap are given in Chapter 41. Model these three objects as solids. Note that the bumper support has a square hole with a chamfer of .02 × .02. Create a box and use the SUBTRACT command to create the hole.

   After you have created the three solids, put them together as they would fit if assembled. Create a full section view using the SECTION command. Use the INTERFERE command to check for interferences among the three pieces.

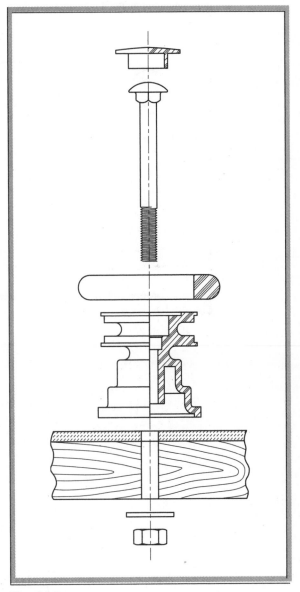

Fig. 44-9

Problem 1 courtesy of Gary J. Hordemann, Gonzaga University

Chapter 44: Downstream Benefits

605

2. A solid model of the rocker arm shown in Fig. 44-10 is needed, along with a full section and a complete set of 2D drawings, for the design and manufacturing departments. Provide the necessary drawings. Save the drawings using names of your choice.

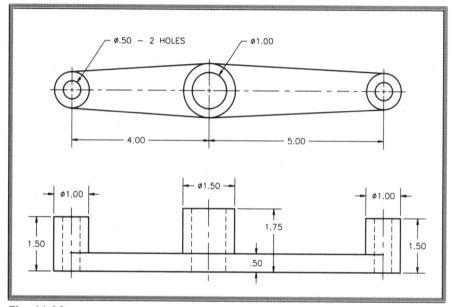

Fig. 44-10

Figure 44-10 prepared by Gary J. Hordemann, Gonzaga University

Chapter 44: Downstream Benefits

Chapter 45 — Documenting Solid Models

Objectives:

- Prepare a solid model for use in creating orthographic and section views
- Produce orthographic and section views from a solid model
- Position and plot orthographic, sectional, and isometric views

Key Term

auxiliary view

AutoCAD enables you to use orthographic projection to lay out multiview and section drawings of solid models. As you may recall, orthographic projection is a method of creating views of an object that are projected at right angles onto adjacent projection planes. The SOLVIEW command uses AutoCAD's paper space layout to establish the viewports and new layers. The SOLDRAW command uses the viewports and layers to place the visible lines, hidden lines, and section hatching for each view.

Preparing the Solid Model

The following steps prepare the pulley model from Chapter 44 for the SOLVIEW and SOLDRAW commands.

1. Open the drawing file named **pulley.dwg**.

2. Using **Save As...**, create a new drawing file named **pulley2.dwg**.

3. Freeze the layers that begin with PH and PV, set the linetype global scale factor to **.5**, and open the Solids toolbar.

PH and PV are the layers that AutoCAD used to store the hidden and visible profile lines.

4. Pick the layout tab named **Profile** to switch to the layout.

5. Click **MODEL** on the status bar so that it reads **PAPER**.

The paper space icon replaces the coordinate system icon.

6. Erase the single viewport.

This deletes the viewport and its contents.

7. Create a new layer named **Vports**, assign the color cyan to it, and make it the current layer.

8. From the **View** pull-down menu, select **Viewports** and **4 Viewports**, and enter **F** (for Fit) at the AutoCAD prompt.

AutoCAD fits four new views of the pulley in the drawing area, as shown in Fig. 45-1.

9. Save your work.

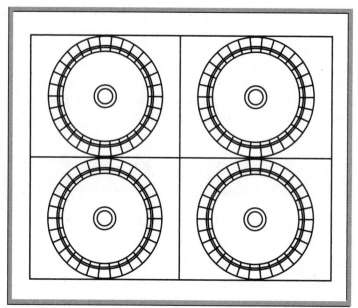

Fig. 45-1

Creating the Views

AutoCAD provides commands that can calculate and display orthographic, auxiliary, and sectional views. The SOLVIEW command performs the calculations and stores the necessary view-specific information. The SOLDRAW command uses that information to produce the final drawing views.

Producing the Top View

Let's produce a top view of the drawing in the upper left viewport.

1. Pick the **Setup View** button from the Solids toolbar.

This enters the SOLVIEW command and displays four options. The Ucs option creates a profile view relative to a user coordinate system. The Ortho option creates an orthographic view from an existing view. Auxiliary creates an auxiliary view from an existing view. An *auxiliary view* is one that is projected onto a plane perpendicular to one of the orthographic views and inclined in the adjacent view. The Section option creates a sectional view, complete with crosshatching.

2. Enter **O** for **Ortho**.

3. In reply to Specify side of viewport to project, pick the lower horizontal line that makes up the upper left viewport, snapping to its midpoint.

4. In reply to Specify view center, pick a point in the center of the upper left viewport, and press **ENTER**.

An orthogonal view of the pulley appears at the center of the viewport.

5. In reply to Specify first corner of viewport, pick a point near but inside one of the four corners that make up the upper left viewport, as shown in Fig. 45-2.

6. In reply to Specify opposite corner of viewport, pick the opposite corner to form a rectangle that is slightly smaller than the viewport, as shown in Fig. 45-2.

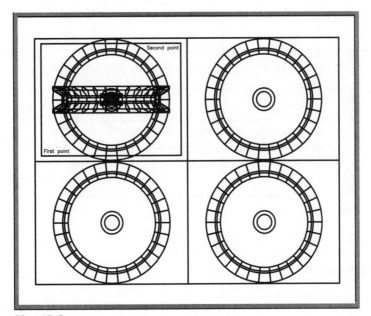

Fig. 45-2

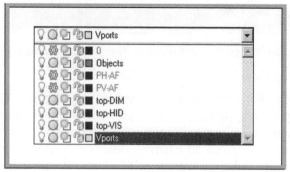

Fig. 45-3

7. In reply to Enter view name, enter **top** and press **ENTER** a second time to terminate the command.

8. Display the list of layers, as shown in Fig. 45-3.

Note the three layers that SOLVIEW created for visible lines, hidden lines, and dimensions.

9. Erase the old upper left viewport. Do not erase the new viewport containing the new orthographic view.

10. Save your work.

Now that SOLVIEW has accumulated the required information, you can use SOLDRAW to generate the final drawing view.

Solids

11. Pick the **Drawing** button from the Solids toolbar.

This enters the SOLDRAW command.

12. Pick the new viewport that you created and press **ENTER**.

AutoCAD creates an orthographic view from the viewport, complete with hidden lines, as shown in Fig. 45-4.

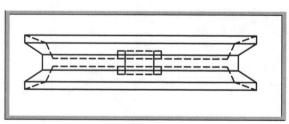

Fig. 45-4

13. On the status bar, change **PAPER** to **MODEL**.

14. Enter **ZOOM** and **.7x**.

15. Save your work.

Creating the Front View

Let's create the front view of the pulley in the bottom left area of the screen.

1. Click inside the lower left viewport to make it the current viewport.

2. View the pulley in this viewport from the front.

 HINT: —————————————————————————

From the View pull-down menu, select 3D Views and Front.

3. Enter **ZOOM** and **.7x**.

Solids

4. Pick the **View** button from the Solids toolbar to enter the SOLVIEW command, and enter the **Ortho** option.

5. Pick the left vertical line that makes up the lower left viewport, snapping to its midpoint.

6. In reply to Specify view center, pick a point in the center of the viewport and press **ENTER**.

7. In reply to Specify first corner of viewport, pick a point inside one of the four corners of the viewport, as you did for the top left viewport.

8. Pick the opposite corner to form a rectangle that is slightly smaller than the viewport.

9. Enter **front** for the view name and press **ENTER** a second time to terminate the command.

10. Save your work.

11. In the status bar, change **MODEL** to **PAPER** and erase the old lower left viewport. Do not erase the new viewport containing the orthographic view.

Solids

12. Pick the **Drawing** button from the Solids toolbar to enter the SOLDRAW command.

13. Pick the newest viewport and press **ENTER**.

AutoCAD creates an orthographic view from the viewport drawing, complete with hidden lines, as shown in Fig. 45-5 on page 612.

14. Save your work.

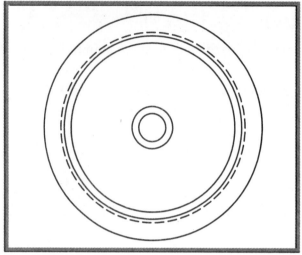

Fig. 45-5

Creating a Full Section View

Let's use the SOLVIEW command's Section option to create a full section.

1. In the status bar, change **PAPER** to **MODEL**.

2. Click in the lower right viewport to make it active.

3. Pick the **View** button from the Solids toolbar and enter **S** for Section.

Solids

4. In reply to Specify first point of cutting plane, snap to the center of the pulley in the lower right viewport.

5. In reply to Specify second point of cutting plane, pick a point a short distance directly above or below the center point.

This defines the cutting plane.

6. In reply to Specify side to view from, pick a point on either side of the center point. (The resulting section is the same either way.)

7. In reply to Enter view scale, press **ENTER** to accept the default value.

8. In reply to Specify view center, pick a point near the center of the viewport and press **ENTER**.

9. For the first and second corners, define a viewport outline as you did before.

10. In reply to View name, enter **section**; press **ENTER** a second time to terminate the command.

The lower right viewport should look similar to Fig. 45-6.

Chapter 45: Documenting Solid Models

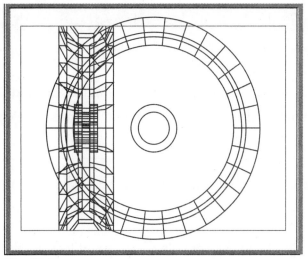

Fig. 45-6

Fig. 45-7

11. Review the list of layers and save your work.

12. Change **MODEL** to **PAPER** in the status bar.

13. Erase the old lower right viewport. Do not erase the new viewport containing the new view.

14. Pick the **Drawing** button from the Solids toolbar.

Solids

15. Pick the newest viewport and press **ENTER**.

AutoCAD creates a full section from the viewport drawing, complete with section lines, as shown in Fig. 45-7.

16. Change **PAPER** to **MODEL** in the status bar, enter **ZOOM** and **.7x**, and center the drawing, if necessary.

17. Save your work.

Creating the Isometric View

Finally, let's add an isometric view of the pulley in the upper right viewport.

1. Click the upper right viewport and enter **ZOOM** and **.7x**.

2. Produce an isometric view of the pulley in this viewport.

HINT:

From the View pull-down menu, select 3D Views and SW Isometric.

The pulley in the upper right viewport should look similar to the one in Fig. 45-8.

Positioning and Plotting the Views

The drawing is a few steps from being complete. All you need to do is position the viewports and turn on hidden line removal in the upper right viewport.

Positioning the Viewports

You can reposition the viewports using the MOVE command. Note that you do not have to resize the viewports. It is okay for the border lines to overlap, as shown in Fig. 45-8.

1. Change **MODEL** to **PAPER** in the status bar.

2. Position the four viewports as you would on a drawing sheet, as shown in Fig. 45-8. (See the following hint.)

HINT:

When you move the left and bottom viewports, move them in pairs with polar tracking or ortho on. For example, move the bottom two viewports upward at the same time. This keeps the views perfectly aligned with one another.

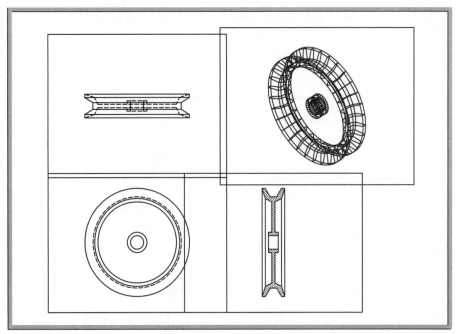

Fig. 45-8

Removing Hidden Lines

1. Select the border of the upper right viewport and then right-click.

2. From the shortcut menu, pick **Shade Plot** and **Hidden**.

This causes AutoCAD to remove hidden lines from this view when plotting.

3. Make **Objects** the current layer and freeze layer **Vports**.

Plotting the Drawing

1. Pick the **Plot** button from the Standard toolbar.

2. In the Plot area, select **Display**.

3. In the Plot scale area, select the **Fit to paper** check box.

Standard

4. Do a full preview, make corrections if necessary, and print the drawing.

The printed sheet should look similar to Fig. 45-9.

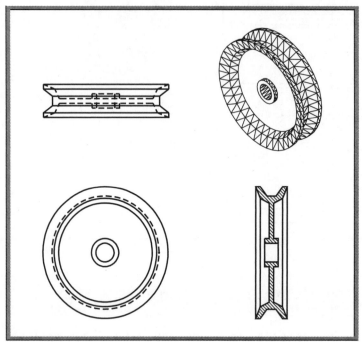

Fig. 45-9

5. Save your work, close the Solids toolbar, and exit AutoCAD.

Review Questions

1. What is the purpose of the SOLVIEW command?

2. Explain the purpose of each of the following SOLVIEW options.
   a. UCS
   b. Ortho
   c. Auxiliary
   d. Section

3. What layers does the SOLVIEW command create?

4. How is the SOLDRAW command used in conjunction with the SOLVIEW command?

5. Is it possible to use SOLDRAW without SOLVIEW? Explain.

Challenge Your Thinking

1. Discuss the Auxiliary option of the SOLVIEW command. Identify a solid object in which an auxiliary view is needed to document the object fully.

2. Under what circumstances might it be helpful to use the Ucs option of SOLVIEW?

Applying AutoCAD Skills

Work the following problems to practice the commands and skills you learned in this chapter.

1. Create a new drawing file named compos2.dwg from the file named compos.dwg (created in Chapter 42). Using SOLVIEW and SOLDRAW, create four new views similar to the ones you produced in this chapter.

2. Open the rocker arm drawing you created in Chapter 44 ("Using Problem-Solving Skills" problem 2). Create the front and top views from the solid, and display them in two viewports. Save the drawing as ch45arm.dwg.

Using Problem-Solving Skills

Complete the following activities using problem-solving skills and your knowledge of AutoCAD.

1. Your supervisor has asked you to create production drawings for the base plate you drew in Chapter 42 ("Using Problem-Solving Skills" problem 1). She wants you to use viewports to display the front view, top view, longitudinal full section, and an isometric view. Open the drawing (or create it, if you haven't already). Save it as ch45base.dwg, and create the required views.

2. Your company's publications department needs some illustrations for its advertising copy. Open the drawing of the height gage you drew in Chapter 42 ("Applying AutoCAD Skills" problem 4). Save the drawing as ch45gage.dwg, and display viewports showing the front view, right-side view, a full section through the V-groove, and an isometric view.

Chapter 45: Documenting Solid Models

Physical Benefits of Solid Modeling

Objectives:

- **Split a solid model in half**
- **Calculate mass properties of a solid model**
- **Create an STL file of a solid model for rapid prototyping**

Key Terms

rapid prototyping (RP)
XYZ octant

Solid modeling offers many benefits in addition to those discussed in previous chapters. One of the benefits is the ease at which a solid allows you to calculate mass properties. Another benefit is the natural link to *rapid prototyping (RP)*. RP is a method of quickly producing physical models and prototype parts from solid model data. Once you have a solid model, it becomes very easy to output the data needed to drive an RP system.

In this chapter, you will split the pulley you created in Chapter 44, as shown in Fig. 46-1. You will use the pulley half to calculate mass properties and create data that can be used to produce an RP part.

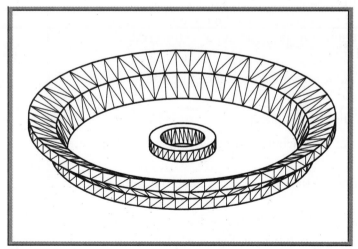

Fig. 46-1

Slicing the Pulley

The SLICE command cuts through a solid and retains either or both parts of it. The command is especially useful if you want to create a mold for half of a symmetrical part, such as a pulley.

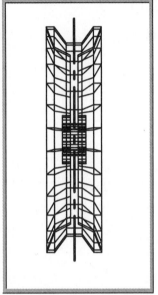

Fig. 46-2

1. Start AutoCAD and use the **pulley.dwg** drawing file to create a new file named **half.dwg**.

2. Freeze all layers except for layer Objects.

3. With snap on, move the pulley to the center of the screen.

4. Enter **UCS** and enter **W** for World.

5. Enter **PLAN** and enter **W** for World.

The view of the pulley should be similar to the one in Fig. 46-2, without the center line.

6. Display the Solids toolbar.

7. From the Solids toolbar, pick the **Slice** button.

This enters the **SLICE** command.

Solids

8. Select the pulley, press **ENTER**, and enter the **YZ** option.

AutoCAD asks you to pick a point on the YZ plane. Try to visualize the orientation of this plane, which is vertical and perpendicular to the screen. "Vertical" is the Y direction and "perpendicular to the screen" is the Z direction.

9. Pick a point anywhere on the center line as shown in Fig. 46-2.

This defines the cutting plane.

10. Pick a point anywhere to the right of the cutting plane.

This tells AutoCAD that you want to keep the right half of the pulley.

AutoCAD cuts the pulley in half and leaves the right half on the screen, as shown in Fig. 46-3. (Ignore the leaders and text.)

11. Save your work.

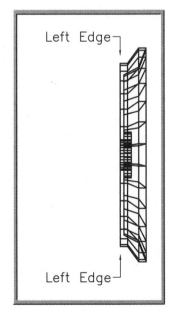

Fig. 46-3

Chapter 46: Physical Benefits of Solid Modeling

Calculating Mass Properties

Because solid models contain volume information, it is possible to obtain meaningful details about the model's physical properties. For example, you can generate mass properties, including the center of gravity, mass, surface area, and moments of inertia.

1. From the **Tools** pull-down menu, pick **Inquiry** and **Region/Mass Properties**.

This enters the MASSPROP command.

2. Select the solid model and press **ENTER**.

AutoCAD generates a report similar to the one shown in Fig. 46-4. Properties such as mass and volume permit engineers to analyze the part for structural integrity.

3. Press **ENTER** to see the entire report.

4. Enter **Yes** in reply to Write analysis to a file? and press **ENTER** or pick the **Save** button to use the default half.mpr file name.

AutoCAD creates an ASCII (text) file, assigning the MPR file extension to it automatically. The MPR file extension stands for "Mass Properties Report." Let's locate and review the file.

5. Close the AutoCAD Text window if you haven't already, and minimize AutoCAD.

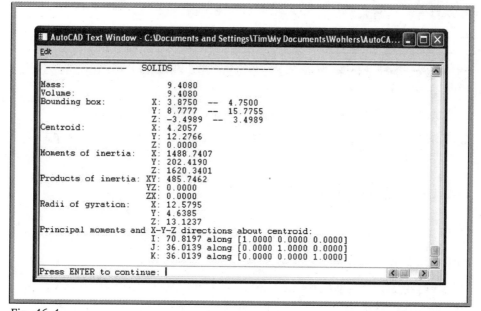

Fig. 46-4

6. Start Notepad.

7. Select **Open...** from the **File** pull-down menu.

8. Locate and open the file named **half.mpr**.

HINT:

Change the file name extension from txt to mpr to list only those files that have the MPR extension.

The contents of the file should be identical to the information displayed by the MASSPROP command.

9. Exit Notepad and maximize AutoCAD.

Outputting Data for Rapid Prototyping

STLOUT creates an STL file from a solid model. STL is the file type required by most rapid prototyping systems, such as stereolithography and fused deposition modeling (FDM). RP systems enable you to create a physical prototype part from a solid model. The part can serve many purposes, such as a pattern for prototype tooling.

Positioning the Solid Model

Before creating the STL file, you must orient the part in the position that is best for the RP system. In most cases, the part should be lying flat on the WCS. You can accomplish this in the current drawing by rotating the part around the Y axis.

1. Enter the **ROTATE3D** command, select the solid, and press **ENTER**.

2. Enter the **Yaxis** option because you want to rotate the part around the Y axis.

3. Pick a point anywhere on the left edge of the part (shown in Fig. 46-3 on page 620) to define the Y axis.

4. Enter **–90** for the rotation angle.

NOTE:

It is important that you rotate the part in the negative direction. A positive rotation positions the part upside-down.

Chapter 46: Physical Benefits of Solid Modeling

5. **ZOOM All** and, if necessary, move the part so that it lies entirely within the drawing area, as reflected by the grid.

If the part lies outside the positive XYZ octant, the STLOUT command will not work. The *XYZ octant* is the area in 3D space where the *x*, *y*, and *z* coordinates are greater than 0.

6. Pick a viewpoint that is slightly above the part to achieve a view similar to the one in Fig. 46-5.

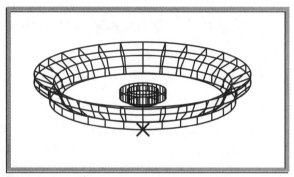

Fig. 46-5

The part should appear to lie on the WCS. Let's check to be sure.

7. Using the **ID** command, pick a point at the bottom of the part (indicated by the x in Fig. 46-5).

HINT:

Use the Nearest object snap to snap to a point at the bottom of the part.

The value of the *z* coordinate should be 0.0000. If it is 0.0000 or a value less than 0, you must move the solid upward into the positive XYZ octant.

8. If necessary, move the solid upward **3** units or until it is above the grid.

Creating the STL File

Now that the pulley is properly positioned, you can create the STL file.

1. Enter the **PLAN** command to view the plan view of the WCS and **ZOOM All**.

2. Enter the **STLOUT** command, pick the solid, and press **ENTER**.

AutoCAD wants to know whether you want to create a binary STL file. Normally, you would create a binary file because they are much smaller than ASCII STL files. However, you cannot view the contents of a binary STL file, and we want to view its contents.

3. Enter **No** to create an ASCII STL file.

4. Pick the **Save** button or press **ENTER** to accept half.stl for the name of the file.

Reviewing the STL File Data

Let's review the contents of the STL file.

1. Minimize AutoCAD and start WordPad.

2. Select **Open...** from the **File** pull-down menu.

3. In the File name box, enter ***.stl**. Be sure to press **ENTER**.

4. Open the file named **half.stl**.

The contents of the file appear. The text listing in Fig. 46-6 shows the first part of the file.

STL files consist mainly of groups of x, y, and z coordinates. Each group of three defines a triangle.

5. After you've reviewed the file, close WordPad and maximize AutoCAD.

6. Close the Solids toolbar, as well as any other toolbars you may have opened.

7. Save your work and exit AutoCAD.

The view shown in Fig. 46-7 was created using a special utility called stlview.lsp. This utility reads and displays the contents of an ASCII STL file.

As you can see, an STL file is indeed made up of triangular facets. You can adjust the size of the triangles using the FACETRES (short for "facet resolution") system variable by changing its value before you create the STL file. A high value creates a more accurate part with a smoother surface finish, but the file size and the time required to process the file increase. You should set it as high as necessary to produce an accurate part. Knowing what this value should be for a given part comes with experience and experimentation.

```
solid AutoCAD
    facet normal 0.0000000e+000  0.0000000e+000  −1.0000000e+000
        outer loop
            vertex 8.4399540e+000  4.8596573e+000  1.0000000e−001
            vertex 8.5000000e+000  4.2500000e+000  1.0000000e−001
            vertex 5.8750000e+000  4.2500000e+000  1.0000000e−001
        endloop
    endfacet
    facet normal 0.0000000e+000  0.0000000e+000  −1.0000000e+000
        outer loop
            vertex 4.8750000e+000  4.2500000e+000  1.0000000e−001
            vertex 2.2500000e+000  4.2500000e+000  1.0000000e−001
            vertex 2.3100460e+000  4.8596573e+000  1.0000000e−001
        endloop
    endfacet
    facet normal 0.0000000e+000 0.0000000e+000  −1.0000000e+000
        outer loop
            vertex 4.9130602e+000  4.4613417e+000  1.0000000e−001
            vertex 4.8750000e+000  4.2500000e+000  1.0000000e−001
            vertex 2.3100460e+000  4.8596573e+000  1.0000000e−001
        endloop
    endfacet
    facet normal 0.0000000e+000  0.0000000e+000  −1.0000000e+000
```

Fig. 46-6

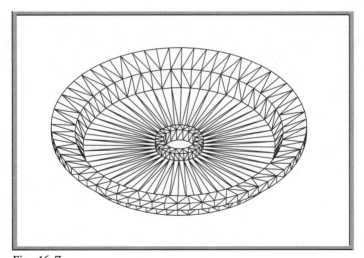

Fig. 46-7

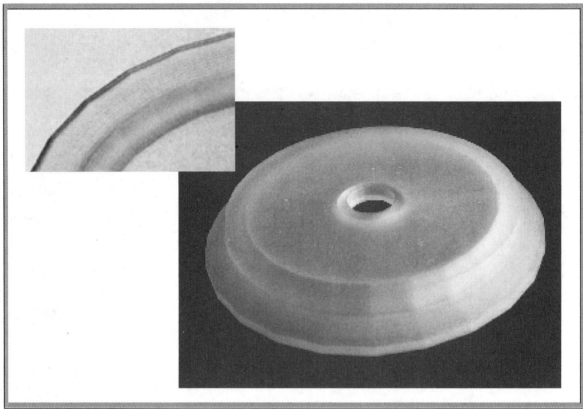

Fig. 46-8   Prototype part courtesy of Laser Prototypes, Inc. of Denville, New Jersey (Photos by Keith M. Berry)

Prototype Part Creation

Hundreds of organizations around the world can create prototype parts from STL files. If you were to send the half.stl file to one of them, that organization could create a part for you, typically for a fee.

Figure 46-8 shows pictures of a completed plastic prototype part. It was created from the half.stl file using a stereolithography system from 3D Systems, Inc. (Valencia, CA). Stereolithography is one of the most widely used RP technologies in the world.

Notice the flat segments that make up the large diameter of the pulley. They are caused by the triangular facets. You could reduce their size by increasing the value of FACETRES before creating the STL file. You could also reduce their visibility by sanding the part, but this could change the accuracy of the part. The part pictured above was built from a version of half.stl with FACETRES set at a relatively low value to show the undesirable effect of large facets.

Chapter 46
Review & Activities

Review Questions

1. Describe the SLICE command.

2. When would the SLICE command be useful?

3. AutoCAD can create files with an MPR file extension. How are these files created, and what information do they provide?

4. Name at least three mass properties for which AutoCAD makes information available about a solid model.

5. Explain the purpose of the STLOUT command.

6. What is the purpose of creating an STL file?

7. What AutoCAD system variable affects the size of the triangles in an STL file?

8. What is the primary advantage of creating a binary STL file? An ASCII STL file?

Challenge Your Thinking

1. Write a paragraph comparing and contrasting the function and purpose of the SECTION and SLICE commands.

2. Describe a situation in which you might need to slice an object using the 3points option.

3. Explore the different rapid prototyping systems that are commercially available. Write an essay that compares each of them.

Applying AutoCAD Skills

Work the following problems to practice the commands and skills you learned in this chapter.

1. Create ASCII and binary STL files from the solid model stored in compos.dwg (created in Chapter 42). Compare the sizes of the two files and then view the contents of the ASCII STL file.

2. Make two slices of the height gage you created in Chapter 42 ("Applying AutoCAD Skills" problem 4). Create one slice through the V groove and the second slice perpendicular to the first and in the same plane.

3. If you know of a rapid prototyping service provider in your area, request a demonstration of its RP system(s). Explore the possibility of having the company create a prototype part for you from an STL file.

4. Open the file named shaft.dwg (created in Chapter 41). Use the SLICE command to split the model down the middle to create a pattern. Use STLOUT to create a binary STL file.

5. Produce the mass properties for shaft.dwg.

Using Problem-Solving Skills

Complete the following activities using problem-solving skills and your knowledge of AutoCAD.

1. The bushing and insert shown in Fig. 46-9 are part of an aerospace project. Weight is critical. Slice the bushing and insert. Determine the volume of each part. Write each of the mass property reports to a file and print the files. Then select a material for each, look up its density, and compute its mass. (To find density values, you may have to go to the reference section of a suitable library or search the Internet.)

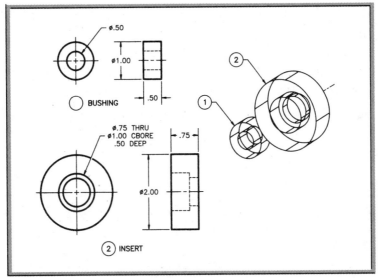

Fig. 46-9

2. Position the bushing and insert from problem 1. Then make an ASCII STL file of the bushing only. View the STL file, but do not print it.

Figure 46-9 prepared by Gary J. Hordemann, Gonzaga University

Part 9 Project

Designing and Modeling a Desk Set

AutoCAD is capable of creating 2D drawings, 3D surface models, and 3D solid models. The differences are meaningful. Although all three types of drawings yield information, 3D models allow more realistic views. Solid modeling, in addition, allows determination of mass properties and the generation of production data such as that used in automated machine tool processes. Familiarity with solid modeling enhances the ability of the drafter/designer to create solutions; it is well worth the effort to learn.

Description

Your company has sold the desk set shown in Fig. P9-1 for several years. Your supervisor has determined that it is time to modernize and streamline the desk set to meet current market demand.

1. Create a solid model of the desk set shown in Fig. P9-1. Include the round space to store pens, pencils, and the like; a large, flat, shallow area for items such as paper clips and rubber bands; and a medium-height section for a note pad, business cards, and similar papers. Wall thickness should range between .1″ and .185″. Assign a plastic material to the model. Save it as desk1.dwg.

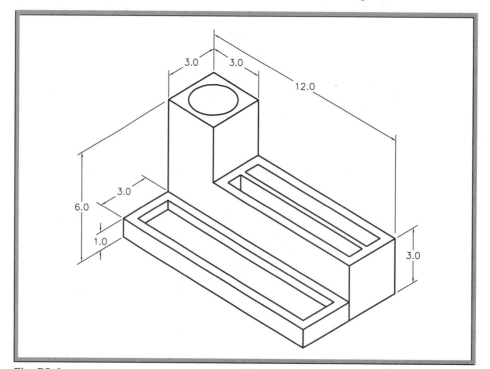

Fig. P9-1

2. Save desk1.dwg with a new name of desk2.dwg. Alter the design to modernize it or to make it more useful. For example, you may decide to provide an inclined surface on the two medium-height holes so that the one in back rises above the one in front of it. You may also want to round the edges to give it a streamlined look, or even add or subdivide compartments.

3. Prepare a presentation for your supervisor. In your presentation, compare and contrast the new and old designs and explain why the new design is superior. Prepare rendered images of both desk sets to support your presentation.

Hints and Suggestions

1. There is often more than one way to construct a solid model. If your approach does not seem to be working, try another.

2. Remember to increase the value of ISOLINES and FACETRES to improve appearance.

3. Create the recesses for the compartments (such as the pencil holder) using the SUBTRACT command.

4. Use the UNION command as necessary so that the final product is a single composite solid.

Summary Questions

Your instructor may direct you to answer these questions orally or in writing. You may wish to compare and exchange ideas with other students to help increase efficiency and productivity.

1. In your desk set design, how could the area actually being used to hold items be calculated?

2. Could the entire desk set design be created via extrusion? Explain how it could or why it could not.

3. What are some advantages of using predefined solid primitives in the creation of more complex solid models?

4. Explain how to use the SUBTRACT command to create a thin-walled cylinder.

5. What is meant by "downstream benefits" of solid modeling?

6. Describe the types of information that are provided by the MASSPROP command.

Careers
Using
AutoCAD

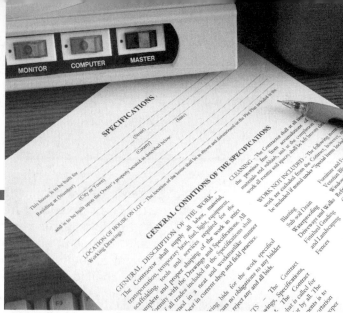
Tim Fuller Photography

Specification Writer

As you walk through your neighborhood, take a look at some of the buildings on your street. How many incorporate brick into the design? How many are covered with vinyl siding or wood? Inside your own home, examine one of the water piping systems (look under the kitchen and bathroom sinks). Are the pipes made of metal or plastic?

While the materials in these situations may appear to be used for cosmetic reasons, most are selected because they are durable and efficient. Trained professionals known as *specification writers* are the people responsible for recommending certain material types for the construction, repair, and installation of products and systems.

Understanding the Job

Specification writers create reports, books, and pamphlets that document the specifications of materials used for a particular job. In order to determine which types of materials are best suited for a project, they draw from their understanding of standard specification codes as well as from architectural notes and drawings.

Many specification writers worked as drafters at one time. This prior experience provides them with a knowledge basis of system and building requirements, skills to interpret blueprints and sketches, and an overall understanding of construction and repair processes.

Multiple Skills Needed

Specification writers should demonstrate computer proficiency. Experience using word-processing programs, spreadsheets, and CAD programs is helpful. Research skills are also important; writers must be familiar with state-of-the-art products and materials in order to make suggestions about new specifications, or to improve old ones.

If a writer encounters a product in which he or she is unfamiliar, the ability to find information about how it works, and about the materials used to build it, is necessary. Also, communicating through writing is a key skill for specification writers. Clearly written documents will help the people who read them to do their jobs more efficiently. Some technical training, preferably thorough college programs, is typically needed. Some specification writing positions may even require applicants to have an engineering degree.

Career Activities

- Research careers for specification writers. Who typically employs them?
- Explain how having prior experience in drafting can help a specification writer.

Additional Problems

Introduction

The following problems provide additional practice with AutoCAD. These problems encompass a variety of disciplines. They will help you expand your knowledge and ability and will offer you new and challenging experiences.

The key to success is to *plan before you begin*. Review the options for setting up a new drawing and consider which commands and features you might use to solve the problem. As you discover new and easier methods of creating drawings, apply these methods to solving the problems. Since there is usually more than one way to complete a drawing, experiment with alternative methods. Discuss these alternatives with other users and create strategies for efficient completion of the problems.

Remember, there is no substitute for practice. The expertise you gain is proportionate to the time you spend on the system. Set aside blocks of time to work with AutoCAD, think through your approach, and enjoy this fascinating technology.

Each problem in this section is preceded by an icon that describes its level of difficulty, as shown below. The level of difficulty assigned to each problem assumes that you have learned the AutoCAD techniques and skills necessary to work the problem successfully. For some of the Level 3 problems, you may be required to expand your experience by combining your knowledge of AutoCAD with knowledge and skills that are not taught in this book.

LEVEL
 Uses basic AutoCAD skills

LEVEL
 Uses intermediate AutoCAD skills

LEVEL
 Uses advanced AutoCAD and problem-solving skills

AdditionalProblems

LEVEL
1

Problem 1

Create and fully dimension the drawing of a link.

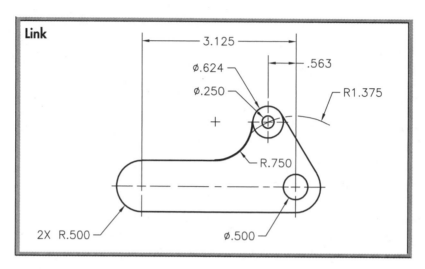

LEVEL
1

Problem 2

Create and fully dimension the drawing of a filler.

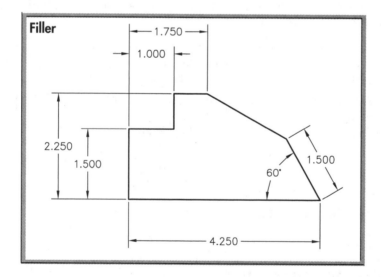

AdditionalProblems

LEVEL
1

Problem 3

Create and fully dimension the drawing of a link.

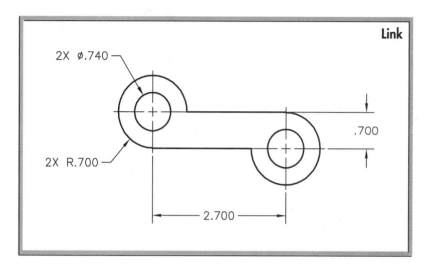

LEVEL
1

Problem 4

Create and fully dimension the drawing of a shaft bracket.

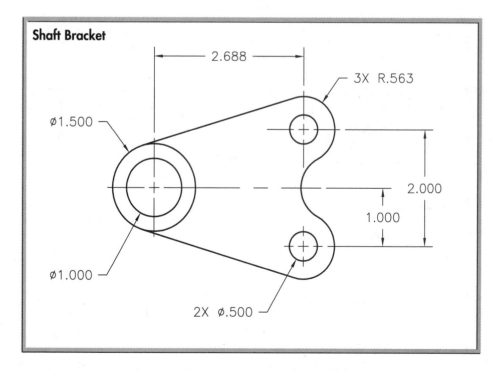

AdditionalProblems

LEVEL 1

Problem 5

Create and fully dimension the drawing of a gage.

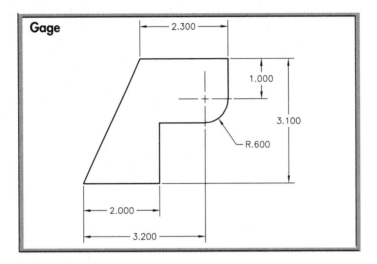

LEVEL 1

Problem 6

Create and dimension the orthographic views of the block.

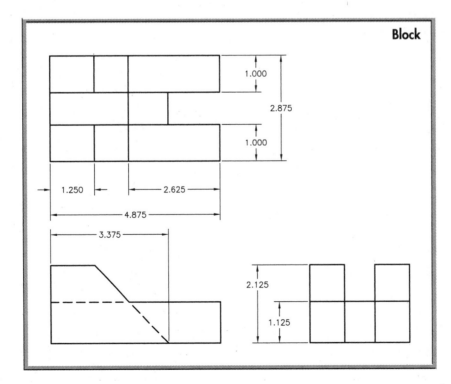

AdditionalProblems

LEVEL 1

Problem 7

Create and dimension the drawing of an end block.

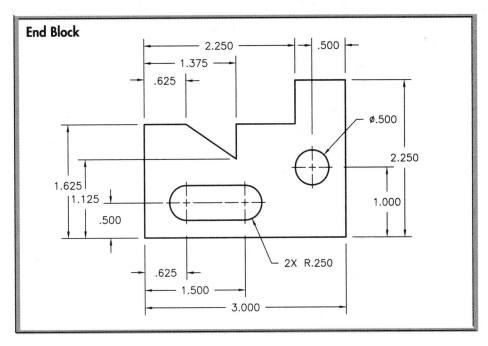

End Block

LEVEL 1

Problem 8

Shown below is an isometric view of an angle block. Create orthographic views of the block.

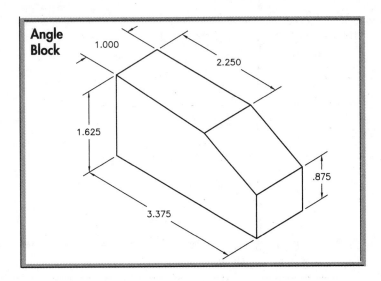

Angle Block

AdditionalProblems

LEVEL
1

Problem 9

Shown at the right is an isometric drawing of a nesting block. Create orthographic views of the block and fully dimension the views.

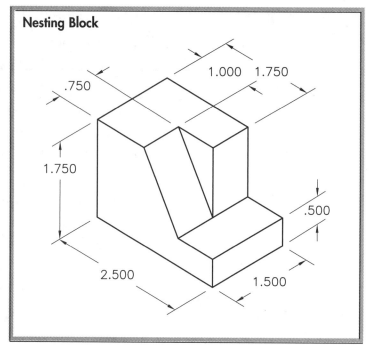

Nesting Block

.750
1.000 1.750
1.750
.500
2.500
1.500

LEVEL
1

Problem 10

Shown at the right is an isometric view of a locator. Create orthographic views of the locator and fully dimension the views.

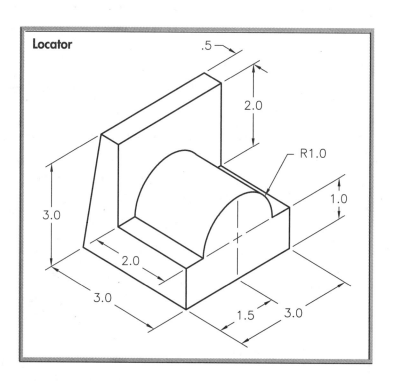

Locator

.5
2.0
R1.0
3.0
1.0
2.0
3.0
1.5 3.0

AdditionalProblems

Problem 11

Shown at the right is an isometric view of a step block. Create orthographic views of the step block and fully dimension the views.

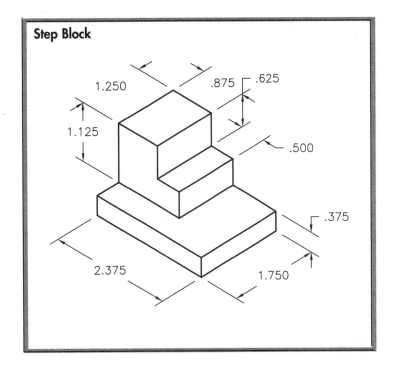

Step Block

1.250 · .875 · .625

1.125

.500

.375

2.375 · 1.750

Problem 12

Create and fully dimension the drawing of a block.

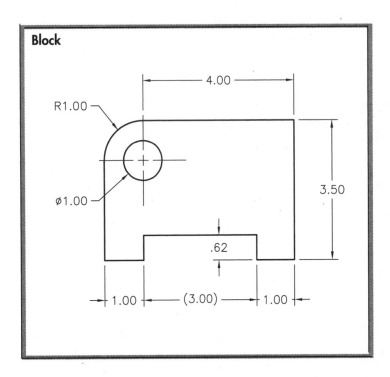

Block

R1.00

4.00

ø1.00

3.50

.62

1.00 · (3.00) · 1.00

AdditionalProblems

Problem 13

Create the drawing of an angle bracket. Use the dimensions shown, but dimension the drawing using decimal units.

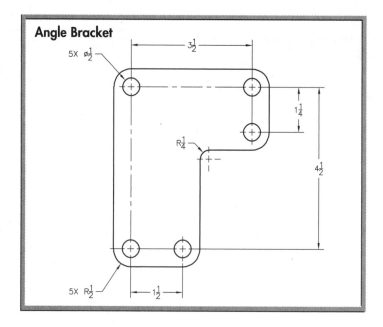

Problem 14

Create and dimension the drawing of a gasket.

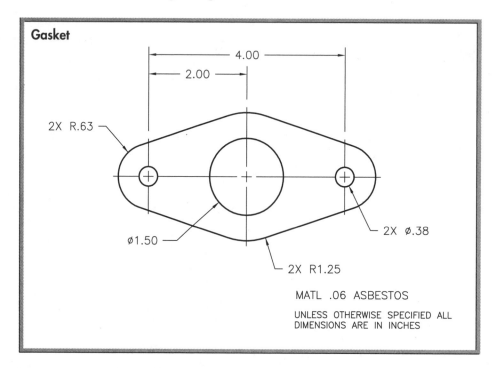

Problem 14 adapted from the textbook Drafting Fundamentals by Scott, Foy, and Schwendau

AdditionalProblems

Problem 15

Create the electrical symbols as shown. Estimate their sizes. Produce a block of each symbol to create a symbol library.

Electrical Symbols

RESISTOR   THERMISTOR   VARIABLE RESISTOR   TRANSISTORS   LOGICAL GATES

NPN   PNP   AND   NAND

DIODE   ZENER   CONNECTION

JFET—N   JFET—P   OR   NOR

CAPACITOR   CRYSTAL   TERMINAL

THYRISTORS   COMMON GROUND   CHASSIS GROUND

INDUCTOR   SWITCH   SPEAKER

Problem 16

Create the following international symbols.

International Symbols

Problem 15 courtesy of Robert Pruse, Fort Wayne Community Schools; Problem 16 courtesy of Joseph K. Yabu, San Jose State University

Problem 17

Create the floor plan drawing of a computer lab. Determine the drawing area according to space requirements. Estimate the dimensions that are not given.

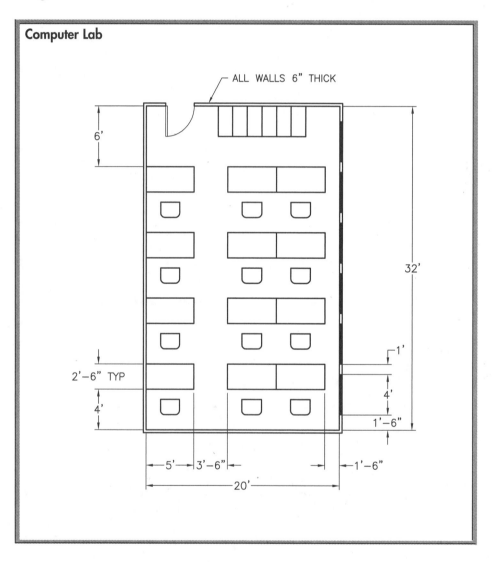

Computer Lab

ALL WALLS 6" THICK

6'

32'

2'-6" TYP

4'

1'

4'

1'-6"

5' 3'-6" 1'-6"

20'

AdditionalProblems

LEVEL
2

Problem 18

Create and fully dimension the drawing of an adjustable link.

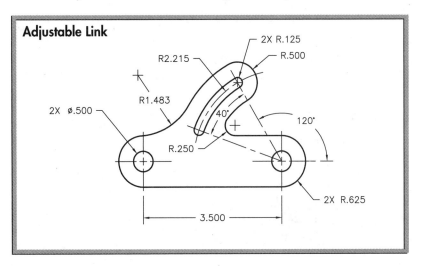

Adjustable Link

- 2X R.125
- R2.215
- R.500
- R1.483
- 2X ⌀.500
- 40°
- 120°
- R.250
- 2X R.625
- 3.500

LEVEL
2

Problem 19

Create and dimension the drawing of an idler plate.

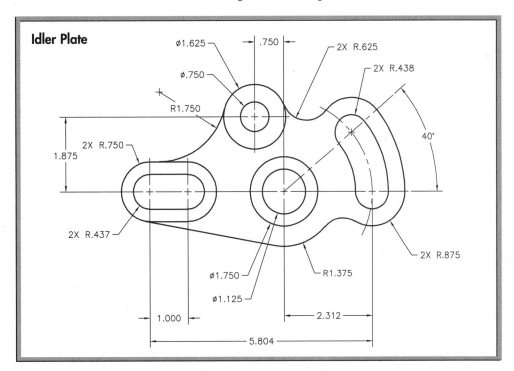

Idler Plate

- ⌀1.625
- .750
- 2X R.625
- ⌀.750
- 2X R.438
- R1.750
- 40°
- 2X R.750
- 1.875
- 2X R.437
- 2X R.875
- ⌀1.750
- R1.375
- ⌀1.125
- 1.000
- 2.312
- 5.804

AdditionalProblems

Problem 20

Create and dimension orthographic views of the support.

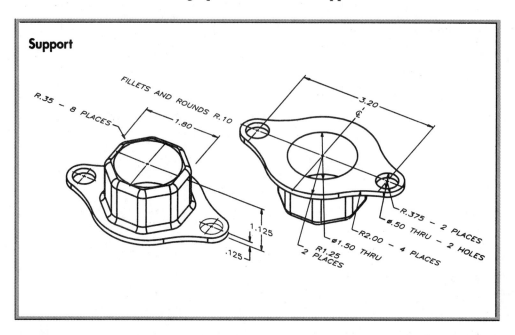

Support

Problem 21

Create and dimension the drawing of a cover plate.

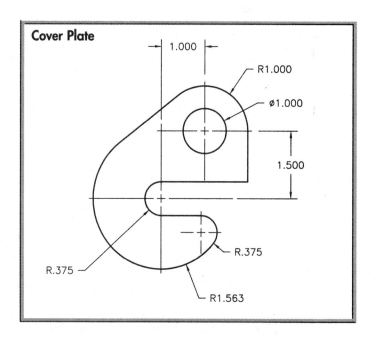

Cover Plate

Problem 20 courtesy of Gary J. Hordemann, Gonzaga University

AdditionalProblems

Problem 22

Shown at the right is an isometric drawing of an angled base. Create orthographic views of the base and fully dimension the views. Use a drawing area of 17 × 11.

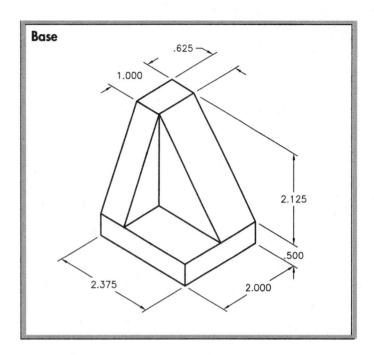

Base

.625

1.000

2.125

.500

2.375

2.000

LEVEL 2

Problem 23

Shown at the right is an isometric drawing of a wedge block. Create orthographic views of the block and fully dimension the views. Use a drawing area of 17 × 11.

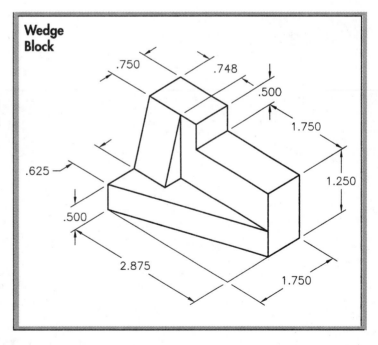

Wedge Block

.750

.748

.500

1.750

.625

1.250

.500

2.875

1.750

LEVEL 2

Problem 24

Shown at the right is an isometric view of an angle block. Create and fully dimension the orthographic views of the angle block.

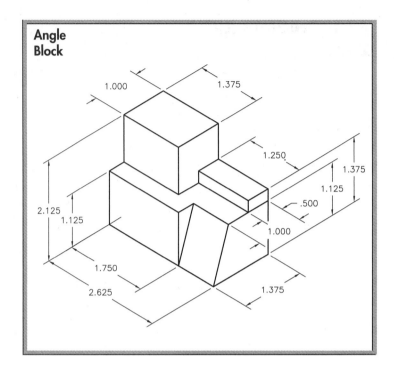

Angle Block

LEVEL 2

Problem 25

Shown at the right is an isometric view of a cradle. Create orthographic views of the cradle and fully dimension the views.

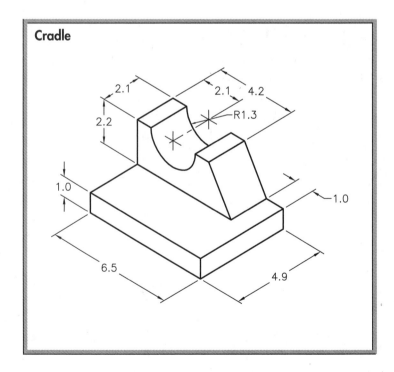

Cradle

AdditionalProblems

LEVEL 2

Problem 26

Draw the hex ratchet. Begin by creating the circles for the ends of the ratchet. Then create the lines and polygon. Use the BREAK command to remove parts of both circles as indicated to finish the drawing. Use a drawing area of 280 × 215 and dimension the drawing in millimeters.

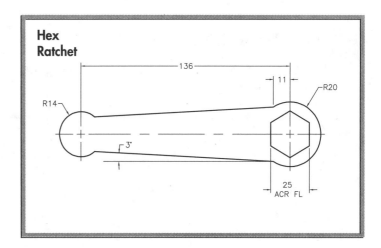

LEVEL 2

Problem 27

The mechanism shown at the right consists of three parts: a driver, a follower, and a link. Use the dimensions shown in the illustration and the CIRCLE, OFFSET, and FILLET commands to draw the front view.

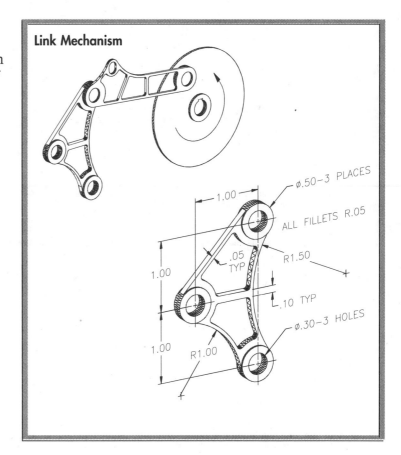

Problem 26 adapted from the textbook *Drafting Fundamentals* by Scott, Foy, and Schwendau; Problem 27 courtesy of Gary J. Hordemann, Gonzaga University

AdditionalProblems

Problem 28

Create and fully dimension the drawing of a master template. Use 22 × 17 for the drawing area.

Template

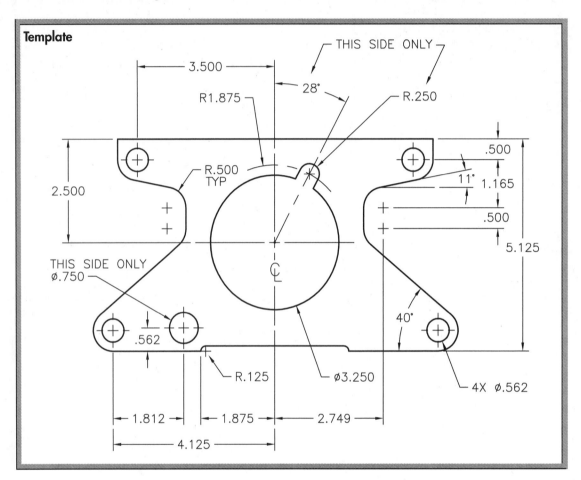

Problem 28 courtesy of Steve Huycke, Lake Michigan College

AdditionalProblems

LEVEL 2

Problem 29

Create and dimension the drawing of a rod support.

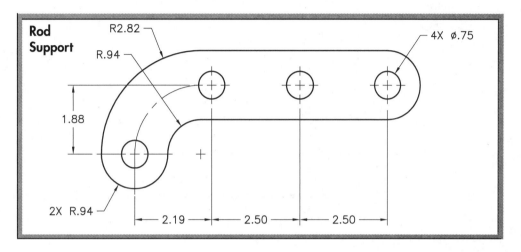

LEVEL 2

Problem 30

Create and dimension the drawing of a steel plate.

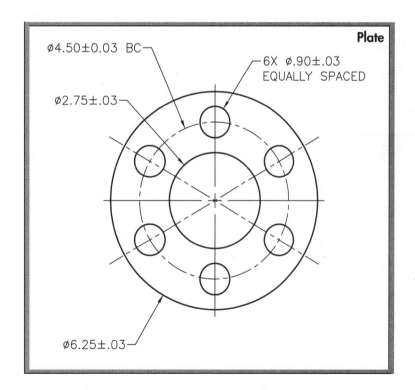

AdditionalProblems

LEVEL
2

Problem 31

Create and dimension the drawing of a rocker arm using decimal units.

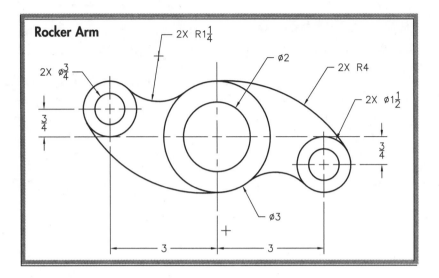

Rocker Arm

2X R1¼
ø2
2X R4
2X ø¾
2X ø1½
¾
¾
ø3
3   3

LEVEL
2

Problem 32

Create and fully dimension the drawing of a Geneva plate. Use a drawing area of 22 × 17.

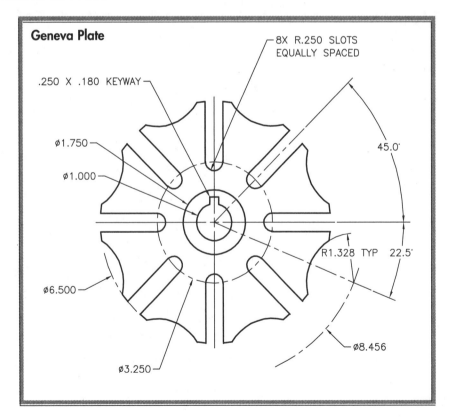

Geneva Plate

8X R.250 SLOTS
EQUALLY SPACED

.250 X .180 KEYWAY

ø1.750

ø1.000

45.0°

R1.328 TYP   22.5°

ø6.500

ø8.456

ø3.250

LEVEL
2

Problem 33

Create and dimension the drawing of a cassette reel. Use a drawing area of 280 × 215.

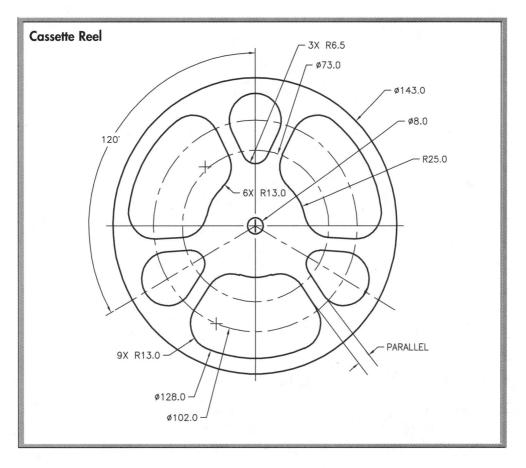

Cassette Reel

3X R6.5
ø73.0
ø143.0
ø8.0
R25.0
120°
6X R13.0
9X R13.0
PARALLEL
ø128.0
ø102.0

Problem 33 adapted from the textbook *Drafting Fundamentals* by Scott, Foy, and Schwendau

AdditionalProblems

LEVEL
2

Problem 34

Create and dimension the drawing of a slotted wheel. Use a drawing area of 17 × 11.

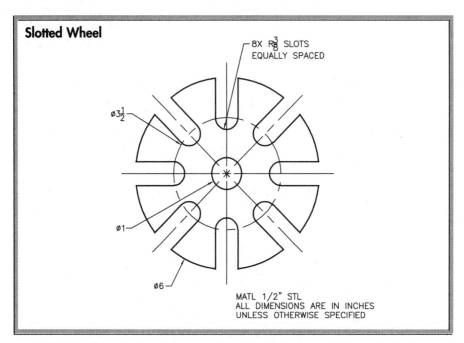

Slotted Wheel

8X R$\frac{3}{8}$ SLOTS
EQUALLY SPACED

ø3$\frac{1}{2}$

ø1

ø6

MATL 1/2" STL
ALL DIMENSIONS ARE IN INCHES
UNLESS OTHERWISE SPECIFIED

LEVEL
2

Problem 35

Create and dimension the drawing of a gasket.

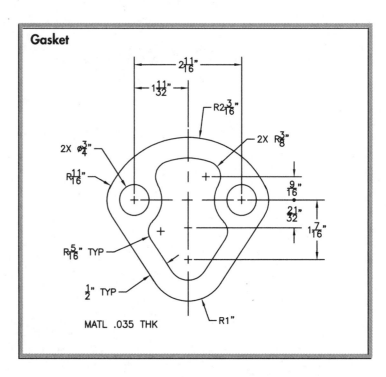

Gasket

2$\frac{11}{16}$"

1$\frac{11}{32}$"

R2$\frac{3}{16}$"

2X R$\frac{3}{8}$"

2X ø$\frac{3}{4}$"

R$\frac{11}{16}$"

$\frac{9}{16}$"

$\frac{21}{32}$"

1$\frac{7}{16}$"

R$\frac{5}{16}$" TYP

$\frac{1}{2}$" TYP

R1"

MATL .035 THK

Problems 34 and 35 adapted from the textbook *Drafting Fundamentals* by Scott, Foy, and Schwendau

Additional Problems

AdditionalProblems

LEVEL 2 Problem 36

Create and dimension the drawing of an 18-tooth cutter. Use a drawing area of 17 × 11.

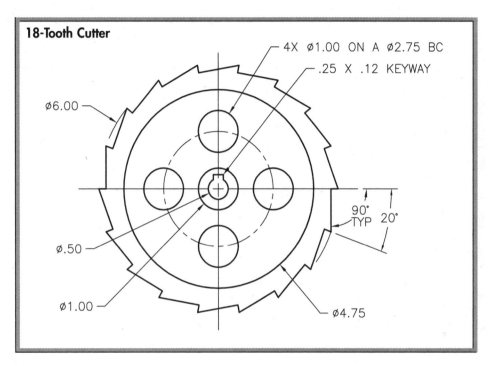

18-Tooth Cutter

4X Ø1.00 ON A Ø2.75 BC

.25 X .12 KEYWAY

Ø6.00

Ø.50

Ø1.00

90° TYP   20°

Ø4.75

AdditionalProblems

Problem 37

Create and dimension the top and front orthographic views of the slotted shaft.

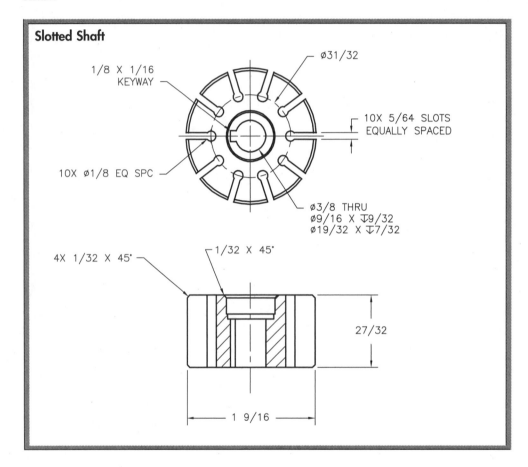

Problem 37 adapted from the textbook *Drafting Fundamentals* by Scott, Foy, and Schwendau

AdditionalProblems

LEVEL 2

Problem 38

Create and dimension the drawing of an adjustable bracket.

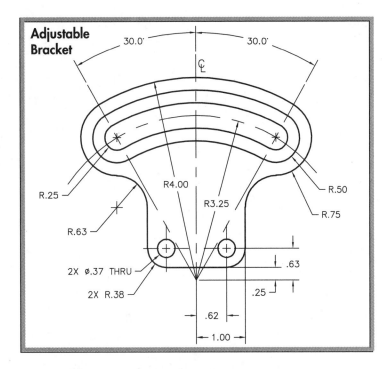

LEVEL 2

Problem 39

Create and dimension the drawing of a basketball backboard. Use a scale of $1'' = 1'$.

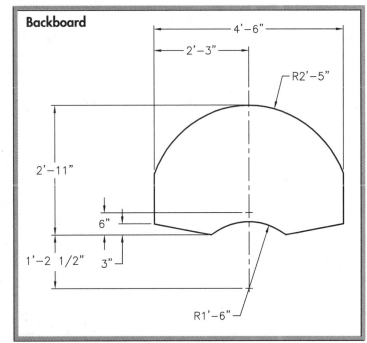

Problems 38 and 39 adapted from the textbook *Drafting Fundamentals* by Scott, Foy, and Schwendau

AdditionalProblems

Problem 40

Create and dimension a complete drawing of the wall plaque.

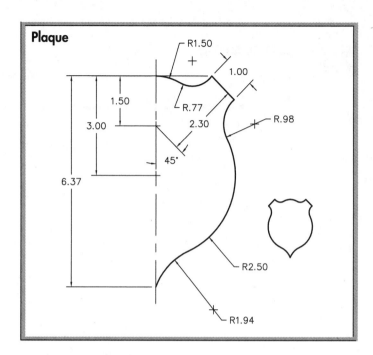

Problem 41

Create orthographic views of the bracket and fully dimension them.

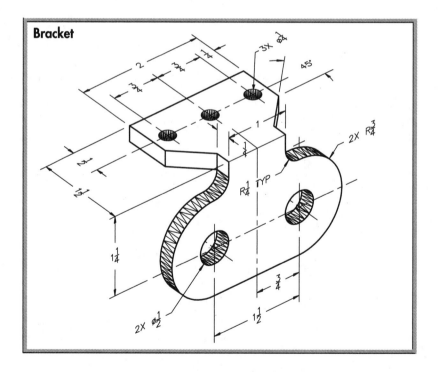

Problem 40 courtesy of Mark Schwendau, Kishwaukee College

Additional Problems

LEVEL
2

Problem 42

Create and dimension the spacer. Use a drawing area of 17 × 11.

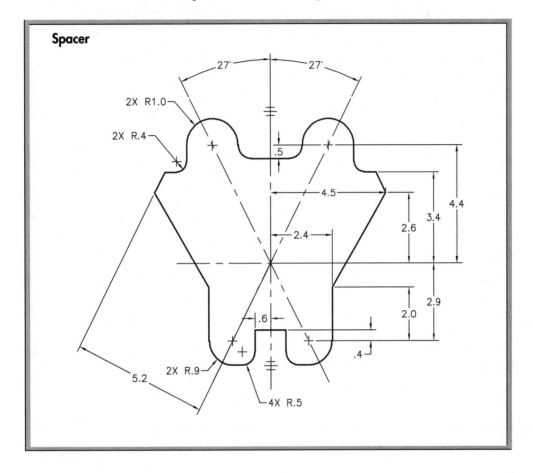

Problem 42 courtesy of Mark Schwendau, Kishwaukee College

AdditionalProblems

Problem 43

Create a single view of the spacer. Dimension the drawing and add a note indicating the thickness of the spacer. Convert the fractions to decimal units.

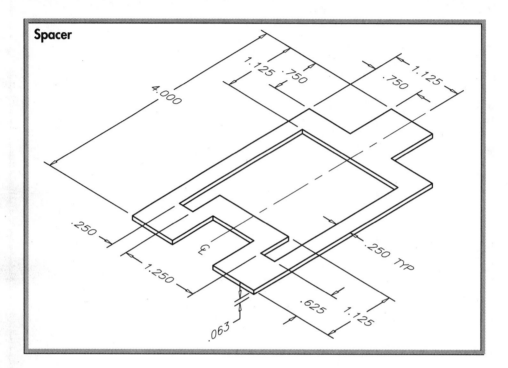

Problem 43 adapted from the textbook *Drafting Fundamentals* by Scott, Foy, and Schwendau

Additional Problems

Problem 44

Create the front and right-side views of the centering bushing. Include dimensions.

Centering Bushing

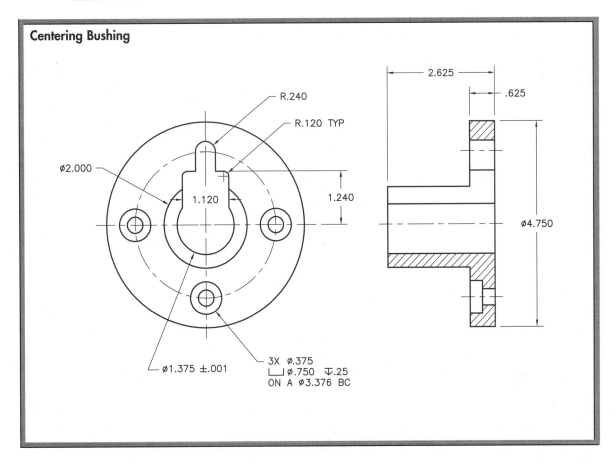

R.240

R.120 TYP

ø2.000

1.120

1.240

2.625

.625

ø4.750

ø1.375 ±.001

3X ø.375
⌴ ø.750 ▽.25
ON A ø3.376 BC

Problem 44 courtesy of Steve Huycke, Lake Michigan College

AdditionalProblems

Problem 45

Create the views of the rod support as shown and fully dimension them. Determine the drawing area on your own, based on space requirements.

Rod Support

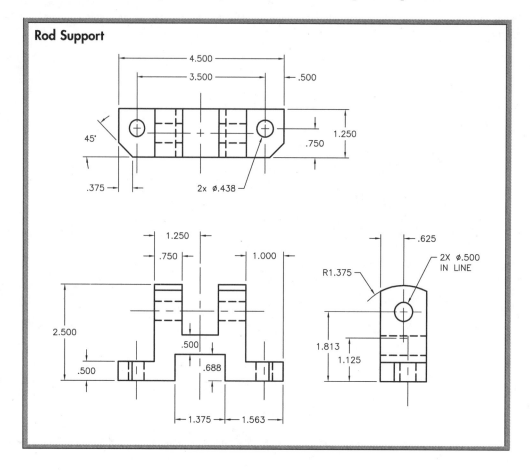

Problem 45 courtesy of Steve Huycke, Lake Michigan College

Additional Problems

LEVEL 2

Problem 46

Create a drawing of the template and dimension it as shown. Determine the drawing area, based on space requirements.

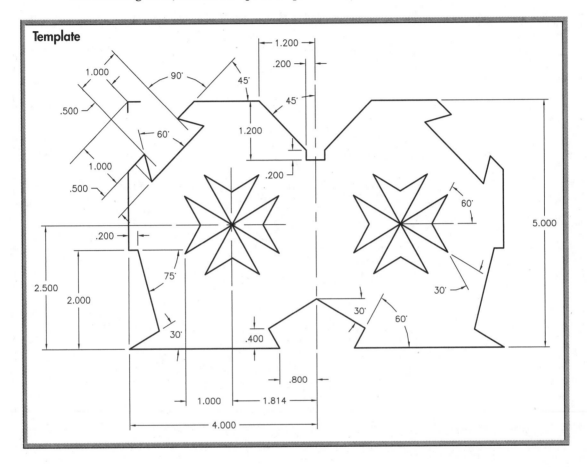

Problem 46 courtesy of Steve Huycke, Lake Michigan College

AdditionalProblems

Problem 47

Create the top and front full section view of the arm. Be sure to include the detail.

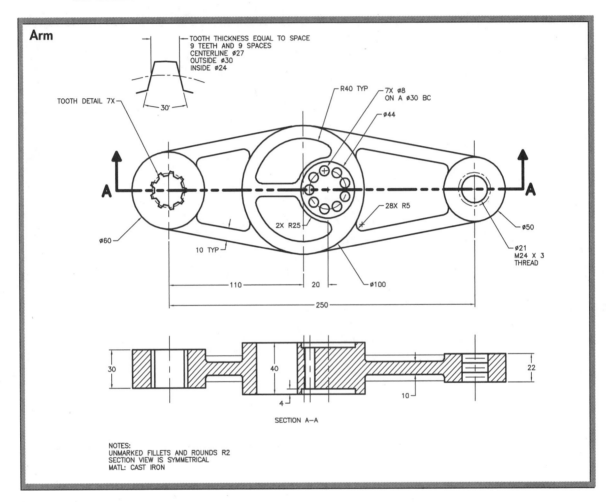

Arm

TOOTH THICKNESS EQUAL TO SPACE
9 TEETH AND 9 SPACES
CENTERLINE ø27
OUTSIDE ø30
INSIDE ø24

TOOTH DETAIL 7X

30'

R40 TYP   7X ø8
ON A ø30 BC

ø44

A

A

28X R5

2X R25

ø60

10 TYP

110

20

ø100

250

ø50

ø21
M24 X 3
THREAD

30

40

22

4

10

SECTION A–A

NOTES:
UNMARKED FILLETS AND ROUNDS R2
SECTION VIEW IS SYMMETRICAL
MATL: CAST IRON

Problem 47 courtesy of Alan Fitzell, Central Peel Secondary School

Problem 48

Create the electrical schematic drawing. Create blocks of each of the components and then insert them as needed.

Electrical Schematic

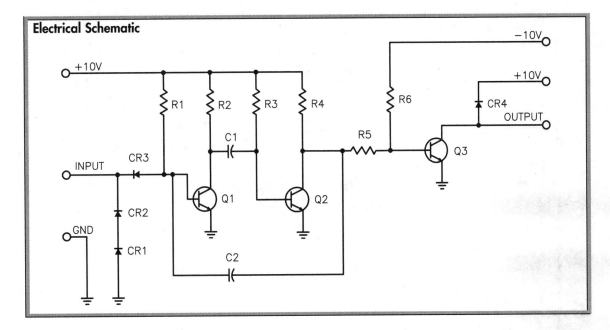

AdditionalProblems

Problem 49

Create and dimension the irregular curve. Determine the drawing area based on space requirements.

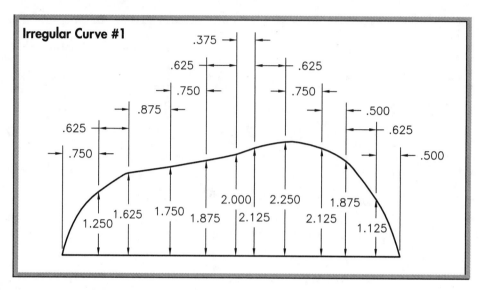

Irregular Curve #1

Problem 50

Create and dimension the irregular curve. Determine the drawing area based on space requirements.

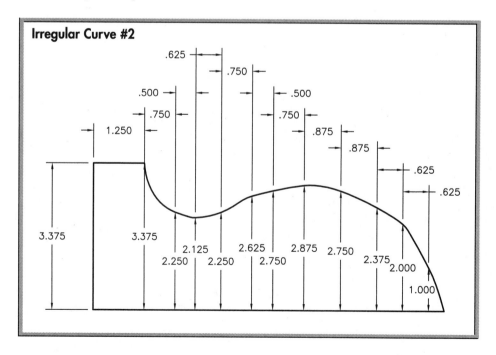

Irregular Curve #2

AdditionalProblems

LEVEL
2

Problem 51

Create and fully dimension the crank handle using decimal inches to three decimal places. Estimate all dimensions.

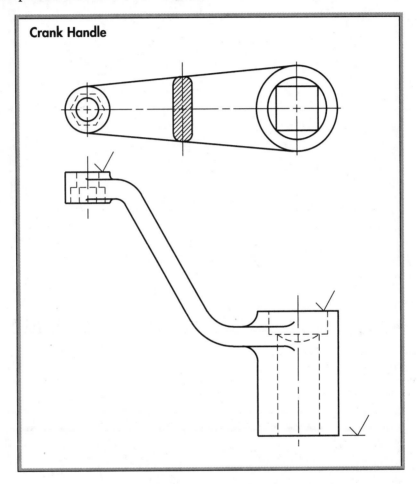

Crank Handle

Problem 51 courtesy of Joseph K. Yabu, San Jose State University

Additional Problems

665

AdditionalProblems

LEVEL 2 Problem 52

Create and dimension the wheel cover as shown below.

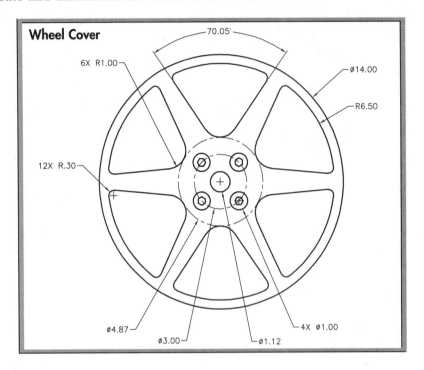

Wheel Cover

70.05°

6X R1.00

Ø14.00

R6.50

12X R.30

Ø4.87

Ø3.00

Ø1.12

4X Ø1.00

LEVEL 2 Problem 53

Create drawings of the wheel covers shown below. Estimate all dimensions.

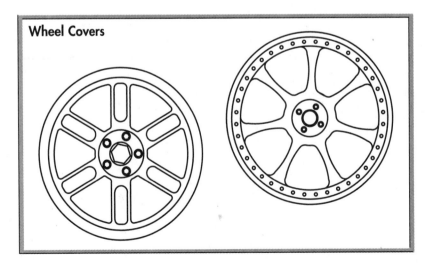

Wheel Covers

Problems 52 and 53 courtesy of Joseph K. Yabu, San Jose State University

Problem 54

In a steel frame building, a bracket is to be connected to a column as shown. The front and right-side views of the structural detail are given below. The W 14 × 26 wide flange beam is to be bolted to a 9 × 5 × 3/8 plate which, in turn, is to be welded to the W 8 × 35 wide flange beam. The dimensions for the two beams are taken from the American Institute of Steel Construction's *Manual of Steel Construction*. Draw the two views.

Structural Assembly

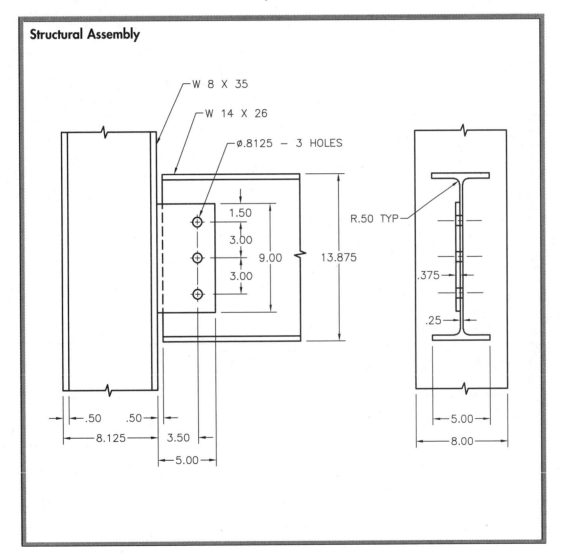

Problem 54 courtesy of Gary J. Hordemann, Gonzaga University

LEVEL
2

Problem 55

Draw the top and front views of the barrier block shown below. Study the top view and figure out how to minimize drawing and maximize the use of replicating commands such as COPY, MIRROR, and ARRAY.

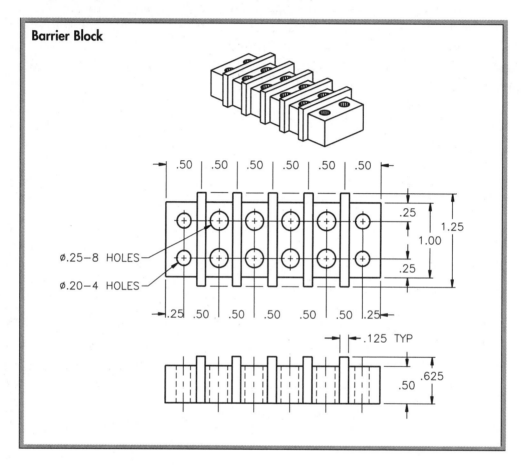

Barrier Block

Ø.25—8 HOLES

Ø.20—4 HOLES

Problem 55 courtesy of Gary J. Hordemann, Gonzaga University

AdditionalProblems

LEVEL 2

Problem 56

The front view of the link shown below is amazingly easy to draw if you draw the large circle first; then draw long horizontal and vertical lines snapped to the center of the circle. Offset the lines to locate the centers of the two smaller circles; if necessary, use a zero-distance chamfer or zero-radius fillet to extend the lines until they meet. Insert the two small circles, snapped to the intersections of the lines. Offset again to construct the lines composing the lower left corner and then fillet that corner. Draw a line snapped tangent to the upper small circle and the large circle; use the FILLET command to insert the two arcs joining the small and large circles. Finally, insert the octagon snapped to the center of the large circle.

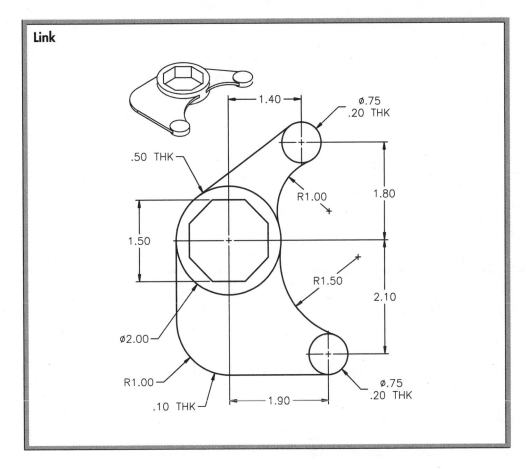

Problem 56 courtesy of Gary J. Hordemann, Gonzaga University

AdditionalProblems

LEVEL 2

Problem 57

Draw the top view of the stanchion shown below.

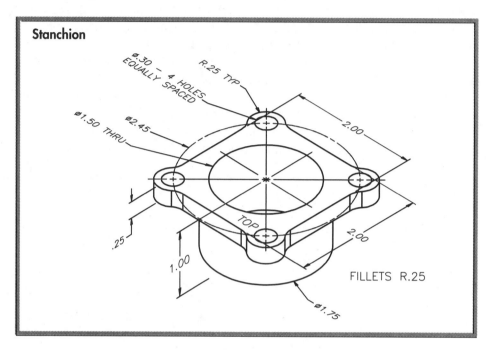

Stanchion

ø.30 — 4 HOLES
EQUALLY SPACED

R.25 TYP

ø1.50 THRU

ø2.45

2.00

.25

1.00

TOP

2.00

FILLETS R.25

ø1.75

LEVEL 2

Problem 58

A drawing of the 20 scale from an engineer's scale is shown below. Using the ARRAY command, draw the 20 scale. Then draw 30 and 50 scales.

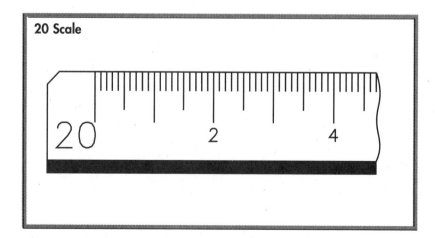

20 Scale

20   2   4

Problems 57 and 58 courtesy of Gary J. Hordemann, Gonzaga University

Additional Problems

LEVEL 2

Problem 59

Draw the front view of the swivel link shown below. Draw half the object using LINE, CIRCLE, OFFSET, and FILLET. Construct the external fillets using circles and lines snapped to tangents. Then use BREAK to trim the circles to arcs. Use MIRROR to complete the other half.

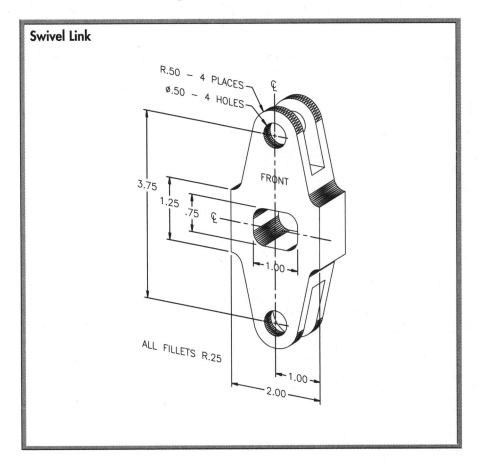

Swivel Link

Problem 59 courtesy of Gary J. Hordemann, Gonzaga University

AdditionalProblems

Problem 60

The footprints for two cordless mice on a mouse docking (recharging) station are shown below. Draw the two footprints *without* using the FILLET command by drawing circles for all the rounded ends. Use the TTR option of the CIRCLE command for the R3.50 arc. Then use TRIM to create the final objects. Would it be easier to use filleted lines? Try it by drawing the outlines again, this time without using the CIRCLE command.

Mouse Footprints

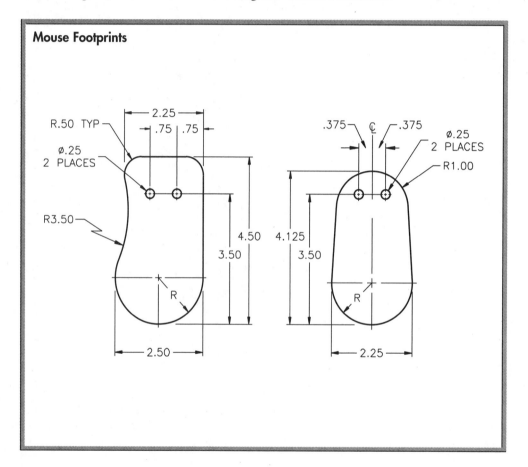

Problem 60 courtesy of Gary J. Hordemann, Gonzaga University

Additional Problems

AdditionalProblems

Problem 61

Draw the front view of the splined separator block shown below. A suggested approach: Draw two circles bounding the spline teeth. Draw a horizontal line through the center and offset it to form the boundary of one tooth. Trim the circles and lines to form one tooth and array it about the center. Using the OFFSET command and horizontal, vertical, and 30° lines through the center, create $1/12$ of the object. Use MIRROR to create $1/6$ of the object, and then array it about the center to finish it. When you have finished the drawing, use SCALE to reduce the spline teeth so that the outer diameter is 12. Use STRETCH to narrow the outer gaps from 8 to 6.

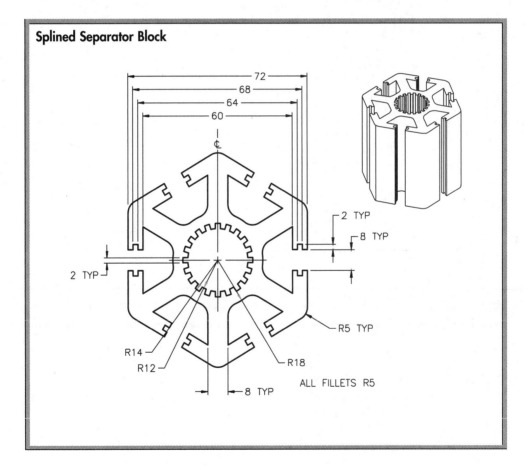

Splined Separator Block

Problem 61 courtesy of Gary J. Hordemann, Gonzaga University

Problem 62

Create and dimension the two views of the cam shown below. The spacing between the dots is .125.

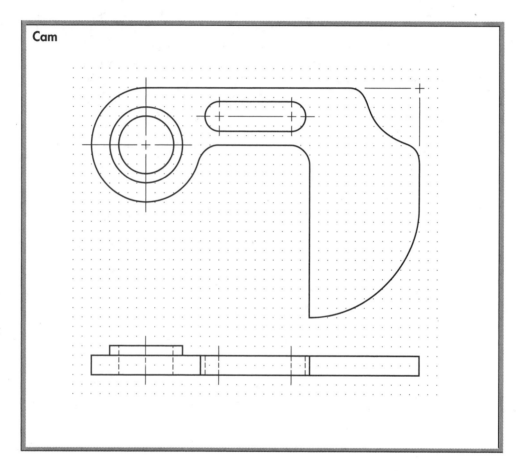

Cam

Problem 62 courtesy of Gary J. Hordemann, Gonzaga University

Problem 63

Create a new text style called Inside with the following settings:

Dimension Line	Spacing: Color:	.30 By Layer
Extension Line	Extension: Origin Offset: Color:	.06 .05 By Layer
Arrow Size	.12	
Center	None	
Fit	Text only	
Units	Precision: Angular Precision: Suppress Leading Zeros	0.00 0
Text	Style: Height: Gap: Color:	ROMANS (font romans.shx) .12 .06 White

Using the new dimension style, draw and dimension the two views of the clock blank shown below. The spacing between dots is .125.

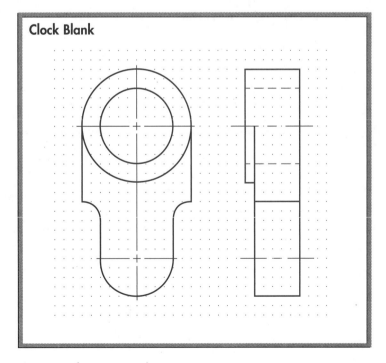

Clock Blank

Problem 63 courtesy of Gary J. Hordemann, Gonzaga University

LEVEL
2

Problem 64

Draw each of the three thread representations shown below and write them to files using the WBLOCK command. You will be able to use them in drawings requiring threads by inserting and scaling them. Although you will usually have to do some editing of the threads to make them work in a particular drawing, the blocks will give you a good starting point. The external threads are most easily drawn using the ARRAY command. The distance between dots in the grid is not important, but .125 should work well.

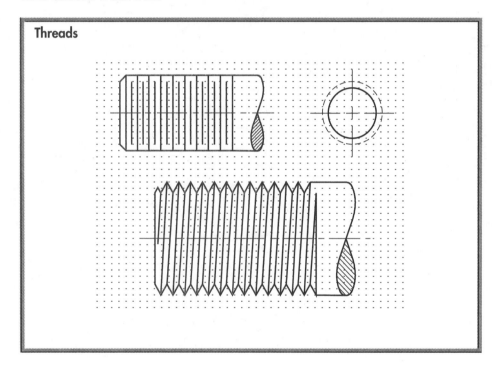

Threads

Problem 64 courtesy of Gary J. Hordemann, Gonzaga University

AdditionalProblems

Problem 65

Imagine an object composed of:

a. A cube measuring 2″ on each side.

b. A four-sided regular pyramid with a height of 1″. The pyramid sits on top of the cube and the four edges of its base align with the four edges of the cube's top surface.

c. A hole, 1½″ in diameter, centered in the front cube face and having a depth of ½″.

d. Another hole, coaxial with the first but 1″ in diameter, extending through the object.

Three ways of drawing this object are shown below. They are called *obliques*. Simply put, they are created by drawing one face at its full size and shape with the receding lines at an angle of 45°. The *cabinet* oblique has half-size receding lines; the *cavalier* oblique has full-size receding lines. The receding lines in *general* obliques vary; those on the one shown here are 0.707-size lines. Draw the three obliques.

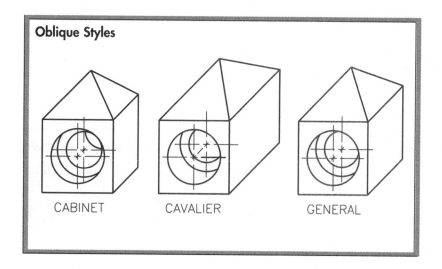

Oblique Styles

CABINET   CAVALIER   GENERAL

Problem 65 courtesy of Gary J. Hordemann, Gonzaga University

AdditionalProblems

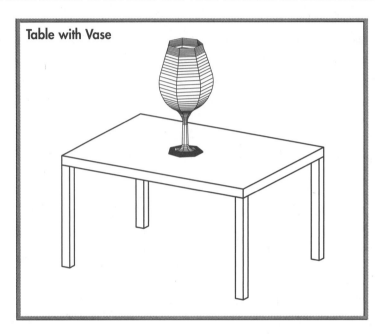

LEVEL 2

Problem 66

Create solid models of the table and vase. Estimate all dimensions.

Table with Vase

LEVEL 2

Problem 67

Create the top and side views of the fighter aircraft. Estimate all dimensions.

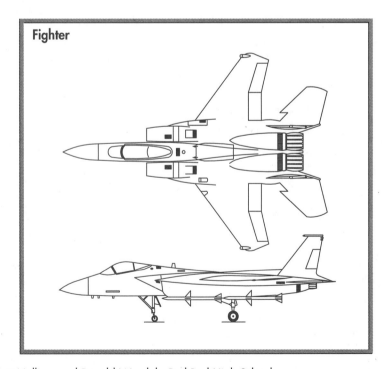

Fighter

Problem 67 courtesy of Matt Melliere and Ronald Weseloh, Red Bud High School

LEVEL 2

Problem 68

Create the landscape drawing as shown. Create blocks of the trees and shrubs and insert them as shown.

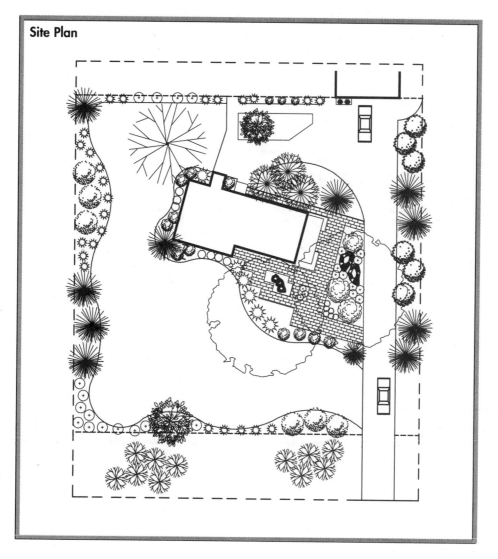

Site Plan

Problem 68 courtesy of Mill Brothers Landscape and Nursery, Inc.

Problem 69

Create the design elevation as shown. Create blocks of the trees and shrubs and insert them as shown. Estimate all dimensions.

Elevation

Problem 69 courtesy of Mill Brothers Landscape and Nursery, Inc.

Additional Problems

LEVEL
2

Problem 70

Create the floor plan for the first story of a residence. Determine the drawing area according to space requirements. Estimate the dimensions that are not given.

Floor Plan

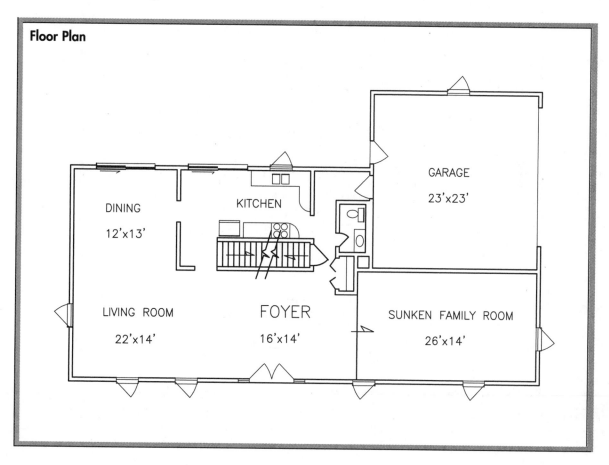

GARAGE
23'x23'

DINING
12'x13'

KITCHEN

LIVING ROOM
22'x14'

FOYER
16'x14'

SUNKEN FAMILY ROOM
26'x14'

Problem 70 courtesy of Mark Schwendau, Kishwaukee College

LEVEL 2

Problem 71

Create the wall section shown below.

Wall Section

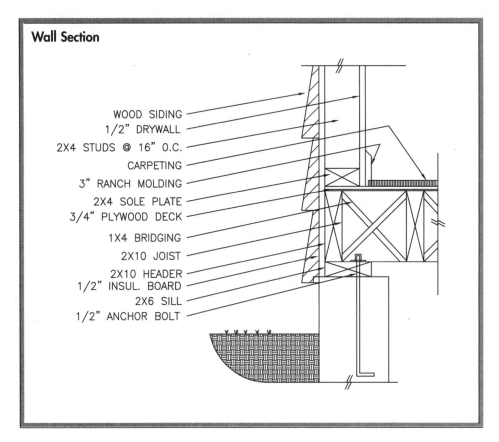

WOOD SIDING
1/2" DRYWALL
2X4 STUDS @ 16" O.C.
CARPETING
3" RANCH MOLDING
2X4 SOLE PLATE
3/4" PLYWOOD DECK
1X4 BRIDGING
2X10 JOIST
2X10 HEADER
1/2" INSUL. BOARD
2X6 SILL
1/2" ANCHOR BOLT

AdditionalProblems

Problem 72

Create a dimension style named Structural that uses feet and inches, oblique arrowheads, and text placed above the dimension lines. Draw and dimension the foundation detail shown below. The hatch patterns are Earth, ANSI37, and AR-CONC.

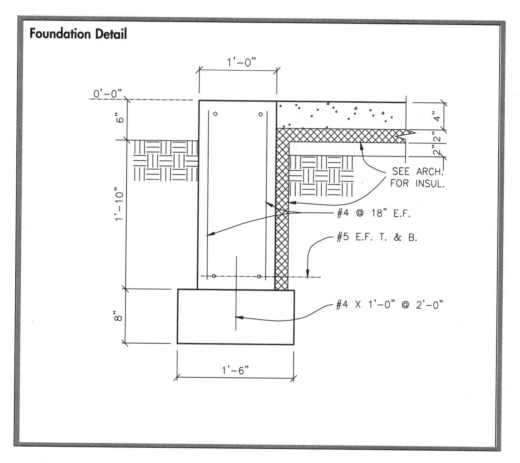

Foundation Detail

Problem 72 courtesy of Gary J. Hordemann, Gonzaga University

AdditionalProblems

Problem 73

Create and dimension
an isometric view of
the drawing.

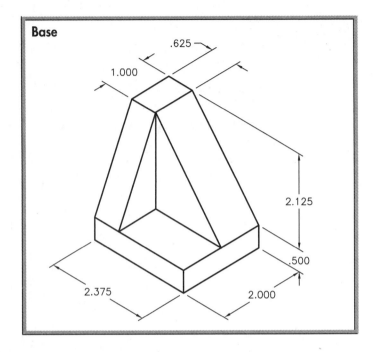

Base

.625
1.000
2.125
.500
2.375
2.000

Problem 74

Create and dimension
an isometric view of
the drawing.

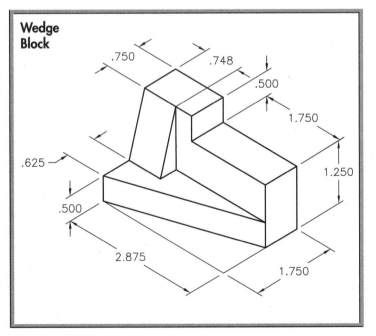

Wedge
Block

.750
.748
.500
1.750
.625
1.250
.500
2.875
1.750

AdditionalProblems

LEVEL 3

Problem 75

Create the isometric view of the 90° link shown at the right.

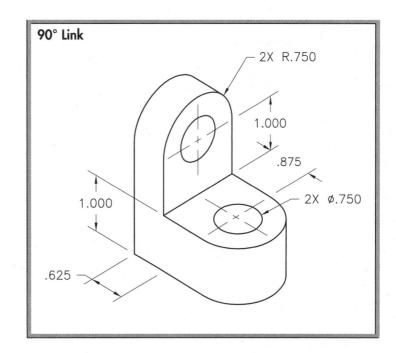

90° Link

2X R.750
1.000
.875
2X ⌀.750
1.000
.625

LEVEL 3

Problem 76

Draw the front and right-side views of the gear as shown. Use the detail drawing to create the teeth accurately. Hatch as indicated using the ANSI31 pattern.

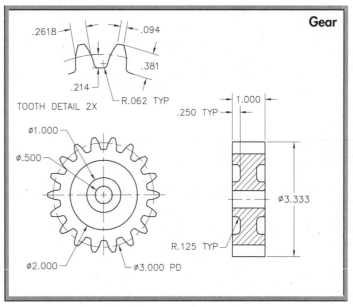

Gear

.2618
.094
.381
.214
R.062 TYP
TOOTH DETAIL 2X
⌀1.000
⌀.500
1.000
.250 TYP
⌀3.333
R.125 TYP
⌀2.000
⌀3.000 PD

Problem 76 courtesy of Steve Huycke, Lake Michigan College

AdditionalProblems

Problem 77

Create and fully dimension the isometric view of the strip block shown below.
Use a drawing area of 17 × 11.

Strip Block

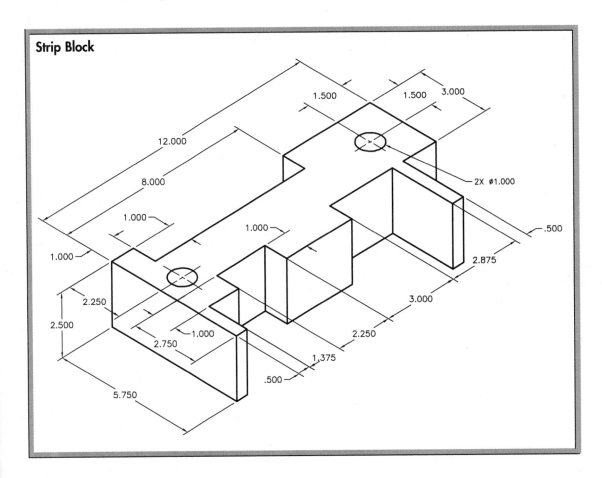

LEVEL
3

Problem 78

Create and dimension the drawing of a wrench. The units are millimeters. Determine the drawing area on your own, based on the dimensions of the wrench.

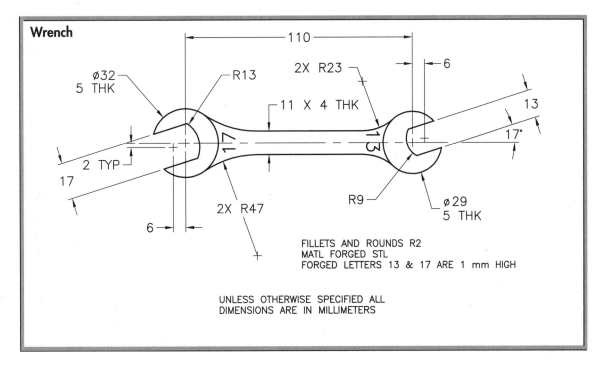

Wrench

110

ø32
5 THK

R13

2X R23

6

11 X 4 THK

13

17

2 TYP

17°

17

2X R47

R9

ø29
5 THK

6

FILLETS AND ROUNDS R2
MATL FORGED STL
FORGED LETTERS 13 & 17 ARE 1 mm HIGH

UNLESS OTHERWISE SPECIFIED ALL
DIMENSIONS ARE IN MILLIMETERS

Problem 78 adapted from the textbook *Drafting Fundamentals* by Scott, Foy, and Schwendau

LEVEL 3

Problem 79

Create the front and right-side views of the fitting and then construct the auxiliary view. Dimension the front and right-side views.

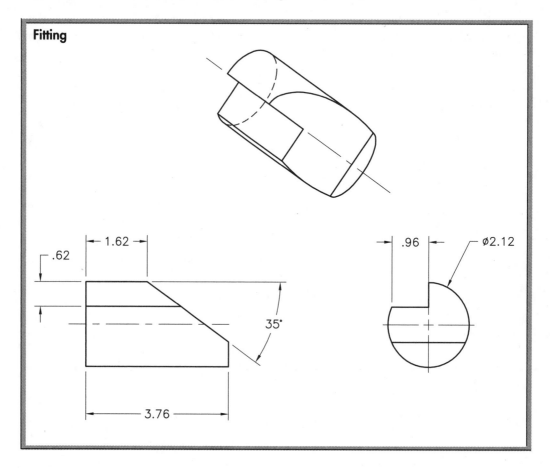

Fitting

Problem 79 courtesy of John F. Kirk, Kirk & Associates

AdditionalProblems

LEVEL 3

Problem 80

Create the front and right-side views of the block and then construct the auxiliary view. Dimension the front and right-side views.

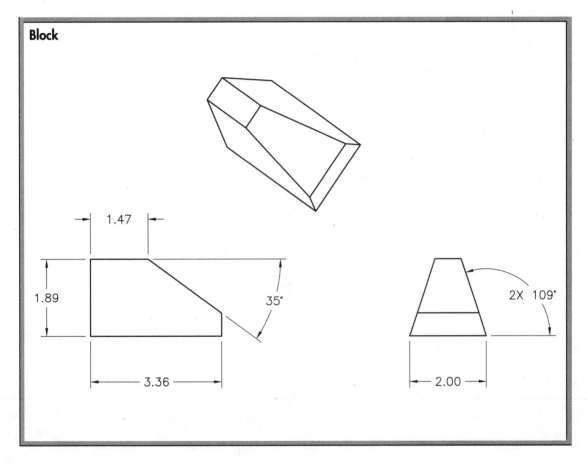

Block

Problem 80 courtesy of John F. Kirk, Kirk & Associates

Additional Problems

AdditionalProblems

LEVEL
3

Problem 81

Create the front and right-side views of the casting. The right-side view should include a half section. Fully dimension both views.

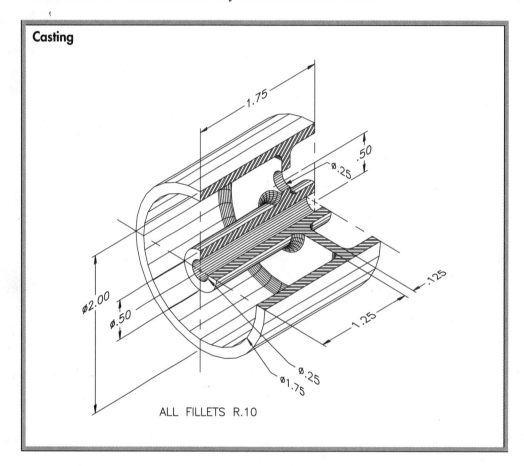

Casting

1.75

.50

Ø.25

.125

Ø2.00

Ø.50

1.25

Ø.25

Ø1.75

ALL FILLETS R.10

Problem 81 courtesy of Gary J. Hordemann, Gonzaga University

Additional Problems

Problem 82

Create the top and front views of the alternator bracket. Use a drawing area of 17 × 11.

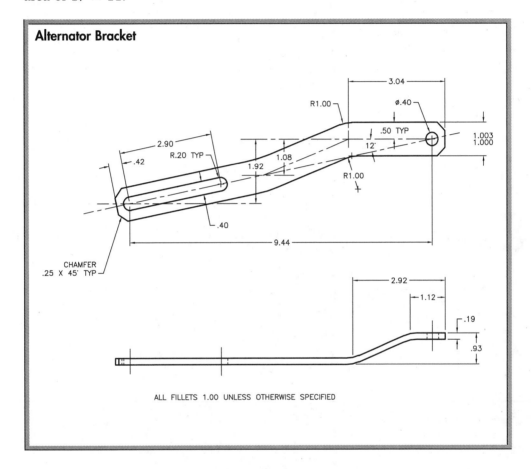

Alternator Bracket

ALL FILLETS 1.00 UNLESS OTHERWISE SPECIFIED

Problem 82 adapted from the textbook *Drafting Fundamentals* by Scott, Foy, and Schwendau

AdditionalProblems

Problem 83

Create the necessary orthographic views of the angle block and fully dimension the views. Use a drawing area of 17 × 11. Add an isometric view in the upper right area of the drawing, but do not dimension it.

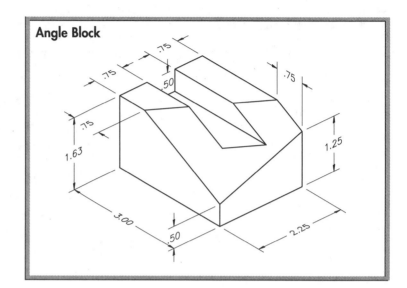

Angle Block

Problem 84

Using AutoCAD's 3D capabilities, create a model of the impeller. Approximate all dimensions.

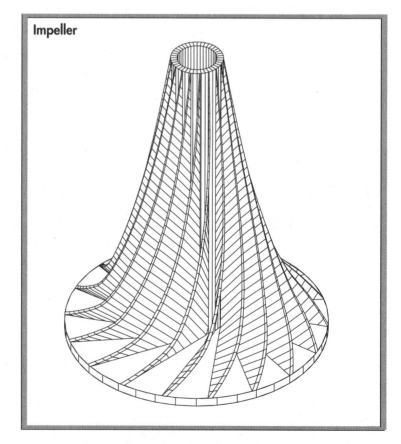

Impeller

Problem 83 adapted from the textbook *Drafting Fundamentals* by Scott, Foy, and Schwendau
Problem 84 courtesy of Gary J. Hordemann, Gonzaga University

AdditionalProblems

Problem 85

Create a solid model of the block.

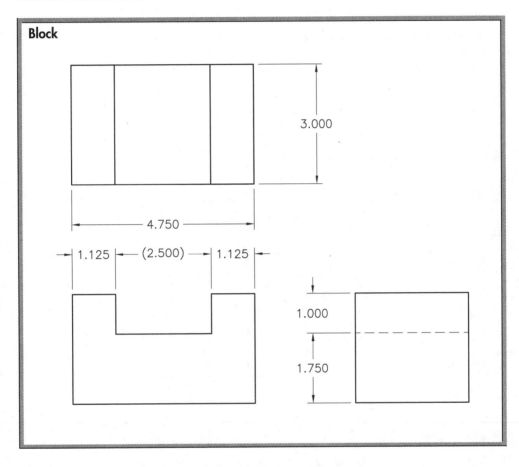

Block

3.000

4.750

1.125 — (2.500) — 1.125

1.000

1.750

LEVEL
3

Problem 86

Create a solid model of the block. Then produce orthographic views from the solid model and dimension the views.

Block

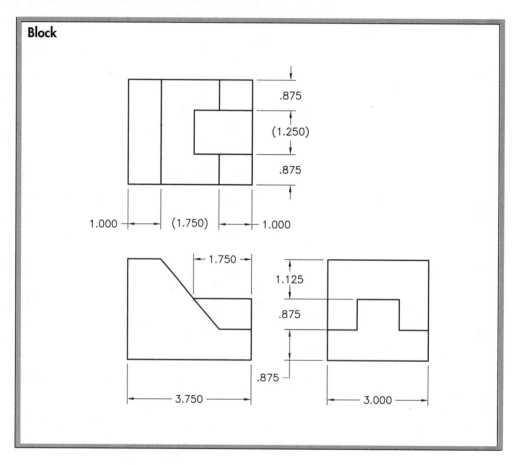

LEVEL 3

Problem 87

Create the views of the 45° elbow as shown and fully dimension them. Determine the drawing area on your own, based on space requirements.

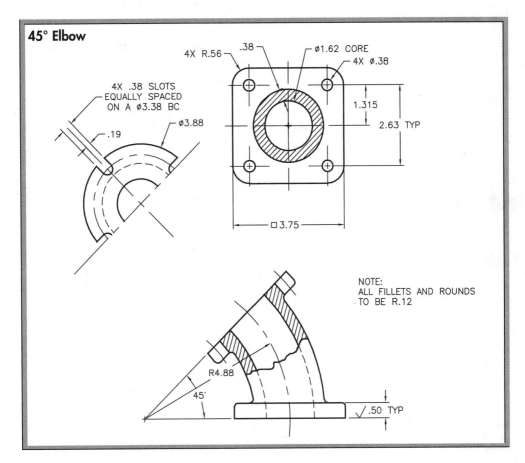

45° Elbow

4X R.56 — .38

Ø1.62 CORE

4X Ø.38

4X .38 SLOTS EQUALLY SPACED ON A Ø3.38 BC

.19

Ø3.88

1.315

2.63 TYP

□3.75

NOTE: ALL FILLETS AND ROUNDS TO BE R.12

R4.88

45°

√ .50 TYP

Problem 87 courtesy of Steve Huycke, Lake Michigan College

Problem 88

Create the front and full section views of the sprocket and fully dimension them. Units are in millimeters. Determine the drawing area on your own.

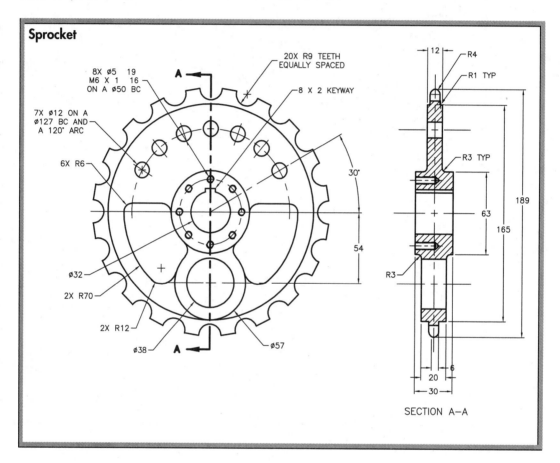

Sprocket

20X R9 TEETH EQUALLY SPACED

8X ∅5   19
M6 X 1   16
ON A ∅50 BC

A

8 X 2 KEYWAY

7X ∅12 ON A
∅127 BC AND
A 120° ARC

6X R6

30°

∅32

54

2X R70

2X R12

∅38   A

∅57

12   R4

R1 TYP

R3 TYP

189

63

165

R3

20   6

30

SECTION A—A

Problem 88 courtesy of Alan Fitzell, Central Peel Secondary School

Problem 89

Create the front and full section views of the tool and dimension them. Also include the detail as shown. Determine the drawing area, based on space requirements. Units are in millimeters.

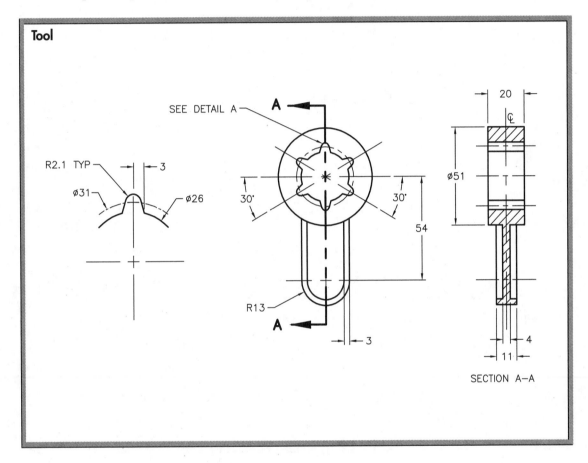

Tool

SEE DETAIL A

R2.1 TYP
ø31
3
ø26
30°
30°
54
R13
3
A
A
20
ø51
4
11
SECTION A–A

Problem 89 courtesy of Julie H. Wickert, Austin Community College

AdditionalProblems

Problem 90

Create the entire set of drawings of the control bracket. Determine the drawing area, based on space requirements.

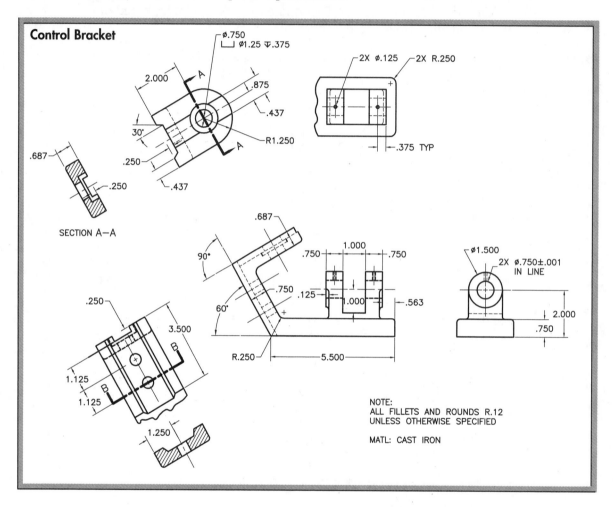

Problem 90 courtesy of Steve Huycke, Lake Michigan College

Additional Problems

Problem 91

Create the drawing of the fork lift as well as the bill of materials. Estimate all dimensions.

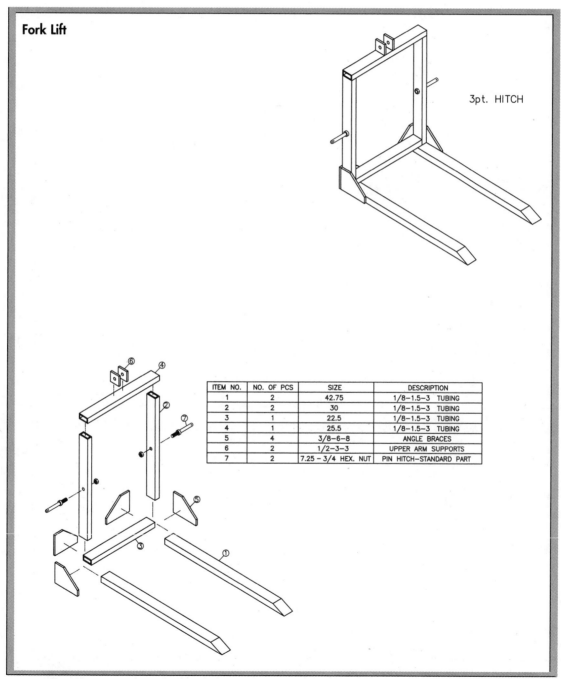

Fork Lift

3pt. HITCH

ITEM NO.	NO. OF PCS	SIZE	DESCRIPTION
1	2	42.75	1/8-1.5-3 TUBING
2	2	30	1/8-1.5-3 TUBING
3	1	22.5	1/8-1.5-3 TUBING
4	1	25.5	1/8-1.5-3 TUBING
5	4	3/8-6-8	ANGLE BRACES
6	2	1/2-3-3	UPPER ARM SUPPORTS
7	2	7.25 - 3/4 HEX. NUT	PIN HITCH—STANDARD PART

Problem 91 courtesy of Craig Pelate and Ron Weseloh, Red Bud High School

Problem 92

Create the top, front, right-side, and isometric views of the picnic table. Estimate all dimensions.

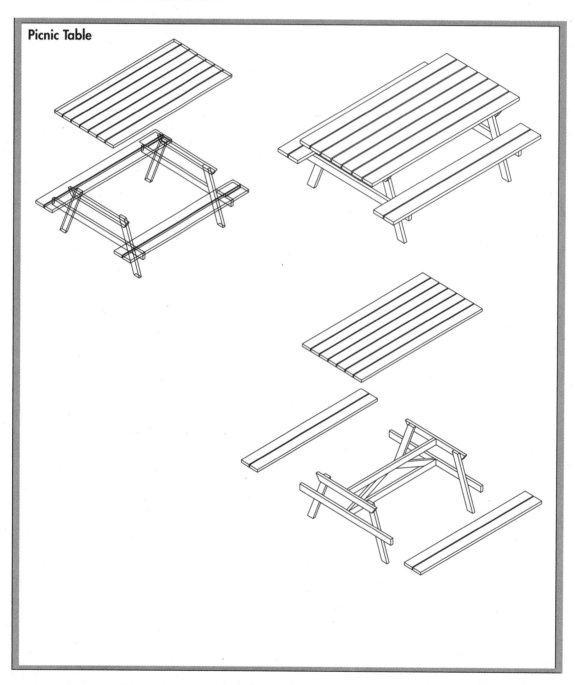

Picnic Table

Problem 92 courtesy of Dan Cowell and Ron Weseloh, Red Bud High School

AdditionalProblems

Problem 93

Create the front and side views of the structural bracket and dimension them as shown. Also, include section and isometric views. Determine the drawing area based on space requirements.

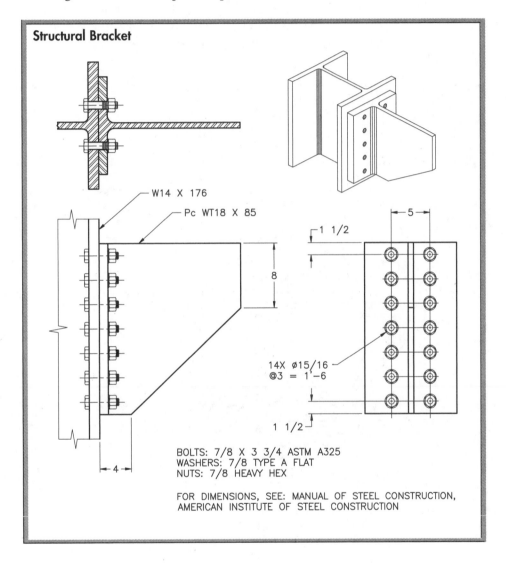

Structural Bracket

W14 X 176

Pc WT18 X 85

8

5

1 1/2

14X ⌀15/16
@3 = 1'-6

1 1/2

4

BOLTS: 7/8 X 3 3/4 ASTM A325
WASHERS: 7/8 TYPE A FLAT
NUTS: 7/8 HEAVY HEX

FOR DIMENSIONS, SEE: MANUAL OF STEEL CONSTRUCTION,
AMERICAN INSTITUTE OF STEEL CONSTRUCTION

Problem 93 courtesy of Gary J. Hordemann, Gonzaga University

AdditionalProblems

Problem 94

Draw the top view of the spacer shown below. Assume that all of the ribs have a width of .2 and are symmetrical about the center lines. Try this approach: Create the hexagon and offset it. Add one set of circles and array them. Draw lines joining the centers of the circles and polygons. Then offset the lines, fillet them, and array them. *Note:* The fillets are between lines and circles and between lines and the outer polygon, *not* between lines.

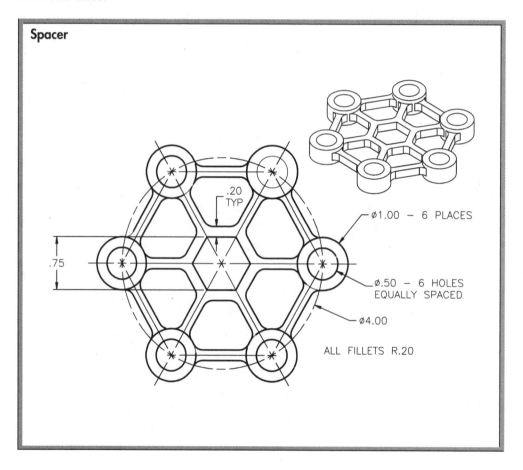

Spacer

.20
TYP

Ø1.00 — 6 PLACES

.75

Ø.50 — 6 HOLES
EQUALLY SPACED

Ø4.00

ALL FILLETS R.20

Problem 94 courtesy of Gary J. Hordemann, Gonzaga University

Problem 95

Create a solid model of the bracket using the dimensions shown in the orthographic views below. Create the orthographic views from the model.

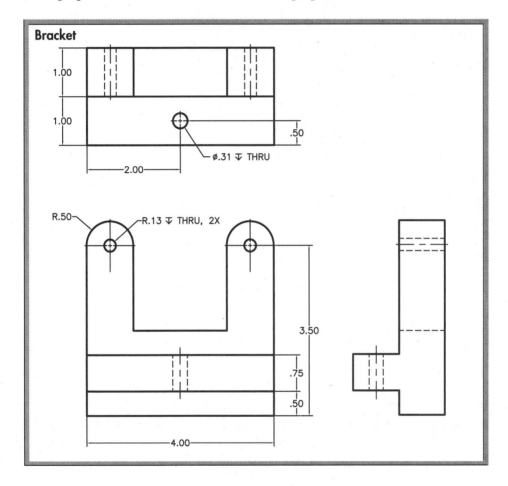

Problem 95 courtesy of Mark R. Stevenson

Problem 96

Create and dimension a solid model of the locking sleeve according to the dimensions shown below.

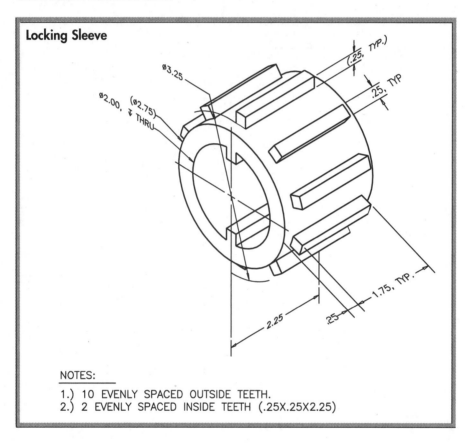

Locking Sleeve

ø3.25

ø2.00, (ø2.75)
↧ THRU

(.25, TYP.)

.25, TYP

1.75, TYP.

2.25

.25

NOTES:
1.) 10 EVENLY SPACED OUTSIDE TEETH.
2.) 2 EVENLY SPACED INSIDE TEETH (.25X.25X2.25)

Problem 96 courtesy of Mark R. Stevenson

Problem 97

Using EXTRUDE, create a solid model of the swivel stop shown below. First draw the profile and the circles; turn them into regions; then use SUBTRACT to obtain the final profile for extruding.

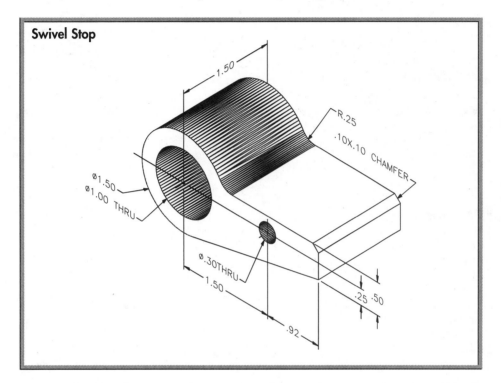

Swivel Stop

1.50

R.25

.10X.10 CHAMFER

Ø1.50
Ø1.00 THRU

Ø.30THRU
1.50

.92

.25 .50

Problem 97 courtesy of Gary J. Hordemann, Gonzaga University

AdditionalProblems

LEVEL 3

Problem 98

Create a dimensioned isometric view of the roller tube block shown at the right. It is sometimes difficult to draw the line tangent to the two ellipses; if both ellipses are the same size, the Quadrant object snap will yield the same result and is more dependable. Note the complex internal shapes caused by the intersecting holes. Do not attempt to draw them—merely place isometric ellipses on the three surfaces.

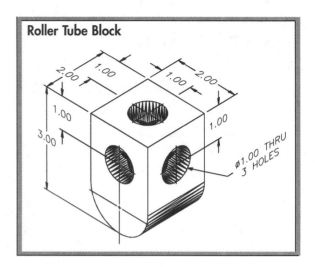

Roller Tube Block

LEVEL 3

Problem 99

Using REVOLVE, create a solid model of the V-belt pulley shown below. Note that by revolving around the axis, you can create a hole without subtracting a cylinder.

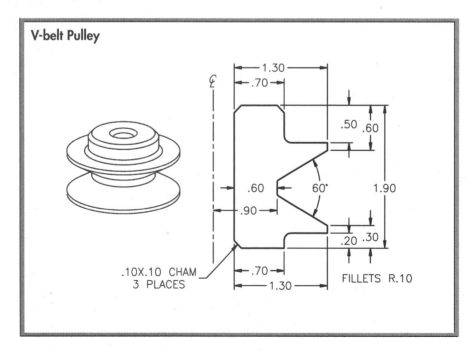

V-belt Pulley

Problems 98 and 99 courtesy of Gary J. Hordemann, Gonzaga University

Additional Problems

Problem 100

Create top, front, and isometric views of the fence. Determine the drawing area, based on space requirements. Dimension the top and front views.

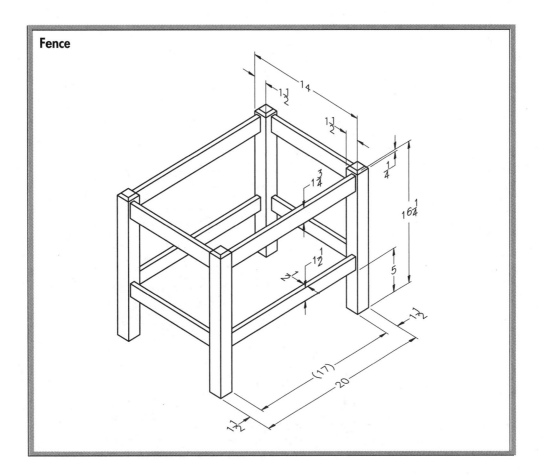

Fence

Problem 100 adapted from the textbook *Drafting Fundamentals* by Scott, Foy, and Schwendau

AdditionalProblems

Problem 101

Create a solid model of the surveyor's transit. Estimate all dimensions.

Transit

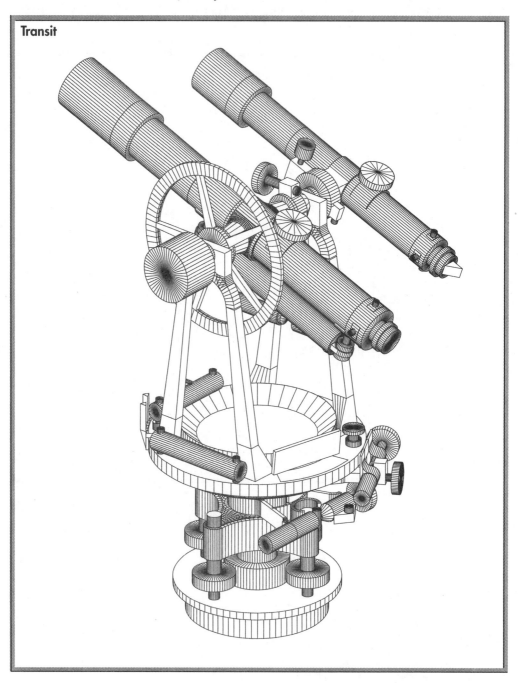

Problem 101 courtesy of Riley Clark

Additional Problems

Problem 102

Create the architectural elevation of the cathedral. Estimate all dimensions.

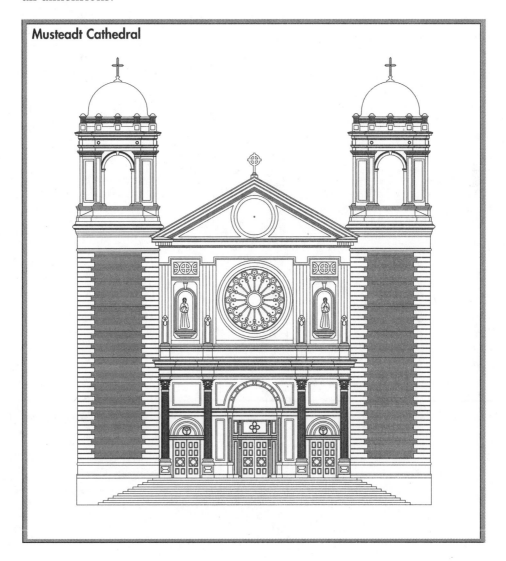

Musteadt Cathedral

Problem 102 courtesy of David Sala, Forsgren Associates, from *World Atlas of Architecture* (G.K. Hall and Company)

LEVEL
3

Problem 103

Create the wall section as shown below.

Wall Section

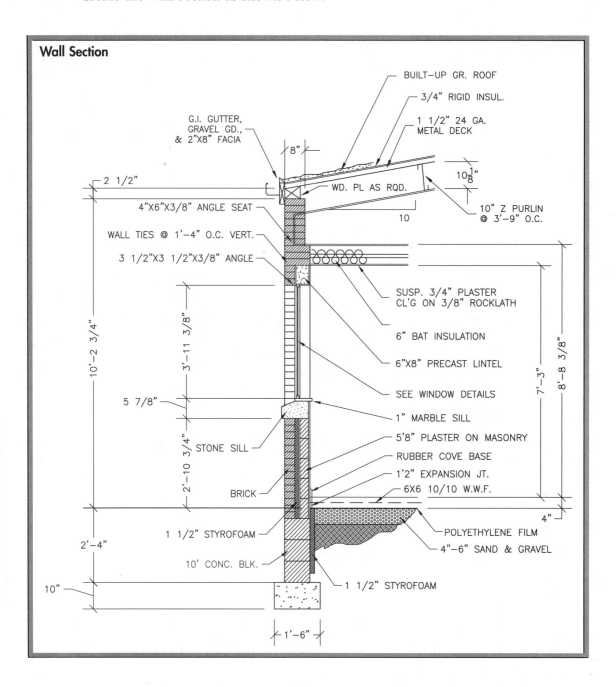

BUILT—UP GR. ROOF

3/4" RIGID INSUL.

1 1/2" 24 GA. METAL DECK

G.I. GUTTER, GRAVEL GD., & 2"X8" FACIA

8"

WD. PL AS RQD.

$10\frac{1}{8}$"

2 1/2"

4"X6"X3/8" ANGLE SEAT

10" Z PURLIN @ 3'—9" O.C.

10

WALL TIES @ 1'—4" O.C. VERT.

3 1/2"X3 1/2"X3/8" ANGLE

SUSP. 3/4" PLASTER CL'G ON 3/8" ROCKLATH

10'—2 3/4"

3'—11 3/8"

6" BAT INSULATION

6"X8" PRECAST LINTEL

SEE WINDOW DETAILS

5 7/8"

8'—8 3/8"

7'—3"

1" MARBLE SILL

STONE SILL

2'—10 3/4"

5'8" PLASTER ON MASONRY

RUBBER COVE BASE

1'2" EXPANSION JT.

BRICK

6X6 10/10 W.W.F.

1 1/2" STYROFOAM

4"

POLYETHYLENE FILM

2'—4"

4"—6" SAND & GRAVEL

10' CONC. BLK.

1 1/2" STYROFOAM

10"

1'—6"

Problem 103 adapted from a drawing by Paul Driscoll

Problem 104

Create the front and left-side architectural elevation drawings. Estimate all dimensions.

Architectural Elevations

6 X 12 RSC TIMBERS
TYP ABOVE WDWS
2X RSC SHUTTERS

FULL BRICK

BRICK CORBEL
BRACKET

12
12

12/12 PITCH

9' RADIUS CUT
SCAB RAFTERS

12/12 PITCH

FULL BRICK
SIDE OPENING
OPTIONAL

Problem 104 courtesy of Rodger A. Brooks, Architect

AdditionalProblems

Problem 105

Create the back and right-side architectural elevation drawings. Estimate all dimensions.

Architectural Elevations

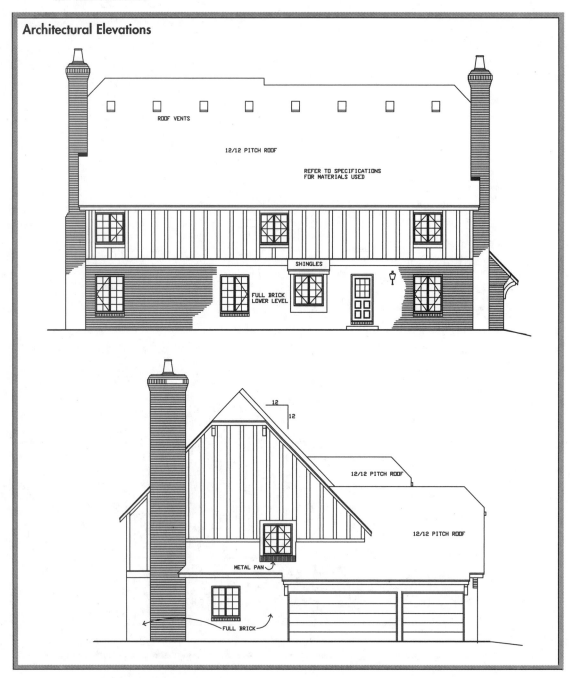

ROOF VENTS

12/12 PITCH ROOF

REFER TO SPECIFICATIONS
FOR MATERIALS USED

SHINGLES

FULL BRICK
LOWER LEVEL

12

12

12/12 PITCH ROOF

12/12 PITCH ROOF

METAL PAN

FULL BRICK

Problem 105 courtesy of Rodger A. Brooks, Architect

Problem 106

Create the architectural elevation drawing. Estimate all dimensions.

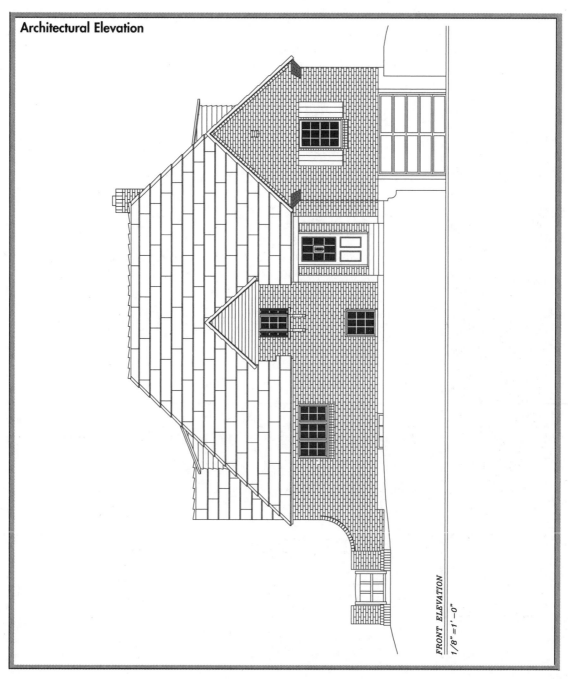

Architectural Elevation

FRONT ELEVATION
1/8" = 1'-0"

Problem 106 courtesy of Gary J. Hordemann, Gonzaga University

Advanced

The advanced projects are designed to help you develop your creativity and problem-solving skills. Each project describes a task or job as you might encounter it in the workplace. In some cases, steps are outlined to help you get started. In others, you must decide for yourself how best to approach the project.

Pepper Mill – page 716

Design, create the plans for, and produce a solid model of a pepper mill.

Drawing Table – page 720

Using a group approach and AutoCAD's external referencing capability, complete the parts of and assemble a solid model of a student drawing table.

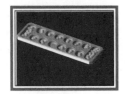

Mancala Game Board – page 722

Design, create the plans for, and produce a solid model of a game board for an ancient African game called Mancala.

Bolted Seat Connection – page 726

Create structural details for the connection between beams in a steel frame building and then create a solid model representing the connection.

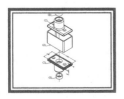

Bushing Assembly – page 730

Create production drawings for a bushing assembly consisting of bushing holders, bushings, and a housing box; then create a solid model of each piece.

Projects

Every project in this section requires considerable thought and planning. These projects are suitable for long-term projects for individuals or groups. They are *not* intended to be short-term projects. Give yourself plenty of time, and do not get discouraged if you do not succeed on your first attempt.

Architectural Plans – page 733

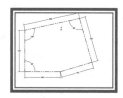

Develop a complete set of residential architectural plans according to the client's specifications.

Buffet Tray – page 734

Design a lap tray for use at a buffet or picnic; document the design, and then create solid models of the pieces and display them as an exploded assembly.

Shackle Assembly – page 737

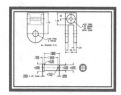

Create models of each piece of the shackle assembly and assemble them; slice the assembly and remove half so that the interior is visible.

Clothing Rack – page 740

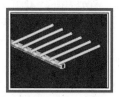

Create solid models of the pieces of this wall-mounted clothing rack and arrange them into an exploded assembly.

Protective Packaging – page 744

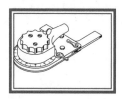

Design and develop plans for protective packaging for shipping a product.

715

Project:
Pepper Mill

Shown in Fig. AV1-1 is a disassembled pepper mill. The complete assembly is composed of the following eight parts:

- 1 plastic top
- 1 plastic barrel to hold the peppercorns
- 1 stainless steel stator
- 1 rotor
- 1 threaded shaft
- 2 nuts
- 1 spring

An exploded assembly view of the eight parts is shown in Fig. AV1-2. The shaft and one nut are permanently affixed to the top. When the top is turned, the shaft and rotor rotate, grinding the peppercorns between the rotor and stator.

Your supervisor has asked you to improve the design by creating a more attractive top and barrel. Do not modify the shaft/rotor/stator/spring/nut assembly, except to adjust the length of the shaft to accommodate the new design. (The dimensions of these parts are given in Fig. AV1-4 on page 718.) Concentrate instead on the shape of the pepper mill and the material from which it is made. Consider ergonomic factors: does your new shape make the pepper mill easier to hold and use?

Fig. AV1-1

AdvancedProjects

■ Part 1

Using the dimensions given, create a new design for the top and barrel. Perform the following tasks:

- Draw the two-dimensional views of all the parts, inserting all hidden and center lines.

- Dimension all of the parts.

- Draw the front view of the assembly with the parts exploded. All of the parts should be drawn as full sections, using the hatch patterns appropriate for the materials. Omit hidden lines, but include the center line.

- Label each part with a circle ("balloon") containing its part number and accompanied by its title and material.

- Add a border and title block.

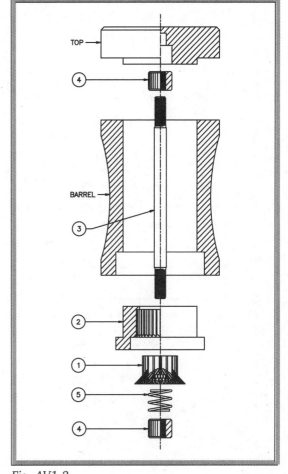

Fig. AV1-2

The complexity of the rotor teeth makes drawing this part especially challenging. A dimensioned single tooth is shown in Fig. AV1-3. While it is a simple object, it is rotated about its axis 10° outward from the center of the rotor, rotated 20° clockwise about the midpoint of its lower edge, and leaned 35° toward the center of the rotor. This twisting and turning results in a drawing complexity that is beyond the scope of this textbook. Consequently, when drawing the rotor, approximate its appearance by following the dimensions given in Fig. AV1-3.

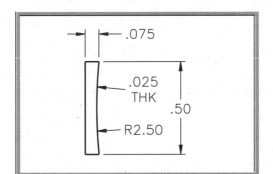

Fig. AV1-3

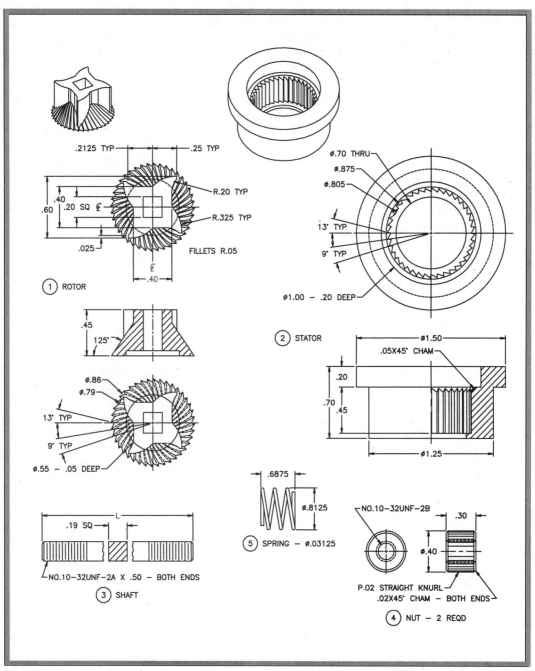

.2125 TYP
.25 TYP

R.20 TYP

R.325 TYP

FILLETS R.05

.40
.60
.20 SQ
.025
.40

① ROTOR

Ø.70 THRU
Ø.875
Ø.805
13° TYP
9° TYP
Ø1.00 – .20 DEEP

② STATOR

Ø1.50
.05X45° CHAM
.20
.70
.45
Ø1.25

.45
125°

Ø.86
Ø.79
13° TYP
9° TYP
Ø.55 – .05 DEEP

.6875
Ø.8125

⑤ SPRING – Ø.03125

L
.19 SQ
NO.10–32UNF–2A X .50 – BOTH ENDS

③ SHAFT

NO.10–32UNF–2B
.30
Ø.40
P.02 STRAIGHT KNURL
.02X45° CHAM – BOTH ENDS

④ NUT – 2 REQD

Fig. AV1-4

AdvancedProjects

After creating the top and barrel (freehand sketches may help), you should establish how you will arrange the views of the objects in the drawing. There are so many parts to this assembly that it will probably pay you to draw rectangles representing the objects and their dimensions, and then use them to lay out the parts and make decisions about scales and how many pages will be required. Given the smallness of the parts, you should plan to plot them to at least double scale.

Most of the dimensions of the parts are "friendly," that is, they have a common multiple. This makes the use of snaps and grids particularly helpful. Some of the parts, such as the rotor and stator, are especially suitable for the ARRAY and MIRROR commands. (You should always examine the parts for symmetries so you can make use of AutoCAD's powerful replication commands.)

With all your preliminary settings completed and the format of your drawings decided (and assuming that you already have a drawing template file containing most of the layers, text style, linetypes, and dimension style), you should outline the steps you will take to create the drawing. The general strategy should be:

1. Draw the needed views of all of the parts. As is often the case with objects that have circular features, it is easiest to begin with the views in which circular features appear as circles or arcs. Then use these views to construct the others.

2. Use copies of the appropriate views to create the exploded assembly.

3. Dimension the orthographic views.

■ Part 2

Make solid models of the pieces of your design. Then perform the following tasks.

- Arrange the solids in an exploded assembly and display them in a view in which you are looking upward at the assembly from below.

- Determine the volumes and the surface areas of all the pieces.

- Look up the densities of the materials of your objects (you may have to go to the reference section of a suitable library or search the Internet); use the densities to determine the weights of the parts.

- Render the assembly using suitable materials.

Pepper Mill project courtesy of Gary J. Hordemann, Gonzaga University

Advanced Project

Project:
Drawing Table

Create the solid model of a student drawing table as shown in Fig. AV2-1. Consider making this a group project. Determine which group member will be responsible for each part of the table. When the individual parts have been created, place the files in a central location. As an optional activity, the group should consider using AutoCAD's external referencing capability to ensure proper fitting of the parts.

As a group project, this project works best with a team of 4 people. Assign each person one of the following pieces of the assembly:

- table top
- table drawer
- frame assembly (legs and feet)
- mounting assembly for the top (tubular support and screwheads)

As with any solid model, there are several ways to create a model of the drawing table. Study the parts and choose the method you think will work best. If it doesn't work as well as you'd like, you may want to consider a different method. The following illustrations and tips may give you some ideas about how to approach each part of the project. For example, one way to create the table frame is to draw a path for the legs and then use the EXTRUDE command. You may want to consider using the SOLREV command to create the feet of the table legs.

Fig. AV2-2 shows a hidden-line view of the table top. Solid primitives of various types may be useful in creating parts such as this.

Fig. AV2-1

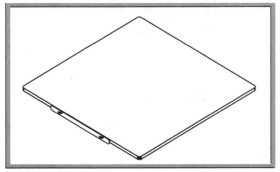

Fig. AV2-2

The drawer handle, hinges, and brackets may also include primitives. Two views of the drawer are provided in Fig. AV2-3. The wireframe view may give you an idea of how the drawer should fit together.

The support assembly, which attaches the table top to the frame, consists of a tubular supporting bracket and holder and Philips-head screws. The tubular bracket is shown in Fig. AV2-4. Fig. AV2-5 shows a closeup of a screw head, and Fig. AV2-6 shows a wireframe view of the assembly in position under the table top.

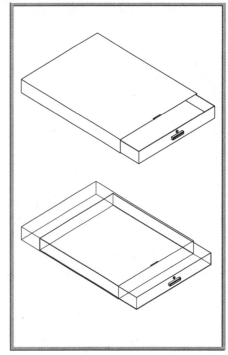

Fig. AV2-3

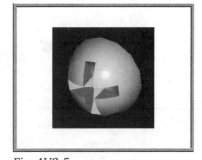

Fig. AV2-5

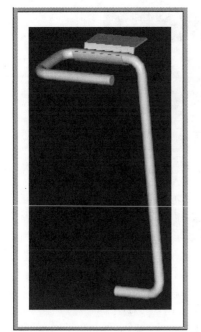

Fig. AV2-4

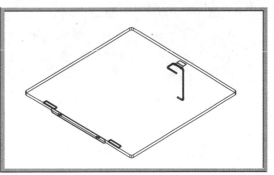

Fig. AV2-6

Drawing Table project courtesy of Bill Fell, AHST High School

Advanced Project

Project:
Mancala Game Board

Mancala is an ancient bean game originating in Africa. It can be played with beans, pebbles, marbles, or any other small objects, which are placed into holes (pockets) dug into the ground or routed out of a board, as shown in Fig. AV3-1.

This project has three parts. In Part 1, you will create all the orthographic views necessary to build a *Mancala* board. In Part 2, you will design your own version of a *Mancala* board. In Part 3, you will create a solid model of the *Mancala* board from Part 1.

■ Part 1

Your task in this part is to create completely dimensioned views of the board. A number of views can be drawn, but the obvious choice is the top and front views. The fillet and depth dimensions of the pockets further suggest that the front view be drawn as a half or full section.

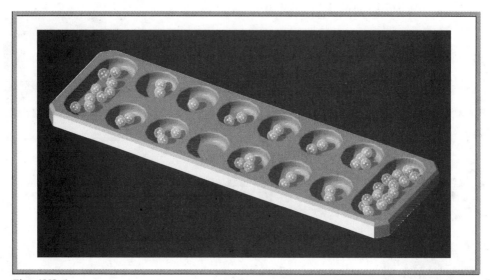

Fig. AV3-1

722

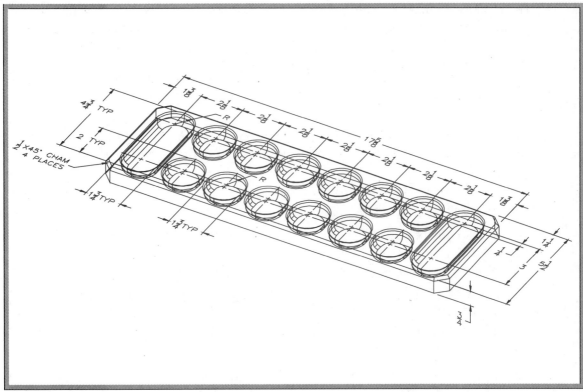

Fig. AV3-2

The board is to have the dimensions shown in Fig. AV3-2. Additional dimensions you will need are:

- Chamfer around the top edge: .125″
- Depth of the pockets: .500″
- Pocket top fillet radius: .062″
- Pocket bottom fillet radius: .250″

The positioning of the pockets on the board is to be symmetrical.

It is probably easiest to begin with the top view and then use it to help construct the front view. You should always begin a drawing by looking for symmetries because AutoCAD is very adept at replication. You can usually save a great deal of time (and boredom!) by taking advantage of commands such as ARRAY, MIRROR, OFFSET, and COPY. The *Mancala* board is particularly suitable for such commands—the entire top view of the pockets could be drawn by replication of just one line and one arc. Another thing you should examine when starting a drawing is the dimensions. Is there a lowest common multiple for at least most of the dimensions? If so, you should use this number for your initial snap and grid settings. What would be good initial snap and grid settings for this project?

AdvancedProjects

The final preliminary considerations for this drawing should be the size of the drawing area and the ultimate size of the sheet on which the drawing will be plotted. Assuming a size A sheet (8.5 × 11), a good drawing limit size for this board is 22 × 17. A final plot scale of 1 = 2 will then give you a nicely sized final plot. Your dimensioning style should then contain a general overall scale setting of 2.

With these settings completed—and assuming that you have begun with a drawing template containing the appropriate layers, text style, linetypes, and dimension style—you could proceed to draw the top view as follows:

- Draw an arc and a line for the large pocket.
- Use MIRROR to finish the pocket.
- Use a copy of the arc, a shortened line, and MIRROR to create a small pocket.
- Make five copies of the small pocket using a rectangular array.
- Draw one-half of the outer edge of the board, using FILLET or snaps to draw the filleted corners.
- Offset these lines to obtain the filleted edges.
- Use MIRROR to obtain the complete top view.
- Construct the front view by drawing one pocket, complete with fillets, and replicating it using a rectangular array.
- Hatch the front view.
- Insert the center lines for all of the circular features, except fillets.
- Dimension the two views.

■ Part 2

Create another, very different *Mancala* board. Include features that you think might be useful to *Mancala* players. See the *Mancala* instructions on the next page for ideas. You may also consider adding the following features:

- handles
- legs
- edges for stacking multiple boards
- a lid
- one or more drawers to store the beans

Create orthographic and isometric views of the board you design.

AdvancedProjects

■ Part 3 (Optional)

Create a solid model of the board from Part 1. Render the board using different materials such as wood, marble, and brass. Add beans, marbles, or other playing pieces (marbles are shown in Fig. AV3-1) and assign appropriate materials to them also. Be sure to create and position lights so that the board is visible and attractive when you render it. (Remember that you will need to render using either the Photo Real or the Photo Raytrace option to display the materials correctly.)

Rules for *Mancala*

Place four beans in each of the twelve smaller pockets. Each of the two players has six of the small pockets and one of the large pockets. The first person to play picks up the four beans in any one of his or her six pockets. Moving counterclockwise, the player then places one bean in each pocket of the twelve and in his or her own large pocket, but not the opponent's.

If the last bean played falls into the player's large pocket, he or she gets another turn. If it falls into an empty pocket on the player's own side of the board, the bean captures all the beans in the opponent's pocket directly across from that pocket. The capturing bean plus all of the captured beans are placed into the player's large pocket.

The game ends when one player runs out of beans in his or her small pockets. When this happens, the other player places all the beans remaining in his or her small pockets into the large one. The player with the most beans in his or her large pocket wins.

Mancala Game Board project courtesy of Gary J. Hordemann, Gonzaga University

Project:
Bolted Seat Connection

When steel frame buildings are designed, a structural engineer prepares *structural details*, which are representations of the connections between beams. The bolted seat connection shown in Fig. AV4-1 is an example of such a connection.

The specifications for structural connections are contained in the *Manual of Steel Construction* published by the American Institute of Steel Construction (AISC). This project uses AISC specifications and is adapted from an example in *Structural Steel Detailing*, also published by AISC.

Bolted seat connections are commonly used to connect grid filler beams to supporting girders (Fig. AV4-2). The filler beam is supported by a seat angle, often merely sitting on the angle without bolts or welds. The other angle is used only for torsional stability.

In its specifications for beams, AISC provides dimensions and physical property data according to beam shape codes. The beams shown here would usually be "W shape," and their identifying codes would be listed in that category.

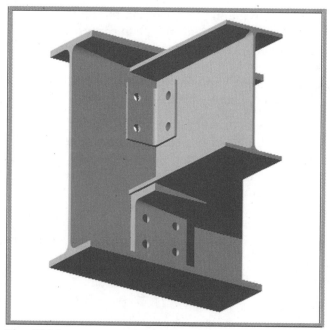

Fig. AV4-1

AdvancedProjects

For the connection in this project, the four members are:

- Girder: W21 × 62 wide flange beam
- Filler: W12 × 26 wide flange beam
- Seat Angle: 6 × 4 × ³/₄ × 6
- Angle: 3¹/₂ × 3¹/₂ × ¹/₄ × 5¹/₂

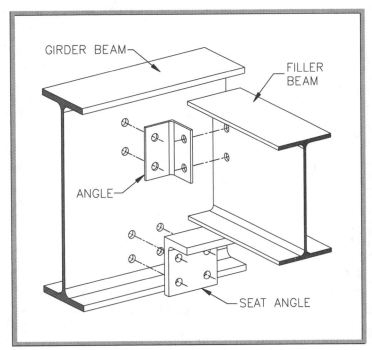

Fig. AV4-2

■ Part 1

Create the front and right-side views of the beam connection. A partially dimensioned front view is shown in Fig. AV4-3 (page 728).

The remaining dimensions, extracted from the *Manual of Steel Construction*, are given in Fig. AV4-4 (page 729).

A few other facts you will need:

- Seat angle fillet radius: .5″
- Angle fillet radius: .375″
- Hole diameters: .8125″
- Clearance between the end of the filler beam and the girder: .5″

AdvancedProjects

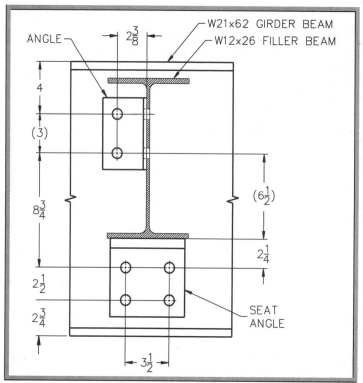

Fig. AV4-3

Structural details typically do not show hidden lines, so leave them out. Center lines should be shown.

This is the type of problem for which snaps and grids are of little use. Furthermore, trying to draw these objects directly, *i.e.*, by using LINE and CIRCLE with coordinates, would be foolhardy. This project is particularly suited to AutoCAD's construction commands OFFSET, EXTEND, and TRIM.

Other preliminary considerations of importance are the size of the drawing area and the size of the sheet on which the drawing will ultimately be plotted. Assuming a size A sheet (8.5 × 11), a good drawing area for this drawing is 44 × 34. A final plot scale of 1 = 4 will then give you a nicely sized final plot.

AdvancedProjects

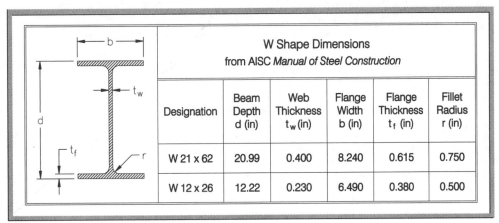

Fig. AV4-4

The W Shape Dimensions table contains:

	W Shape Dimensions from AISC *Manual of Steel Construction*					
Designation	Beam Depth d (in)	Web Thickness t_w (in)	Flange Width b (in)	Flange Thickness t_f (in)	Fillet Radius r (in)
W 21 x 62	20.99	0.400	8.240	0.615	0.750
W 12 x 26	12.22	0.230	6.490	0.380	0.500

Assuming that you have begun with the proper layers, linetypes, and other routine settings, you could create this drawing as follows:

- Start with the front view.
- Draw a vertical line representing the center line of the front view.
- Draw a horizontal line representing the top edge of the girder.
- Use OFFSET to create all the other lines, including center lines for the holes.
- To avoid getting confused by having too many offset construction lines, pause occasionally to trim the lines to their final lengths.
- Use the front view to construct the right-side view, again making extensive use of the OFFSET command.

■ Part 2

If you have access to a copy of the AISC *Manual of Steel Construction*, draw the two views of the bolted column and bracket shown in Problem 96 in the "Additional Problems" section of this book. Leave out the fasteners; they usually are not shown.

■ Part 3

Create solid models of the connection members from Part 1. As an optional activity, render the models using materials that make the members resemble steel.

Bolted Seat Connection project courtesy of Gary J. Hordemann, Gonzaga University

Advanced Project

Project:
Bushing Assembly

The assembly shown in Fig. AV5-1 with nominal dimensions is composed of five parts: two bushing holders, two bushings, and a box.

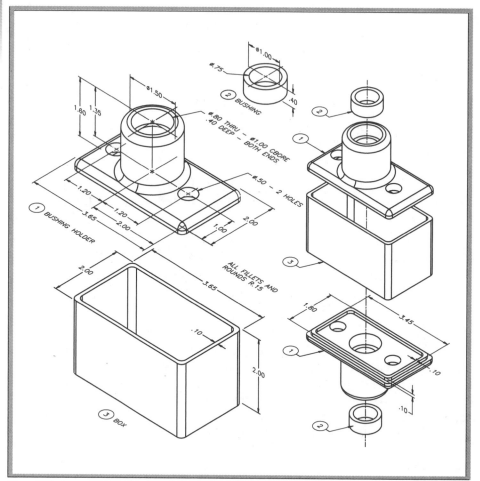

Fig. AV5-1

AdvancedProjects

■ Part 1

Create a full set of drawings to describe the assembly, as follows:

- Draw the necessary two-dimensional views of all the parts, inserting all hidden and center lines.
- Dimension the objects, including tolerances, as specified below.
- Draw the front view of the assembly with all of the parts put together. The front view should be drawn as a full section. Assume the holders, box, and bushings to be made of steel, cast iron, and bronze, respectively. Because hidden lines are usually not included in simple sections, leave them out. Include all center lines.
- Add a border and title block.
- Add notes and the properly toleranced dimensions to specify:
  - Overall tolerance of $\pm.020''$ on all dimensions unless otherwise specified
  - Bushing outside diameter: $1.000'' - 1.0008''$
  - Holder inside diameter for bushing: $1.0014'' - 1.0019''$
  - Bushing inside diameter: $.7500'' - .7512''$
  - Perpendicularity tolerance of holder bushing center line to the base: $.002''$

Assuming that some sort of shaft is to be supported by the bushings, what tolerances would you add to ensure that the two bushings align properly?

The three different objects are most easily drawn by beginning with their top views. You should first determine how many views of each object will be required to describe it completely.

Plan how you will arrange the views and the assembly on the sheet. Will you need more than one sheet? What plot scale(s) will you use? Although it would probably be overkill with so few parts, it often helps to draw rectangles representing the outside dimensions of the views and move them around on the drawing to get a preliminary idea of the layout. In planning the layout, be sure to allow plenty of room for dimensions.

When you have finished the dimensioning, ask yourself whether you could redraw it using only your dimensions; this will help you discover any missing dimensions. Since such redrawing usually goes very quickly, you might even take the time to redraw it. (It helps to let a day or so lapse between dimensioning and redrawing). As a practical matter, you should end up with at least as many dimensions as on the original problem statement.

These objects have "friendly" dimensions and therefore should be drawn using grids and snaps. Determine the lowest common multiple of the dimensions and use that for your snap setting. Use twice that for your grid setting.

With all of your preliminary settings completed and the format of your drawings decided (and assuming that you have begun with a prototype drawing containing all of the appropriate layers, text style, linetypes, and dimension style), you may wish to proceed as follows:

- Draw the top view of the holder and use it to construct any other view(s) you need.
- Draw the box and bushing views, starting with the top views.
- Use copies of the appropriate views of the three objects to construct the sectioned assembly.
- Dimension the orthographic views of the three pieces.
- Place a circle containing a part number (balloon) with each piece; use a circle four times larger than the number and insert the number using the Middle text option and the Center object snap.
- Use the QLEADER command to add "balloons" to the assembly.

■ Part 2

Create solid models of all the pieces in this assembly. Use the models to create an exploded assembly drawing similar to that shown on the right in Fig. AV5-1.

Bushing Assembly project courtesy of Gary J. Hordemann, Gonzaga University

Advanced Project

Project:
Architectural Plans

You are a senior designer for an architectural firm. A client wants to see design proposals for a new home to be built on property she already owns.

A copy of the plot plan is shown in Fig. AV6-1. The client has given you the following list of specifications:

- two-story residence
- 4 bedrooms, 2 baths
- walk-in closets in all bedrooms
- brick construction or brick facing
- attached 2-car garage
- noise level as low as possible in the bedroom area(s)
- total square footage not to exceed 2800 square feet

Design a residence that meets these specifications. Then create a complete set of plans, including (but not necessarily limited to) a site plan, floor plans for each floor, elevations, and a landscaping plan. Include all the information the client needs to make a decision about whether to accept this design. Note that the dimensions shown in the plot plan have been rounded to the nearest foot.

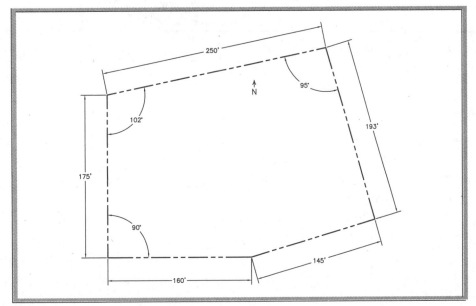

Fig. AV6-1

Project:
Buffet Tray

Two examples of a lap tray for use at a buffet or picnic are shown in Fig. AV7-1. Note that each example has handles and other adaptations to make it easier for the user to control various dishes and implements.

Fig. AV7-1

■ Part 1

Design your own buffet/picnic tray. In doing so, nearly everything is left up to you. The only constraints are:

- The tray is to contain depressions and/or holders for items such as drink containers, plates, silverware, etc.
- The tray is to have handles that are separate pieces; they may or may not be removable or rotatable.

AdvancedProjects

Document your creation by drawing completely dimensioned orthographic views. The thinness of the object, plus the need for dimensions on the depressions and their fillets, make section views appropriate. The nature of the tray also suggests the use of offset sections. Thus, you should:

- Draw complete orthographic views of the tray body, the handles, and any other objects that make up your tray.
- Use section views as needed.
- Dimension the objects.
- In preparing the orthographic views, show each piece individually. Identify each piece with a name and an associated part number inside a circle (balloon).
- Draw a section view of the assembly with the connection between the tray and handle magnified.

An example set of dimensioned drawings for the tray shown in the top of Fig. AV7-1 is given in Fig. AV7-2 (page 736). Use this example as a guide in preparing the drawings.

After you have thought out the design (a pencil sketch may help), prepare to draw it by first considering its dimensions. Is there a common multiple of the dimensions that will make it convenient to use grid and snap? What will be the ultimate size of the sheet on which the drawing will be plotted? Will you need more than one sheet? Set the drawing area and dimension scale accordingly.

■ Part 2

Prepare solid models of the pieces of the tray you designed or the tray shown in the sample design.

After creating the solids, perform the following tasks:

- Display the solids as an exploded assembly displayed in a southeast isometric view.
- Determine the volumes and surface areas of the tray and handles.
- Select your material(s) and look up the density of each. You may have to go to the reference section of a suitable library or search the Internet. Use the density to determine the weights of the parts.

The modeling should probably be started in the top view, and EXTRUDE will likely be the "workhorse" command. Tubular handles like those shown in the bottom tray in Fig. AV7-1 can be created by extruding a circle along a polyline.

Try rendering the model with different materials, such as wood or plastic. Be sure to consider the weight of the tray built with each material or combination of materials. Which material(s) will be the most practical to use for the tray you designed?

AdvancedProjects

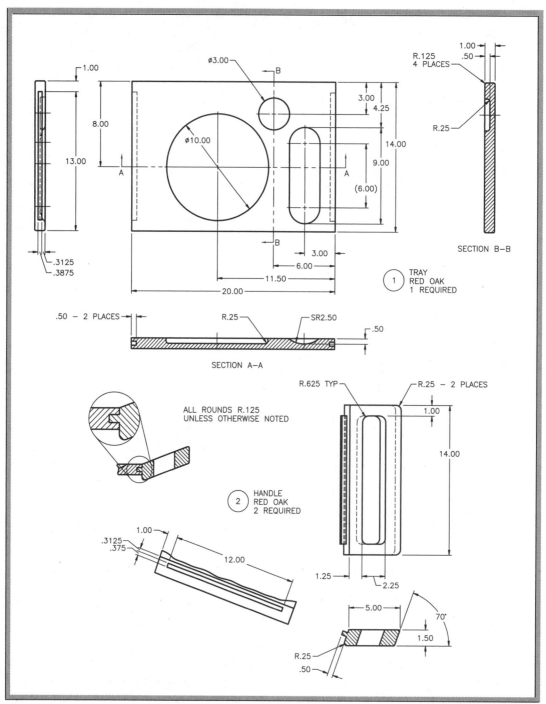

Fig. AV7-2

Buffet Tray project courtesy of Gary J. Hordemann, Gonzaga University

Project:
Shackle Assembly

The shackle assembly shown here is composed of seven parts. Five of the parts (the chain, two yokes, and two clevis pins that secure the chain to the yokes) are shown in Fig. AV8-1. Not shown are two cotter pins that, when inserted into the holes in the pins, keep the pins from working loose.

The dimensions of the yoke and clevis pin are given in the drawing in Fig. AV8-2 (page 739). The clevis pin dimensions are those specified for a nominal 1/2″ standard clevis pin by the American National Standards Institute (ANSI) and published as *ANSI B18.8.1-1972, R1983* by the American Society of Mechanical Engineers (ASME).

Each link of the chain has a diameter of .375″ and outside measurements of 2.125″ × 1.375″.

Using these dimensions, perform the following tasks:

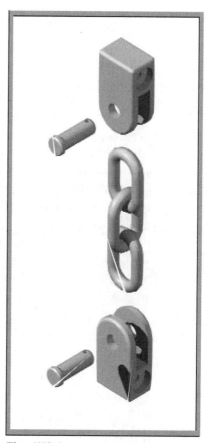

Fig. AV8-1

- Create solid models of the five components shown and arrange them into an exploded assembly similar to the one shown in Fig. AV8-1.

- Determine the volumes of all pieces, including the chain.

- Assuming the pieces to be steel, with a nominal density of .28 lb/cu in., determine their weights.

- Determine the surface area of a yoke.

- Put the solids together and slice the assembly with a cutting plane oriented along the axes of the pins; remove half so the inside of the assembly can be seen.

- Repeat the slice and remove operations with the cutting plane perpendicular to the axes of the pins.

AdvancedProjects

When creating the solids, it is prudent first to choose the starting view. You should select a view such that the world coordinate system standard orthographic views of top, front, right-side, etc., correspond to those same views of the object. This will minimize the confusion over what you are looking at when you switch views. Assume that this assembly is meant to hang vertically, so it has a natural top view. Make the view facing the heads of the clevis pins the front view. The dimensioned drawing of the yoke shown in Fig. AV8-2 shows its front and right-side views.

After settling on the orientation of the solid, take a little time to consider how you are going to create it. There are usually several ways to build a solid, and some may be very difficult and time-consuming.

One way to create the yoke is to follow this procedure:

- Draw the front view profile, making it a closed polyline.
- Extrude the polyline to the full depth of 1″.
- Insert and subtract a cylinder.
- In the right-side view, insert and subtract cylinders to create the counterbored hole.
- Insert and subtract a block (or extruded rectangle) to create the cavity between the arms.
- Fillet the outside edges.

Some variations on this procedure include:

- Extrude the profile to a depth of .25″ to create one arm; then use a duplicate for the other arm, joining the two arms with a box.
- Make the counterbored hole by drawing and revolving its profile in the front view.
- Draw and extrude the yoke profile in the right-side view. Then use FILLET to obtain the geometry common to both extrusions. (This approach is perhaps the most elegant.)
- Create the space between the arms by making three slices and unions. (This approach, however, is not recommended.)
- You could even create the yoke without using either EXTRUDE or REVOLVE by unioning and subtracting a series of cylinders and boxes. This approach, too, is impractical in reality.

Clearly, there are several practical ways to create this assembly, as well as several very impractical ways.

You should also think through the sequence of operations. The most common error resulting from an incorrect sequence is a hole that does not go all the way through the object.

Finally, always start a drawing at a known and easy-to-remember point such as 0,0,0 or 2,2,2. This is particularly critical when you are working in 3D space.

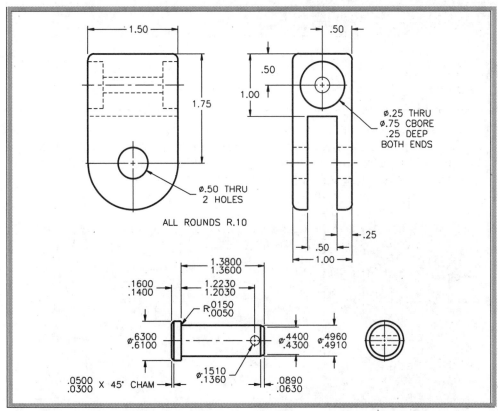

Fig. AV8-2

If you wish to proceed with "elegance," create the yoke as follows:

- Draw the profile of the front view of the yoke and extrude it to 1".
- Insert a cylinder and subtract it.
- In the front view, draw the half-profile of the counterbored hole and revolve it.
- In the right-side view, draw and extrude the profile, without the holes.
- Fillet the edges of both objects.
- After making sure the two extrusions are coincident, obtain the final object using INTERSECT.

You may create the clevis pins from unioned cylinders or from a revolved profile. The chain link can be made from a circle extruded around a polyline (consider using multiple viewports) or from a sliced torus unioned with two cylinders.

You can model the two cotter pins by extruding a half-circle along a polyline. Select the appropriate pin from a table of ANSI standard cotter pins. Often, you can find a table such as this in the appendix of a graphics or drafting textbook.

Shackle Assembly project courtesy of Gary J. Hordemann, Gonzaga University

Project:
Clothing Rack

Shown in Fig. AV9-1 is a wall-mounted rack for drying clothing. It is
composed of six rods inserted into a large cylinder, which is housed in
a square tube. The tube has cutouts that allow the rack to be folded
downward when the rods and cylinder assembly are slid to the right. In
this view, we are looking upward from below the object.

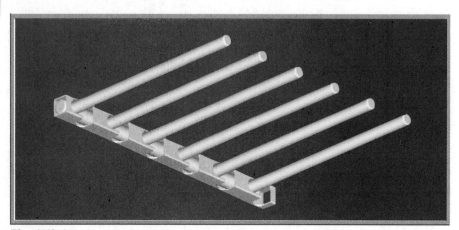

Fig. AV9-1

■ Part 1

The dimensions of the components are shown in Fig. AV9-2. Using these
dimensions, perform the following tasks:

- Create solid models of the eight pieces and arrange them into an
  exploded assembly.
- Determine the volumes and surface areas of all eight pieces.
- Assuming the tube to be aluminum with a density of 160 lb/cu ft,
  determine its weight.
- Assuming the rod assembly to be oak with a density of 45 lb/cu ft,
  determine its weight.

AdvancedProjects

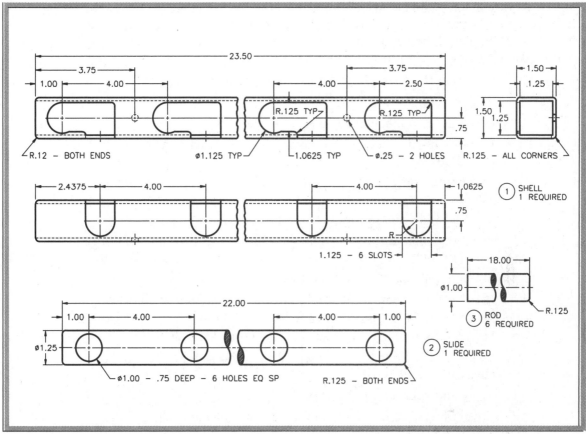

Fig. AV9-2

When creating three-dimensional objects, you will find it least confusing to select the proper view before you begin. Try to align the object with the world coordinate system standard orthographic views of top, front, right-side, etc. The standard views will then make sense when you switch from view to view. Many objects—such as this one—have a natural top and front, and you should plan to orient the object accordingly. Following this orientation, the dimensioned drawing in Fig. AV9-2 shows the front, bottom, and right-side views.

Before starting the model, consider the different ways to carry out this project. Always look for symmetries so that you can take advantage of AutoCAD's replication commands. This object is ideal for use of the ARRAY command. The modeling of this assembly is straightforward, being primarily a collection of cylinders. The square tube at first glance appears complex, but if you realize that cavities usually are best modeled by creating the positive of the cavity and then subtracting it, the creation of this object becomes much easier.

AdvancedProjects

After you have laid out your approach to the modeling, begin the drawing at some known and easy-to-remember point, such as 0,0,0 or 1,1,1. The least difficult approach to this model is probably to begin with the rods. Doing so, you could proceed as follows:

- Switch to the front view, change the UCS, and insert a cylinder.
- Fillet one end of the cylinder.
- Use ARRAY to make five copies on 4″ centers.
- In the right-side view, begin the slide by inserting a cylinder; fillet both ends.
- In the front view, insert and array a set of small cylinders for the six holes.
- Subtract the small cylinders from the slide.
- Begin the shell in the right-side view, inserting two squares; fillet the outer one.
- Extrude and subtract the squares.
- In the top view, create the profile of the cavity, joining all of the pieces with the Join option of the PEDIT command.
- Extrude the profile to at least the thickness of the shell.
- Switch to the front view and repeat the process for the profile of the cavity in that view.
- Array and subtract the two profiles.
- Create the two mounting holes in the back of the shell.
- Arrange the objects in an exploded assembly.

■ Part 2

Using AutoCAD's isometric mode, create an isometric exploded assembly of the rack. Using a leader, tag each part with a balloon. A *balloon* is a circle containing the part number of the piece. The circle (balloon) diameter should be four times the height of the numeral.

Drawing the shell in isometric mode will be a bit of a challenge. Sometimes you can create the isometric view more easily by first drawing an orthographic view such as the front view. Then switch to one of the four standard isometric views to complete the drawing. When switching views, you will want to stay in the world coordinate system. Because different parts of the drawing may end up on different planes, you will discover that TRIM and EXTEND may not work properly. However, TRIM can usually be made to work if you change the coordinate system so that the current view is an XY plane. Once created, the isometric may be transferred into any other view by cutting and pasting. (Do not forget always to change the UCS to the current view before cutting and again before pasting.)

■ Part 3

The rack as specified in Part 1 has a few shortcomings. First, the mounting holes are not located properly—they should be in line with the cavities so the installer can use a tool on whatever fastener is in the hole. Second, if the mounting fasteners have hex or round heads, the heads will interfere with the movement of the rods/slide assembly. The fasteners should have flat heads, but then the holes should be countersunk. Third, the rectangular tubing might be too expensive.

Address these problems by creating a new design. Use a round tube for the shell. Make the rods and slideout of 3/4" PVC pipe. Use 3/4" PVC pipe elbows and Ts to connect the pipe pieces, and 3/4" PVC pipe caps on the ends of the rods. A trip to the local hardware store will provide you with the outside dimensions of these pipe fittings.

Using the drawings on page 741 as a guide, create a complete set of dimensioned drawings of your design. Remember, when you are drawing orthographic views, draw only as many views as necessary to specify the object completely, and only draw one copy of each part. Tag each part with a balloon specifying the part number, material, and how many are required in the full assembly.

In a drawing with so many parts, you should first plan how you will arrange the parts on the paper. This will help you figure out how many sheets to use and what the plot scales will be. Be sure to allow sufficient room for the dimensions.

■ Part 4

Using AutoCAD's isometric mode, create an isometric exploded assembly of the rack you designed in Part 3. Using leaders, tag each part with a balloon. Create the isometric drawing as a separate drawing file. Then insert the isometric into the drawing from Part 3.

■ Part 5 (Optional)

Make solid models of the pieces of your design. Then perform the following tasks:

- Arrange the solids in an exploded assembly and display them in a view in which you are looking upward at the assembly from below.
- Determine the volumes and surface areas of all the pieces.
- Select a material for the shell and look up its density and that of PVC; use the densities to determine the weight of the shell and of the rods/slide assembly.
- Render the rack using suitable materials.

Clothing Rack project courtesy of Gary J. Hordemann, Gonzaga University

Project:
Protective Packaging

Designers for packaging companies typically design custom packaging to protect various products. The packaging company's clients depend on the expertise of these designers to provide safe, effective shipping containers and other packaging. In this project, you will develop protective packaging for a client's new product, shown in Fig. AV11-1.

The device is made of a titanium alloy to reduce weight. The client has specified that the packaging must:

- protect the device from sudden impacts that may result from being dropped or shifting in an airplane cargo hold.

- provide an attractive appearance that will be appealing to the client's potential customers.

Design custom packaging for this instrument and create drawings to show how the packaging will protect it. Also, create assembly drawings as necessary to show the client how to assemble the packaging and how to pack the instrument properly.

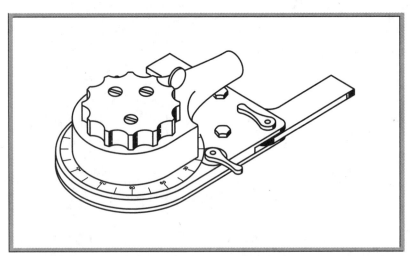

Fig. AV11-1

Appendix A: Options Dialog Box

AutoCAD's Options dialog box permits you to customize many of AutoCAD's settings. Also, the dialog box enables you to add pointing devices and printers to your AutoCAD system. Select Options... from the Tools pull-down menu to display the Options dialog box, as shown in Fig. A-1.

■ Files

The Files tab specifies the folders in which AutoCAD searches for text fonts, drawings to insert, linetypes, hatch patterns, menus, plug-ins, and other files that are not located in the current folder. It also specifies user-defined settings such as the custom dictionary to use for spell checking.

■ Display

This tab customizes the AutoCAD window. For instance, you can specify the display of scroll bars in the window and indicate the number of lines of text to show in the docked Command Line window. It allows you to set the colors for the model and layout tabs, Command line background and text, and AutoTrack® vector. You can also control the size of the crosshairs and the resolution of arcs and circles.

Fig. A-1

Appendix A: Options Dialog Box (Continued)

■ Open and Save

This tab controls the options related to opening and saving files. Examples include the default file type when saving an AutoCAD drawing and whether a thumbnail preview image appears in the Select File dialog box. The settings on this tab also allow you to set the number of minutes between automatic saves.

■ Plot and Publish

The Plot and Publish tab deals with settings and controls related to plotting and printing devices. You can set the default output device and add and configure a new device. You can also add or edit plot style tables.

■ System

Use this tab to control miscellaneous system settings, such as the display of 3D graphics and whether AutoCAD displays the Startup dialog box. This is also where you change the current pointing device.

■ User Preferences

This tab helps you optimize the way you work in AutoCAD. For example, you can customize how the right-click features in AutoCAD behave. Also, you can indicate the source and target drawing units when dragging and dropping drawing content.

■ Drafting

The Drafting tab controls many editing options. For instance, you can make adjustments to the AutoSnap features, such as changing the color and size of the AutoSnap marker. Here, you can also change AutoTrack® settings, including whether to display polar tracking vectors.

■ Selection

This controls settings related to methods of object selection. Examples include turning noun/verb selection on or off and adjusting the size of the pickbox and grip boxes. You can also set the color of selected and unselected grip boxes.

■ Profiles

This tab, shown in Fig. A-2 (page 733), controls the use of profiles. In this context, a *profile* is a user-defined configuration that allows each user to set up and save various personal preferences, such as which toolbars are present and where they appear on the screen. Profiles allow several people to share a single computer without having to change the settings individually each time they use it.

Appendix A: Options Dialog Box (Continued)

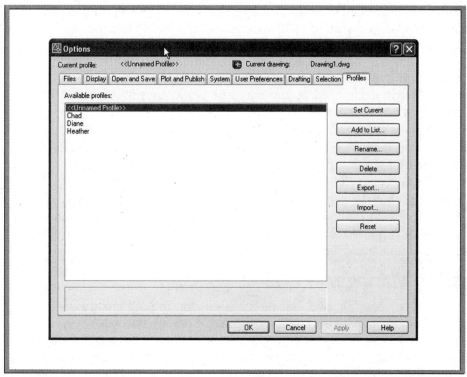

Fig. A-2

If no profiles have been defined for a computer, you can define one by picking the Add to List... button. A dialog box appears that allows you to name the profile and give it a description. Note that creating a new profile does not automatically select it as the current profile. You must pick the Set Current button to set the active profile.

When more than one profile exists, any changes you make to AutoCAD's settings are automatically saved to the current profile. To change the current profile, display the Options dialog box, pick the Profiles tab, select the profile you want to make current, and pick the Set Current button.

Appendix B: Managing AutoCAD (Continued)

Organizing an AutoCAD installation requires several considerations. For example, you should store files in specific folders. Name the files and folders using standard naming conventions. If you don't, you may not be able to find the files later. It is also very important that you back up regularly. If you overlook these and other system management tasks, your productivity will decline. Also, you may lose work, causing problems that you could otherwise avoid.

AutoCAD users and managers alike will find the following information helpful. If you manage an AutoCAD system with care, you should never have to create the same object twice. This comes only with careful planning and cooperation among those who will be using the system.

Bear in mind that the process of implementing, managing, and expanding an AutoCAD system evolves over time. As you and others become more familiar with AutoCAD and the importance of managing files and folders, refer back to this appendix.

■ System Manager

One person (or possibly two, but no more) within the organization should have the responsibility for managing the AutoCAD systems and overseeing their use. This person should be the resident CAD champion and should answer questions and provide directions to other users of the systems. The manager should oversee the software, documentation, and hardware and should work with the users to establish procedural standards for use with the systems.

■ Management Considerations

The system manager should consider questions such as those listed below when installing and organizing AutoCAD. The questions are intended to help guide your thinking, from a management perspective, as you become familiar with the various parts of AutoCAD.

- How can I best categorize the files so that each folder does not grow to more than 200 files total? Folders with a larger number of files become difficult to navigate.

- If I plan to install three or more AutoCAD stations, should I centralize the storage of user-created files and plotting by using a network and file server? A network and a centralized file system are recommended.

After AutoCAD is in place and you are familiar with it, you will create many new files. The following questions address the efficiency with which you create and store these files.

- Are there template files (or existing drawings) on file that may serve as a starting point for new drawings? It is a good practice to benefit from past work.

- Where should new drawing files be stored, what should they be named, and how can users easily locate them? Develop organizational standards that are clear and easy to follow.

- Are there predefined libraries of symbols and details that I can use while I develop a drawing? Do any of the symbol libraries that come with AutoCAD meet my needs?

- Is a custom pull-down menu or toolbar available that lends itself to my drawing application?

- Are AutoLISP routines or other programs available that would help me perform certain drafting operations more easily?

If you feel uncomfortable about your answers to these questions, there is probably room for improvement. The following discussion is provided to help you organize and manage your CAD systems more effectively.

 NOTE:

Generally, the following discussion applies to almost all AutoCAD users and files. However, there are inevitable differences among users (backgrounds and interests, for example), drawing applications, and the specific hardware and software. Take these differences into consideration.

Software/Documentation

The AutoCAD system manager spends considerable time organizing and documenting files, establishing rules and guidelines, and tracking new software and hardware developments.

File Management—Know where files are located and the purpose of each. Understand which ones are AutoCAD system files and which are not. Create a system for backing up files and folders, and back up regularly. Emphasize this to all users. Delete "junk" files.

Template Files—Create an easy-to-use system for the development, storage, and retrieval of AutoCAD template files. Allow for ongoing development and improvement of each template. Store the template files in a folder dedicated to templates so that they are accessible by other users.

Document the contents of each template file by printing the drawing status information, layers, text styles, linetype scale, and other relevant information. On the first page of this information, write the name of the template file, its location, sheet size, and plot scale. Keep this information in a three-ring binder for future reference to other users.

User Drawing Files—Store these in separate folders. Place drawing objects on the proper layers. Assign standard colors, linetypes, and lineweights to the standard layers. Make a backup copy of each drawing and store it on a separate disk or backup system. Plot the drawings most likely to be used by others and store them in a three-ring binder for future reference.

Symbol Libraries—Develop a system for ongoing library development. (See Chapter 34 for details on creating symbol libraries.) Plot each symbol library drawing file, and place the library drawings on the wall near the system(s) or in a binder. Encourage all users to contribute to the libraries.

Menu Files—Develop, set up, and make available custom toolbars and pull-down menus. Store the menus in a folder dedicated to menus so that they will be accessible to others.

AutoCAD Upgrades—Handle the acquisition and installation of AutoCAD software upgrades. Inform users of the new features and changes contained in the new software. Coordinate training.

AutoCAD Third-Party Software—Handle the acquisition and installation of third-party software developed for specific applications and utility purposes. Inform users of its availability and use.

■ Hardware

Oversee the use and maintenance of the hardware components that make up the system. Consider hardware upgrades as user and software requirements change.

■ Procedural Standards

Develop clear, practical standards in the organization to minimize inconsistency and confusion. Each template file should have a standard set of drawing layers, with a specific color and linetype dedicated to each layer. For example, you may reserve a layer called Dimension, with color green, a continuous linetype, and a lineweight of .3 mm, for all dimensions. This will avoid confusion and improve consistency within your organization. Develop similar standards for other AutoCAD-related practices.

■ AutoCAD Standards Checker

AutoCAD offers a method of checking arrowhead size, center marks, zero suppression, text height, and other items for organizational standards. The following steps outline this method.

1. Open a drawing file that you would consider a model example of how drawings should appear at your organization.

2. Using Save As..., save the drawing in AutoCAD Drawing Standard (DWS) format.

3. Open or create a new drawing and then open the CAD Standards toolbar.

4. Pick the Configure Standards button.

5. In the Configure Standards dialog box, pick the Add Standards File button (the one containing the plus sign) and open the DWS file you saved in Step 2.

6. Pick the Check Standards... button.

AutoCAD audits the current drawing against the DWS file and displays the results in the Check Standards dialog box.

7. Pick the Settings... button, check the Automatically fix non-standards properties check box, and pick OK.

8. Pick the Next button; pick the Next button again and again until the Checking Complete box appears.

Similar to a spell checker, AutoCAD fixes the current drawing to conform to the standards contained in the DWS file.

9. Close the Checking Complete and Check Standards dialog boxes.

In summary, take seriously the standards and management of your AutoCAD systems. Encourage users to experiment and to be creative by making software and hardware available to them. Make a team effort out of learning, developing, and managing AutoCAD so that everyone can learn and benefit from its power and capabilities.

Appendix C: Drawing Area Guidelines

	Sheet Size	Approximate Plotting Area	Scale	Drawing Area
Architect's Scale	A: 12″ × 9″ B: 18″ × 12″ C: 24″ × 18″ D: 36″ × 24″ E: 48″ × 36″	10″ × 8″ 16″ × 11″ 22″ × 16″ 34″ × 22″ 46″ × 34″	⅛″ = 1′ ½″ = 1′ ¼″ = 1′ 3′ = 1′ 3′ = 1′	96′ × 72′ 36′ × 24′ 96′ × 72′ 12′ × 8′ 16′ × 12′
Civil Engineer's Scale	A: 12″ × 9″ B: 18″ × 12″ C: 24″ × 18″ D: 36″ × 24″ E: 48″ × 36″	10″ × 8″ 16″ × 11″ 22″ × 16″ 34″ × 22″ 46″ × 34″	1″ = 200′ 1″ = 50′ 1″ = 10′ 1″ = 300′ 1″ = 20′	2400′ × 1800′ 900′ × 600′ 240′ × 180′ 10,800′ × 7200′ 960′ × 720′
Mechanical Engineer's Scale	A: 11″ × 8″ B: 17″ × 11″ C: 22″ × 17″ D: 34″ × 22″ E: 44″ × 34″	9″ × 7″ 15″ × 10″ 20″ × 15″ 32″ × 20″ 42″ × 32″	1″ = 2″ 2″ = 1″ 1″ = 1″ 1″ = 1.5″ 4″ = 1″	22″ × 17″ 8.5″ × 5.5″ 22″ × 17″ 51″ × 33″ 11″ × 8.5″
Metric Scale	A: 279 mm × 216 mm (11″ × 8½″) B: 432 mm × 279 mm (17″ × 11″) C: 55.9 cm × 43.2 cm (22″ × 17″) D: 86.4 cm × 55.9 cm (34″ × 22″) E: 111.8 cm × 86.4 cm (44″ × 34″)	229 mm × 178 mm (9″ × 7″) 381 mm × 254 mm (15″ × 10″) 50.8 cm × 38.1 cm (20″ × 15″) 81.3 cm × 50.8 cm (32″ × 20″) 106.7 cm × 81.3 cm (42″ × 32″)	1 mm = 5 mm 1 mm = 20 mm 1 cm = 10 cm 2 cm = 1 cm 1 cm = 2 cm	1395 × 1080 8640 × 5580 559 × 432 43.2 × 27.95 237.6 × 172.8

Note: 1 = 25.4 mm

Appendix D: Dimensioning Symbols

Geometric Characteristic Symbols

Type of Tolerance	Symbol	Name
Location	◎	Concentricity
	⊕	Position
	═	Symmetry
Orientation	∠	Angularity
	//	Parallelism
	⊥	Perpendicularity
Form	⌭	Cylindricity
	▱	Flatness
	○	Circularity (roundness)
	—	Straightness
Profile	⌒	Profile of a line
	⌓	Profile of a surface
Runout	↗	Circular runout
	↗↗	Total runout
Supplementary	Ⓜ	Maximum material condition (MMC)
	Ⓛ	Least material condition (LMC)
	Ⓟ	Projected tolerance zone

Basic Dimensioning Symbols

Symbol	Type of Dimension	Symbol	Type of Dimension
⌀	Diameter	∨	Countersink
R	Radius	⊔	Counterbore/Spotface
SR	Spherical radius (ISO name)	▼	Deep
S⌀	Spherical diameter (ISO name)	X	Places, times, or by
( )	Reference		

Appendix E: AutoCAD Fonts

AutoCAD provides several standard fonts, which have file extensions of SHX. You can use the STYLE command to apply expansion, compression, or obliquing to any of these fonts, thereby tailoring the characters to your needs. (See Chapter 18 for details on the STYLE command.) You can draw characters of any desired height using any of the fonts.

Examples of some of the fonts supplied with AutoCAD are listed in this appendix, along with samples of their appearance. With the exception of monotxt.shx (not included in the samples below), each font's characters are proportionately spaced. Hence, the space needed for the letter "i," for example, is narrower than that needed for the letter "m."

Each font resides in a separate file. This is the "compiled" form of the font, for direct use by AutoCAD. Examples of standard and TrueType fonts are shown in this appendix.

Standard Fonts

AutoCAD's standard SHX fonts include both text and symbol files that have been created as AutoCAD shapes. You can change the appearance of these fonts by expanding, compressing, or slanting the characters.

Name	Description	Appearance
txt.shx	Basic AutoCAD font; this font is-very simple and generates quickly on the screen	ABCDEFGHIJKLMNOPQRSTUVWXYZ abcdefghijklmnopqrstuvwxyz
romans.shx	A "simplex" roman font drawn by means of many short line segments; produces smoother characters than txt.shx	ABCDEFGHIJKLMNOPQRSTUVWXYZ abcdefghijklmnopqrstuvwxyz
romand.shx	Similar to romans.shx, but instead of a single stroke, it uses a double stroke technique to produce darker, thicker lines	ABCDEFGHIJKLMNOPQRSTUVWXYZ abcdefghijklmnopqrstuvwxyz0123456789
itallicc.shx	Complex italic font using double stroke and serifs.	ABCDEFGHIJKLMNOPQRSTUVWXYZ abcdefghijklmnopqrstuvwxyz0123456789
scripts.shx	A single-stroke (simplex) script font	ABCDEFGHIJKLMNOPQRSTUVWXYZ abcdefghijklmnopqrstuvwxyz 0123456789
gothice.shx	Gothic English font	ABCDEFGHIJKLMNOPQRSTUVWXYZ abcdefghijklmnopqrstuvwxyz0123456789
syastro.shx	A symbol font that includes common astronomical symbols	☉♀♁⊕♂♃♄♅♆♇ЕВС☾♃*♌♉♐♈♅♍♏♎♒♑♏♌♐♈≈ ✳"∪⊃∩∈→↑←↓∇~‿ˇ˘×§†‡∫£®©
symusic.shx	A symbol font that includes common music symbols	‧↗○○○♦♯♭♭—–×↴𝄞⊙:‖∘‧‧‧—⌐╼▽ ‧↗○○○♦♯♭♭—↗↴𝄢♭:♫☉♀♁⊕♂♃♄♅ΨＰ

Appendix E: AutoCAD Fonts (Continued)

TrueType Fonts

AutoCAD uses several TrueType fonts and font "families" (groups of related fonts). In the TrueType fonts, each font in a family has its own font file. The examples below are shown in outline form, as they appear by default in AutoCAD. To display solid characters, set the TEXTFILL system variable to 1. You may also change the print quality by changing the value of the TEXTQLTY system variable.

Name	Description
Arial Narrow	ABCDEFGHIJKLMNOPQRSTUVWXYZ abcdefghijklmnopqrstuvwxyz0123456789
Dutch801 Rm BT	*ABCDEFGHIJKLMNOPQRSTUVWXYZ* *abcdefghijklmnopqrstuvwxyz0123456789*
Lucida Console	ABCDEFGHIJKLMNOPQRSTUVWXYZ abcdefghijklmnopqrstuvwxyz0123456789
Swis721 BdOul BT	ABCDEFGHIJKLMNOPQRSTUVWXYZ abcdefghijklmnopqrstuvwxyz0123456789
UniversalMath1 BT	ΑΒΨΔΕΦΓΗΙΞΚΛΜΝΟΠΘΡΣΤΘΩбXYZ αβψδεφγηιξκλμνοπϑρστθωφχυζ″ + − × ÷ = ± ∓ °′
Wingdings	(symbol characters)

Glossary of Terms

A

absolute points Specific coordinate points entered directly (without using relative positioning).

alignment grid A non-printing grid that can be set at any desired spacing in Standard or Isometric drawing mode.

alignment paths Temporary, non-printing lines that appear at predefined angles to help you create objects more accurately.

alternate units A second set of distances shown inside brackets on some dimensioned drawings.

aperture box A small box that defines an area around the center of the crosshairs within which an object or point will be selected when you press the pick button.

arcball An object that displays when 3D Orbit is active and allows the user to manipulate the drawing view in realtime.

area The number of square units needed to cover an enclosed 2D shape or surface.

array An orderly arrangement of objects.

aspect ratio The ratio of height to width of a rectangular object or region.

associative dimensions Dimensions that update automatically when you change the drawing by stretching, scaling, etc.

associative hatch A hatch object that updates automatically when its boundaries are changed; a hatch created with the BHATCH command.

attribute extraction Gathering information stored in attributes and placing it into an AutoCAD table or external file.

attributes Text data stored in blocks.

attribute tag A variable that identifies each occurrence of an attribute in a drawing.

attribute values The information stored by attribute tags in specific instances of an attribute.

audit In AutoCAD, to examine the validity of a drawing file.

auxiliary view A view that is projected onto a plane perpendicular to one of the orthographic views and inclined in the adjacent view.

B

baseline dimensions A set of dimensions that have a common baseline to reduce measurement error that can accumulate from "stacking" dimensions.

basic dimension A dimension to which allowances and tolerances are added to obtain limits.

block A collection of objects that you can group to form a single object.

Boolean logic (See *Boolean operation*.)

Boolean mathematics (See *Boolean operation*.)

Boolean operations In CAD, combining two solid objects, subtracting one from another, or determining intersecting areas or volumes between overlapping solid objects.

B-spline A type of spline. (See *splines*.)

buttons In AutoCAD, the small areas in toolbars that, when pressed, enter an associated command or function.

C

camera The point of view that moves around the target in 3D Orbit.

Cartesian coordinate system A coordinate system used in geometry that assigns a specific coordinate pair (x,y) or triplet (x,y,z) to points in space so that each point has a unique identifier.

cascading menus Submenus in a pull-down menu that offer further choices.

cells The individual boxes within a table's grid.

center lines Imaginary lines that mark the exact center of an object or feature; shown by a series of alternating long and short line segments.

chamfers Beveled edges.

circular array (See polar array.)

circumference The distance around a circle.

clipping plane Plane that slices through one or more 3D objects, causing the part of the object on one side of the plane to be omitted.

colinear Lying on the same infinite line.

comma-delimited file (CSV) A file format that allows a user to write attributes to an ASCII text file. Entries in this file format are separated by commas.

command alias Command abbreviation that usually consists of the first one to three letters of a command name.

Command line A text area in AutoCAD where you can key in commands and options and view information.

compass When the VPOINT command is active, the compass is a globe representation that allows the user to orient the drawing in 3D space.

composite solids Solids composed of two or more solid objects.

concentric Sharing a common center point.

constrain Assign limits or values that can be used for parameters in dynamic blocks.

construction lines Line objects in AutoCAD that extend infinitely in both directions.

context-sensitive AutoCAD "remembers" the last command used and displays it at the top of right-click shortcut menus.

context-sensitive help Help related to the currently entered command.

control points Points on or near a curve that exert a "pull," influencing its shape.

coordinate pair A combination of one x value and one y value that specifies a unique point in the coordinate system.

coordinates Sets of x,y (for 2D drawings) or x,y,z (for 3D drawings) values used to describe specific locations in space.

copy and paste A Windows-standard copy and paste feature that allows you to copy objects from one drawing and insert them into another.

crosshairs A special cursor that consists of two intersecting lines and a pickbox; used to select objects and pick points in the drawing area.

crosshatch (See hatch.)

cutting plane The plane through a part from which a sectional view is taken.

D

datum A surface, edge, or point that is assumed to be exact.

datum dimensions (See ordinate dimensions.)

delta In mathematics, a Greek symbol used to signify change.

deviation tolerancing A method of tolerancing that allows the user to set the upper and lower tolerances separately (to different values).

Glossary of Terms (Continued)

dialog box A window that provides information and permits you to make selections and enter information.

diameter The length of a line that extends from one side of a circle to the other and passes through its center.

dimensioning A system of describing the size and location of features on a drawing.

dimension line The part of a dimension that typically contains arrowheads at its ends and shows the extent of the area being dimensioned.

dimension styles Saved sets of dimension settings that control the format and appearance of dimensions in a drawing.

direct distance method A method of entering points in which the user aims using the pointing device and enters a number to specify the length of a line.

docked toolbar A toolbar that is attached to the top or side of the drawing area.

donuts Thick-walled or solid circles created using the DONUT command.

double-clicking Positioning the pointer on an object or button and pressing the pick button twice in rapid succession.

drawing area The part of the AutoCAD screen in which the drawing appears. Also refers to the boundaries for constructing the drawing, which should correspond to both the drawing scale and the sheet size.

drawing database Information associated with each AutoCAD file that defines the objects in the drawing and contains specific numerical information about individual objects and their locations.

dynamic blocks Blocks created with certain changeable characteristics, such as length, width, rotation, or location.

dynamic input Entering commands and information directly at the cursor.

E

ellipse A regular oval shape that has two centers of equal radius.

entity An individual predefined element in AutoCAD; the smallest element that you can add to or erase from a drawing. Also called an *object*.

equator The inner ring of the globe representation used in the VPOINT command.

extension lines The lines in a dimension that extend from the object to the dimension line.

extrusion thickness Thickness in the Z axis; can be set using the THICKNESS variable or by applying the EXTRUDE command.

F

facet shading (See *flat shading*.)

feature control frames The frames used to hold geometric characteristic symbols and their corresponding tolerances.

file attributes The characteristics of a file, such as its size, type, and the date and time it was last modified.

fillets Rounded inside corners.

filtering Including or excluding items from a selection set using specific criteria.

fixed attributes Attributes whose values you define when you first create a block.

flat shading Method of rendering a polygonal model that fills each polygon with a single shade of color to give the model a faceted look. Also called *facet shading*.

floating toolbar A toolbar that displays as a window in the drawing area.

Glossary of Terms (Continued)

font Distinctive set of characters consisting of letters, numbers, punctuation marks, and symbols.

formula An expression that performs an automatic mathematical equation in an AutoCAD table.

freeze In AutoCAD, to remove a layer from view without turning it off or deleting it.

full section A section that extends all the way through a part or object.

G

geometric characteristic symbols Symbols that are used to specify form and position tolerances on drawings.

geometric dimensioning and tolerancing (GD&T) Standardized format for dimensioning and showing variations (called tolerances) in the manufacturing process. Tolerances specify the largest variation allowable for a given dimension. GD&T uses geometric characteristic symbols and feature control frames that follow industry standard practices. GD&T standards are defined in the American National Standards Institute (ANSI) Y14.5M Dimensioning/Tolerancing handbook.

Gouraud shading Method of smooth rendering of polygonal models by interpolating (averaging and blending) adjacent color intensities. The Gouraud method applies light to each vertex of a polygon face and interpolates the results to produce a realistic-looking model.

grips Small boxes that appear at key points on an object when you select the object without first entering a command. Grips allow you to perform several basic operations such as moving, copying, or changing the shape or size of an object.

group A named set of objects. A circle with a line through it, for instance, could become a group.

H

hatch A repetitive pattern of lines or symbols that shows a related area of a drawing.

hidden line removal The deletion of tessellation lines that would not be visible if you were viewing the 3D model as a real object.

hidden lines Lines that would not be visible if you were looking at an object without "see-through" capability; shown as dashed or broken lines.

I

icons The small pictures on buttons in AutoCAD that help identify the function of the buttons.

image tiles Selectable squares displayed in some dialog boxes that contain thumbnail images from which the user can choose.

in-place text editor The text editor that allows you to edit dynamic text and mtext in place on the screen.

integer Any positive or negative whole number or 0.

interference The overlap of two objects that occupy the same area in 3D space.

interference solids Solids created in AutoCAD that show the overlapping portions of two solids that interfere in 3D space.

intersection A method of creating a solid from the overlapping portion of two intersecting solid objects.

Glossary of Terms (Continued)

island An area within an object, such as a hole, that is not hatched.

isometric drawing A type of 2D drawing that gives a 3D appearance.

isometric planes Three imaginary drawing planes spaced at 120° from each other; used to draw isometric views.

J

justify Align multiple lines of text at a specific point (left, right, center, etc.).

L

landscape A paper orientation in which the wider edge of the paper is at the top and bottom of the sheet.

layers Similar to transparent overlays in manual drawing; allow AutoCAD users to display and hide information to clarify the drawing.

layout An arrangement of one or more views of an object on a single sheet.

leaders Lines with an arrowhead at one end that point out a particular feature in a drawing.

limits (See *drawing area.*)

limits tolerancing A method of tolerancing in which only the upper and lower limits of variation for a dimension are shown.

linear dimensions Vertical and horizontal dimensions, as well as straight-line dimensions that are aligned with an edge of the object being dimensioned.

M

material condition symbols Symbols used in geometric dimensioning and tolerancing to modify the geometric

tolerance in relation to the produced size or location of the feature.

Maximum Material Condition (MMC) A notation specifying that a feature such as a hole or shaft is at its maximum size or contains its maximum amount of material.

mline (See *multiline.*)

model space The space, or drawing mode, in AutoCAD in which most drawing is done.

mtext A multiple-line text object created in AutoCAD using a text editor.

multiline A type of line object in AutoCAD that consists of up to 16 parallel lines created simultaneously and viewed by AutoCAD as a single object.

multiview drawing A drawing that describes a three-dimensional object completely using two or more two-dimensional views.

N

nested block A block that contains one or more other blocks.

notes Text that refers to an entire drawing, rather than one specific feature. (See also *specifications.*)

noun/verb selection An object selection method in which you select the object first, then enter a command (verb) to perform an operation on the object.

O

object An individual predefined element in AutoCAD; the smallest element that you can add to or erase from a drawing. Also called an *entity*.

Glossary of Terms (Continued)

object snap A magnet-like feature in AutoCAD that allows the user to "snap" to endpoints, midpoints, and other specific points easily and accurately.

object snap tracking A feature that, when used in conjunction with object snap, provides alignment paths that help the user produce objects at precise positions and angles.

offsetting Creating a new object (such as a line or circle) at a specific distance from an existing line or circle.

ordinate dimensions Dimensions that use a datum, or reference dimension, to help avoid confusion that could result from "stacking" dimensions; also called *datum dimensioning*.

origin The point at which the axes cross in the world coordinate system.

ortho A mode in AutoCAD that forces all lines to be perfectly vertical or horizontal.

orthogonal Drawn at right angles.

orthographic projection A projection of views at right angles to each other in a multiview drawing.

P

palette In AutoCAD, a selection of colors, similar to an artist's palette of colors.

panning A feature that allows the user to move the viewing window around the drawing without changing the zoom magnification.

paper space The space, or drawing mode, in AutoCAD from which multiple views can be arranged and plotted on the same sheet.

parallel projection The standard projection mode in AutoCAD, in which objects are shown in parallel regardless of perspective.

parameters Characteristics in a dynamic block that can be changed after the block is inserted into a drawing.

parametric shape editing Editing an object's shape by activating grips and entering a new size in the text box that appears.

perimeter The distance around a two-dimensional shape or surface.

perpendicular Lines, polylines, or other objects that meet at a precise 90° angle.

perspective projection A type of projection in which objects that are farther away appear smaller, as they would in a photograph.

phantom lines Lines drawn using a thick lineweight, alternating two short dashes with one long dash; used for cutting planes in sectional views.

pickbox A cursor that consists of a small box that allows you to pick objects on the screen.

pictorial representation A drawing of an object or part that offers a realistic view, compared to orthographic projections.

plane An imaginary flat surface that can be defined by any three points in space or by a coplanar object.

plot scale The scale at which a drawing is plotted to fit on a drawing sheet.

plot style A collection of property settings saved in a plot style table.

plotter Originally, an output device that could handle very large sheets of paper; now used interchangeably with printer.

point filtering The process of using the pointing device to enter any two of the

Glossary of Terms (Continued)

3D coordinates that define a point and then entering the third coordinate from the keyboard.

polar array An array in which objects are arranged radially around a center point.

polar method A relative method of entering points in which the user can produce lines at specific angles around a central point.

polar tracking method A method of entering points in which AutoCAD displays alignment paths at prespecified angles to help the user place the points at precise positions and angles.

polyline A type of line object in AutoCAD that consists of a connected sequence of line and arc segments and is treated by AutoCAD as a single object.

portrait A paper orientation in which the narrower edge of the paper is at the top and bottom of the sheet.

primary units The units that appear by default when you add dimensions to a drawing in AutoCAD.

printer Originally, a small, tabletop output device; now used interchangeably with *plotter*.

profile An outline or contour, such as a side view, of an object.

projected tolerance zone Controls the height of the extended portion of a perpendicular part.

pull-down menus Menus whose names appear across the top of the screen, from which you can perform many drawing, editing, and file manipulation tasks.

purge In AutoCAD, selectively deleting blocks or other elements that are not currently in use in a drawing.

Q

quadrant A quarter of a circle, defined by the points at exactly 0°, 90°, 180°, and 270°.

quadrant points The points at 0°, 90°, 180°, and 270° on a circle.

R

radial Describes a feature on which every point is the same distance from an imaginary center point.

radius The length of a line extending from the center to the periphery of an arc or circle.

rapid prototyping (RP) Refers to a class of machines used for producing physical models and prototype parts from 3D computer model data such as CAD. Unlike CNC machines, which subtract material, RP processes join together liquid, powder, and sheet materials to form parts. Layer by layer, these machines fabricate plastic, wood, ceramic, and metal objects from thin horizontal cross-sections taken from the 3D computer model. Stereolithography and fused deposition modeling are examples of RP processes.

ray A type of line object in AutoCAD that extends infinitely in one direction from a specified point.

realtime zooming A method of changing the level of magnification by picking the Realtime Zoom button and moving the pointing device.

rectangular array An array in which the objects are arranged in rows and columns.

regular polygon A polygon in which all the sides are of equal length.

Glossary of Terms (Continued)

relative method A method of entering points that allows the user to base the new points on the position of a point that has already been defined.

rendered Shaded to look more realistic.

resolution The number of pixels per inch on a display system; this determines the amount of detail you can see on the screen.

revolve To turn around an axis.

roll In 3D Orbit, moving the view around an axis at the center of the arcball.

rotating Changing the angle of an object in the drawing without changing its scale.

rounds Curved or rounded outside corners on an object.

rubber-band effect The stretching effect when a drawing command is active that lets you see the line, circle, etc., "grow" as you move the pointing device.

running object snap modes/running object snaps Object snaps that have been preset to function automatically.

S

scale Increasing or decreasing the overall size of an object or assembly without changing the proportions of the parts to each other.

screen regeneration Recalculation of each vector, or line segment, in a drawing.

section drawings Drawings used to show the interior of a part.

selectable In AutoCAD groups, selectable means that when you select one of the group's members (*i.e.*, one of the circles), AutoCAD selects all members in the group.

selection previewing AutoCAD's automatic highlighting of objects as you run the cursor over them.

selection set All the objects that are currently highlighted, or selected, for a command or operation in AutoCAD.

selection window A window created by picking diagonally opposite corners of an imaginary box at the Select objects prompt.

shelling A method in AutoCAD that allows the user to remove material from an object to create a shell or hollow object.

shortcut menu A context-sensitive menu that appears when you right-click; shortcut menus contain several of the commands and functions the user is most likely to need in the current drawing situation.

F9

snap grid An invisible, user-controlled grid that restricts cursor movement to points that lie on the grid.

solid models Three-dimensional models defined using solid modeling techniques, which are somewhat similar to using physical materials such as wood or clay to produce shapes. CAD solid modeling programs that take advantage of Constructive Solid Geometry (CSG) techniques use primitives, such as cylinders, cones, and spheres, and Boolean operations to construct shapes. CAD programs that use the Boundary Representation (B-Rep) solid modeling technique store geometry directly using a mathematical representation of each surface boundary. Some CAD programs use both CSG and B-Rep modeling techniques.

solid primitives Basic solid shapes, such as cylinders or spheres, usually used to produce more complex shapes.

Glossary of Terms (Continued)

special characters Characters that are not normally available on the keyboard or in a basic text font.

specifications Text that provides information about sizes, shapes, and surface finishes that apply to specific portions of an object or part.

spline leader Similar to a regular leader, except that the leader line consists of a spline curve, offering more flexibility with its shape.

splines Curves that use sampling points to approximate mathematical functions that define a complex curve.

status bar The line at the bottom of the AutoCAD screen that displays various types of status information about modes and settings, including snap, grid, the ortho mode, polar tracking, object snap, object tracking, lineweight, and the current space (model or paper).

surface controls Feature control frames, not associated with a specific dimension, that refer to a surface specification regardless of feature size.

surface models Three-dimensional models defined by surfaces. Each surface consists of a mesh of polygons, such as triangles, or a mathematical description such as a Bezier B-spline surface or non-uniform rational B-spline (NURBS) surface.

symbol library Drawing file that contains a series of blocks for use in other drawings.

symmetrical deviation (See *symmetrical tolerancing*.)

symmetrical tolerancing A method of tolerancing in which the upper and lower limits of a plus/minus tolerance are equal.

system variable Similar to a command, except that it holds temporary settings and values instead of performing a specific function.

T

table A set of rows and columns that contain text.

tangent Lines, circles, or other objects that meet at a single point.

tangential edges Transition lines that occur in a solid model when a curved face meets a flat face.

template file A file that contains predefined characteristics and is used to create new drawings with those characteristics. Also a file that defines text format for attribute extraction.

tessellation Lines that describe a curved surface or 3D model.

3D face A planar surface.

thumbnail A small preview image of the currently selected drawing in the Select File dialog box.

title block A portion of a drawing that is set aside to give important information about the drawing, the drafter, the company, and so on.

tolerance A designation that specifies the largest variation allowable for a given dimension.

toolbar A strip of related buttons that can be displayed or hidden by the user.

tooltips One or more words that appear when you position the pointer over a button for a second or longer to help you understand the purpose or function of the button.

Glossary of Terms (Continued)

traces Thick lines commonly used on printed circuit boards; also, lines drawn using AutoCAD's TRACE command.

transparent zooms Zooms that the user forces to occur while another command is in progress.

two-dimensional (2D) representation A single profile view of an object seen typically from the top, front, or side.

U

UCS icon An icon, placed by default in the lower left corner of the drawing area, that shows the orientation of the axes in the current coordinate system.

user coordinate systems (UCSs) Custom coordinate systems that a user can define for use in creating 3D objects and models.

V

value set The set of specified values to which a constrained parameter is limited in dynamic blocks.

variable attributes Attributes whose value you can change as you insert the block.

vector Line segment defined by its endpoints or by a starting point and direction in 3D space.

verb/noun selection The traditional method of entering a command in AutoCAD, in which the user enters a command and then selects the objects on which the command should operate.

vertex Any endpoint of an individual line or arc segment in a polyline, polygon, or other segmented object.

view box The window that appears in Aerial View to define the current screen magnification and viewing window.

viewports Portions of the drawing area that a user can define to show a specific view of a drawing.

view resolution The accuracy with which curved lines appear on the screen.

W

wireframe A "stick" representation of a 3D model whose shape is defined by a series of lines and curves.

wizard A series of dialog boxes that steps you through a sequence.

world coordinate system (WCS) AutoCAD's name for the Cartesian coordinate system. (See *Cartesian coordinate system*.)

X

xlines A type of construction lines created by the XLINE command that extends infinitely in both directions.

xrefs Ordinary drawing files that have been attached to the current (base) drawing for reference.

XYZ octant The area in 3D space where the x, y, and z coordinates are greater than 0.

X/Y/Z point filtering The process of using the pointing device to enter any two of the 3D coordinates that define a point and then entering the third coordinate from the keyboard.

Z

Z axis The third axis in a 3D coordinate system, in which depth is measured.

Index

Index

Index

Index

Index

Index

Index

Index

Index

Index

Index

Index

Index

Index

Index

user coordinate system, 535-541, 765
  aligning with an object, 538-539
user interface, AutoCAD, 10

V

value set, 461, 765
variable attributes, 483-484, 765
vectors, 422, 765
  evaluation of, with CAL command, 422
  format of, 159
  regeneration of, 179
verb/noun selection, 62, 765
vertex, 180, 765
vertical construction lines, 141
view box, 169, 765
View button, on UCS toolbar, 539
VIEW command, 167-168
viewing interior of 3D objects, 524
viewing options, in 3D space, 517-524
  predefined viewpoint buttons, 518
  3D Orbit, 519-524
  Viewpoint Presets dialog box, 518-519
  View toolbar, 518
viewpoint, manipulating, 520-521
View pull-down menu, 12
Viewpoint Presets dialog box, 518-519
  Set to Plan View option of, 519
viewports, 332, 337, 765
  current, 339
  floating, 338
  in model space, 337-340, 345
  multiple viewports, in paper space, 346-347
  in paper space, 341-343, 346

objects in, 343-345
plotting, 614-616
polygonal, 343
positioning, in paper space, 347, 614-616
VIEWPORTS command, 338. *See also* VPORTS command.
View pull-down menu, 12, 514
VIEWRES command, 148
view resolution, 765
  controlling, 158
views
  aerial, 169
  capturing, 167-168
  isometric, 518
  plan, 514
  3D, 514-516
VPOINT command, 514-515
VPORTS command, 338-340

W

WBLOCK command, 448
WCS (world coordinate system), 68, 535
Welding Fixture Model.dwg, 4
Welding Fixture-1.dwg, 7, 8, 170
wheel.dwg, 218
window.dwg, 483, 485
window selection, of objects, 53
wireframe, 333, 765
wizard, 102, 765
  Add-A-Plotter, 334
  Advanced Setup, 299-301
  Quick Setup, 101-102
word processor, importing text from, 265
workshop.dwg, 473
world coordinate system (WCS), 68, 765
World UCS button, 538, 540
WPolygon object selection option, 54-55
Write Block dialog box, 448

X

X axis, 68, 513
x coordinate, 531
Xdatum option, of ordinate dimensions, 363
XLINE command, 140-141
xlines, 140-142, 765. *See also* construction lines.
XLS (Microsoft Excel®) file, 493
xref, 491, 765
XYZ octant, 623, 765
X/Y/Z point filtering, 532-533, 765
X/Y/Z point filters, 531

Y

Y axis, 68, 513
y coordinate, 531
Ydatum option, of ordinate dimensions, 363-364

Z

Z axis, 69, 513, 765
Z axis vector, to create UCS, 539
z coordinate, 302, 531
zero, leading, 406
zero suppression, 394
Zoom All button, 156
ZOOM command, 74, 153-158
zoom.dwg, 154, 338, 339
Zoom Extents button, 156
Zoom In button, 155
zooming, 4-6, 153-158
  with 3DORBIT command, 521
Zoom Object button, 156
Zoom Previous button, 6, 156
Zoom Realtime button, 157
Zoom Window button, 4, 6, 154